SULFHYDRYL AND DISULFIDE GROUPS OF PROTEINS

STUDIES IN SOVIET SCIENCE

LIFE SCIENCES

1973

Motile Muscle and Cell Models
N. I. Arronet

Pathological Effects of Radio Waves
M. S. Tolgskaya and Z. V. Gordon

Central Regulation of the Pituitary-Adrenal Complex
E. V. Naumenko

1974

Sulfhydryl and Disulfide Groups of Proteins
Yu. M. Torchinskii

Mechanisms of Genetic Recombination
V. V. Kushev

A Continuation Order Plan is available for this series. A continuation order will bring delivery of each new volume immediately upon publication. Volumes are billed only upon actual shipment. For further information please contact the publisher.

STUDIES IN SOVIET SCIENCE

SULFHYDRYL AND DISULFIDE GROUPS OF PROTEINS

Yu. M. Torchinskii

Institute of Molecular Biology
Academy of Sciences of the USSR
Moscow, USSR

Translated from Russian by

H. B. F. Dixon

Department of Biochemistry
University of Cambridge
Cambridge, England

Revised by Yu. M. Torchinskii and H. B. F. Dixon

CONSULTANTS BUREAU • NEW YORK AND LONDON

Library of Congress Cataloging in Publication Data

Torchinskiĭ, Iŭriĭ Moiseevich.
Sulfhydryl and disulfide groups of proteins.

(Studies in Soviet science)
Translation of Sul'fgidril'nye i disul'fidnye gruppy belkov.
1. Proteins. 2. Thiols. 3. Sulphides. I. Title. II. Series. [DNLM: 1. Cystine. 2. Sulfhydryl compounds. QU60 T676s 1974]
QP551.T6913 574.1'9245 73-83903
ISBN 0-306-10888-7

The original Russian text, published by Nauka Press in Moscow in 1971, has been corrected by the author for the present edition. This translation is published under an agreement with the Copyright Agency of the USSR (VAAP).

Сульфгидрильные и дисульфидные группы белков
Ю. М. Торчинский
SUL'FGIDRIL'NYE I DISUL'FIDNYE GRUPPY BELKOV
Yu. M. Torchinskii

A Division of Plenum Publishing Corporation
227 West 17th Street, New York, N.Y. 10011

United Kingdom edition published by Consultants Bureau, London
A Division of Plenum Publishing Company, Ltd.
4a Lower John Street, London W1R 3PD, England

Printed in the United States of America

Contents

Preface

(Written for the original Russian edition)

Out of all the functional groups of protein molecules that belong to the side chains of amino acid residues, the sulfhydryl or thiol (SH) groups of cysteine residues and the disulfide (S–S) groups of cystine have for long attracted particular attention from chemists and biochemists. This is due, on the one hand, to the high chemical reactivity of these groups, which enter into many, and often very characteristic and selective, reactions with several types of compound. On the other hand, this attention is connected with the great significance of SH and S–S groups for the specific functions of a number of enzymes, hormones, and other biologically active proteins, and consequently for the normal course of many physiological processes.

The varied functions of sulfhydryl and disulfide groups in enzymes are particularly interesting; their role can consist of binding substrates and cofactors, in direct participation in the catalytic act, or in maintaining (in ways not so far sufficiently studied) the native, catalytically active conformation of the protein.

It is entirely natural that the detailed study of the chemical behavior and biological role of these sulfur-containing groups in proteins (and in such low-molecular cofactors as glutathione, coenzyme A, lipoic acid, etc.) has, in addition to its academic, general biological interest, great significance for many applied branches of medicine and biology (toxicology, radiobiology, diagnostics and therapeutics, biochemical engineering, plant growing, etc.).

The vast and rapidly growing literature on this subject is almost boundless. It is surveyed in many review articles published over the years and in the proceedings of specialist symposia and conferences. Nevertheless there is no single monograph, published in the last decade, either in Russian or in a foreign language, in which this valuable information is systematized and subjected to critical review.

The present book successfully fills this gap, and is presented for the attention of a wide range of biochemists and experimental biologists of many types. It is written both concisely and at the highest contemporary level. Chapters 1-3 describe the chemical properties of sulfhydryl and disulfide groups and review the experimental methods for their study. Chapters 4-7 competently review the vast factual material on the biochemistry of these groups and critically discuss modern ideas on their role in enzymes and in other biologically active proteins.

The author of the book, Yu. M. Torchinskii, has long personal experience of studying the SH and S–S groups of muscle proteins and of various enzymes; he has published many experimental papers and some authoritative reviews on this subject.

A. E. Braunstein

Introduction

Introduction

The sulfhydryl and disulfide groups of proteins belong respectively to the residues of cysteine and cystine. Cystine was first discovered by Wollaston (1810) in stones of the bladder, whence it obtained its name.* Only many years later was it found in proteins (Mörner, 1899) and its structure established (Neuberg, 1902; Friedmann, 1903). In 1907 Heffter (1907) discovered SH groups in proteins, but only the discovery of glutathione and the study of its role (Hopkins, 1921) excited wide interest in the investigation of SH groups.

This interest grew after Lundsgaard (1930a,b) discovered that iodoacetate inhibited glycolysis in muscle and alcoholic fermentation in yeast.† By that time it was already known that iodoacetate reacts with SH groups. Later it was established that the suppression of glycolysis by iodoacetate is caused by the inhibition of glyceraldehyde-3-phosphate dehydrogenase (Green et al., 1937; Adler et al., 1938). At about the same time, Hopkins and Morgan (1938) observed the inhibition of succinate dehydrogenase under the influence of oxidized glutathione. Somewhat earlier the significance of SH groups for the action of pyruvate oxidase (Peters et al., 1935; Peters, 1936) and of papain (Hellerman and Perkins, 1934; Bersin, 1933, 1935) had been demonstrated.

---

* The word "cystic" means "of the bladder."

† The inhibition of glycolysis by bromoacetate was first observed by Schwartz and Oschmann (1925), but their discovery remained unnoticed for long.

Since the thirties SH groups have attracted more and more attention from enzymologists; a more important role has been ascribed to them than to the many other, less reactive, functional groups of proteins. The appearance of new reagents specific for SH groups and the working out of exact methods for their quantitative determination has helped the development of investigations in this field. Among the various reagents for SH groups one should note especially p-chloromercuribenzoate (Hellerman, 1937), o-iodosobenzoate (Hellerman et al., 1941), methylmercury iodide (Hughes, 1950), N-ethylmaleimide (Gregory, 1955), and 5,5'-dithiobis(2-nitrobenzoate) (Ellman, 1959).* Barron (1953), in a review article published in 1953, already mentioned 17 reagents for SH groups in proteins, and listed 42 of the so-called thiol, or sulfhydryl, enzymes, i.e., enzymes whose activity is inhibited by blocking their SH groups. The counts made by Boyer (1959) bear witness to the subsequent scale of investigations in this field; in a period of about two and a half years (1955 to the beginning of 1957) 134 papers were published throughout the world on the influence of thiol reagents on the activity of isolated enzymes and enzyme systems; 87% of these noted inhibition of activity on blocking of SH groups.

Among the thiol enzymes now known there are representatives of all classes of enzymes, catalyzing the most varied chemical changes. The majority of intracellular enzymes, including almost all the known dehydrogenases, many other oxidative enzymes, and the enzymes of amino acid, carbohydrate, and fat metabolism, and also of protein biosynthesis, are inhibited by reagents for SH groups.

In parallel with the discovery of the significance of SH groups for the action of various enzymes, data on the important role of SH groups in a number of physiological and biochemical processes have accumulated in the literature; these processes include muscular contraction, nerve activity, cell division, regulation of the permeability of mitochondrial membranes, oxidative phosphorylation, and photosynthesis. Thiol groups are also important in the mechanism of radiation damage and in the action of certain toxic substances. It is clear that the influence of thiol reagents on these physiological processes is caused by the blocking of SH groups of

* The authors given in parentheses are those who first used the stated reagent for studying SH groups in proteins.

enzymes and other proteins, and also of low-molecular, functionally important thiols (coenzyme A, 4'-phosphopantetheine, lipoic acid, and glutathione), which act as cofactors or prosthetic groups in various enzyme systems. The successful application of the dithiols BAL (2,3-dimercaptopropanol) and unithiol (sodium 2,3-dimercaptopropanesulfonate) to the treatment of poisoning by compounds of arsenic and of heavy metals is widely known, and also the effective use of aminothiols (e.g., cysteamine) as protective agents against radiation damage. The facts listed explain the great interest in the study of SH groups not only from enzymologists and protein chemists, but also from physiologists, pathologists, pharmacologists, toxicologists, radiobiologists, and cyto- and histochemists.

The last few years have been marked by an impetuous development of the chemistry of proteins and enzymes. Great advances have also been made in the study of the other functional groups of proteins. Data have been obtained about the participation in various enzymic reactions of the imidazole group of histidine, the hydroxyl group of serine, the ε-amino group of lysine, the ω-carboxyl groups of aspartic and glutamic acids, the phenolic group of tyrosine, etc. These findings, however, have not diminished interest in the study of SH groups, which are distinguished from the other functional groups of proteins by their high reactivity and by the unique diversity of the chemical reactions into which they enter: alkylation, acylation, oxidation, thiol–disulfide exchange, reactions with sulfenyl halides, and the formation of mercaptides, hemimercaptals, mercaptols, and, possibly, charge-transfer complexes. As Chinard and Hellerman (1954) justly remarked, the majority of reagents for the various functional groups of proteins, with few exceptions, react fastest of all with free SH groups.

The reactivity of mercaptide ions toward α,β-unsaturated compounds – acrylonitrile, for example – is 280 times greater than that of unprotonated amino groups (Friedman et al., 1965). The rate of reaction of SH groups with 1-fluoro-2,4-dinitrobenzene at pH 8.4 is about 90 times greater than that of phenolic groups and 166 times greater than that of amino groups (Wallenfels and Streffer, 1966). Even those SH groups that are markedly ionized at physiological pH values (i.e., those characterized by a comparatively low pK) are effective nucleophiles. According to Ogilvie et al. (1964), an SH group with a pK of 7 is about four times more powerful as a nucleophile in reacting with p-nitrophenyl acetate than an imidazole with a pK of 7. The high nucleophilic power of mercaptide ions is given by the particular electronic structure of the sulfur atom, its high polarizability and weak solvation.

In comparison with SH groups, disulfide groups have a more limited reactivity: the resistance of these groups to various influences and the stability of the S–S bond correspond well with the main function of disulfide groups in proteins. This function consists of stabilizing the macromolecular structure of the proteins.

Investigations of the properties of SH and S–S groups and of the role of these groups in enzymes and other proteins are rapidly expanding. There is no doubt that these investigations will provide in the not too distant future more precise answers to some of the questions discussed below. Nevertheless the vast experimental material already accumulated in this field deserves, in our opinion, extensive analysis and review, an attempt at which is undertaken in this book.

The general review by Jocelyn (1972) of the biochemistry of the SH group appeared after the publication of the Russian edition of this book in 1971, and there is naturally some overlap between the two books. Nevertheless the two approaches prove to be largely complementary. The contents of this English edition has been revised up to mid-1973. In translation the whole text has been reviewed, and some sections largely rewritten.

PART I

THE CHEMISTRY OF SH AND S–S GROUPS AND METHODS FOR THEIR DETERMINATION

Chapter 1

The Chemical Properties of SH Groups Sulfhydryl Reagents

This chapter describes the chemical properties of SH groups of low-molecular thiols and of proteins, and also describes reagents for these groups. Investigations of low-molecular thiols have played an important part in the development of the chemistry of the SH groups of proteins. One should note, however, that the SH groups of proteins sometimes differ in reactivity and chemical behavior both from those of low-molecular thiols and from one another (see Chapter 4).

In describing sulfhydryl reagents we have concentrated on the problem of their specificity of action, and also on the conditions for using them to determine and modify the SH groups of proteins.

1. The Ionization of SH Groups

In order to evaluate the reactivivity of SH groups in proteins it is important to know the pH at which ionization of these groups takes place. SH groups take part in most reactions in the form of the mercaptide ion RS^-. We can roughly assess the degree of ionization of the SH groups of proteins on the basis of data obtained by studying cysteine and other low-molecular thiols. Cysteine contains three dissociable protons. On titration with alkali three "macroscopic" ionization constants K_1, K_2, and K_3, are obtained; the pK values that correspond with them are, at zero ionic strength, 1.71, 8.37, and 10.70 (Borsook et al., 1937; Coates et al., 1969). The constant K_1 is clearly related to the ionization of the carboxyl group. The ownership of the constants K_2 and K_3 is not so obvious. Cohn and Edsall (1943) ascribed K_3 to the SH group. Later, however, Edsall expressed the opinion that the dissociations of the SH and NH_3^+ groups occur at very close pH values and that consequently the constants K_2 and K_3 are mixed; they presented the following scheme of ionization of these groups in the cysteine molecule:*

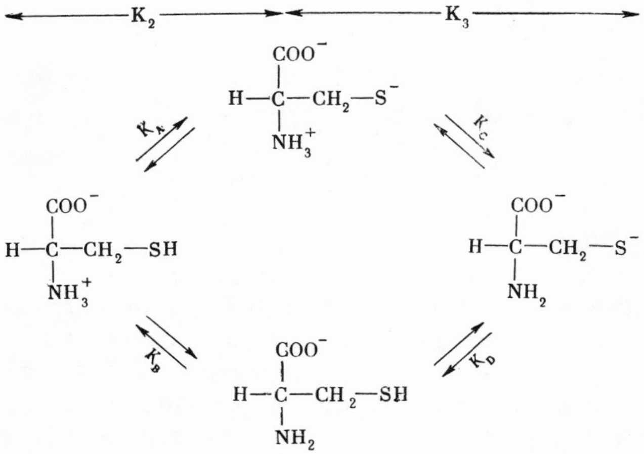

$\longleftarrow K_2 \longrightarrow \longleftarrow K_3 \longrightarrow$

$H-C(COO^-)(NH_3^+)-CH_2-SH \overset{K_A}{\rightleftharpoons} H-C(COO^-)(NH_3^+)-CH_2-S^- \overset{K_C}{\rightleftharpoons} H-C(COO^-)(NH_2)-CH_2-S^-$

$H-C(COO^-)(NH_3^+)-CH_2-SH \overset{K_B}{\rightleftharpoons} H-C(COO^-)(NH_2)-CH_2-SH \overset{K_D}{\rightleftharpoons} H-C(COO^-)(NH_2)-CH_2-S^-$

The symbols K_A, K_B, K_C, and K_D in the scheme represent the individual or "microscopic" ionization constants of the SH and NH_3^+ groups. They are linked with the titration, "macroscopic" constants K_2 and K_3 by the following equations:

$$K_2 = K_A + K_B$$

$$1/K_3 = 1/K_C + 1/K_D$$

$$K_2K_3 = K_AK_C = K_BK_D$$

The individual ionization constants of the SH and NH_3^+ groups of cysteine and related compounds were calculated by Benesch and Benesch (1955) on the basis of data obtained on spectrophotometric

* Edsall's proposal and the scheme proposed for the ionization of cysteine were first presented in a paper by Ryklan and Schmidt (1944) with a reference to a personal communication from Edsall. De Dekken et al. (1956) proposed another scheme for the ionization of cysteine; according to their scheme a proton first dissociates from the NH_3^+ group and then a hydrogen bond forms between the SH and NH_2 groups, after which the SH group ionizes; this scheme does not correspond well with experimental findings (Edsall, 1965).

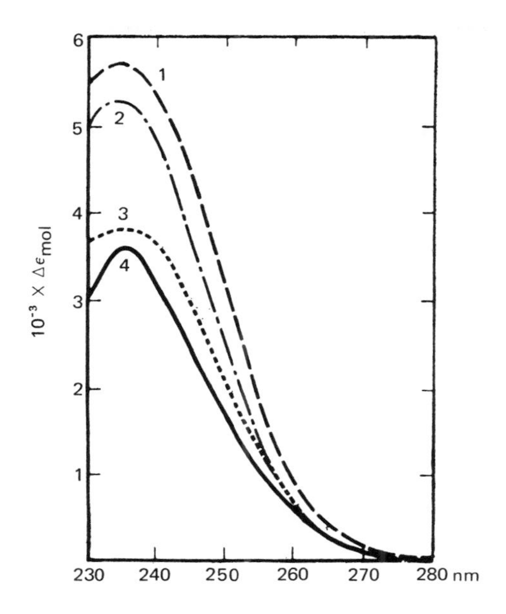

Fig. 1. Difference spectra of ionization of SH groups (Donovan, 1964). The pH of comparison solutions was between 6 and 7. The spectra were measured at concentrations of mercaptoethanol, cysteine, and aldolase of 150 μM, 100 μM, and 1.6 μM, respectively. 1) β-Mercaptoethanol in water, pH 12.2; 2) as 1, but in 4 M urea, pH 12.9; 3) cysteine in 4 M urea, pH 12.3; 4) aldolase in 4 M urea.

titration of these compounds (the titration is based on the difference in absorbance in the region 230–240 nm between RS^- ions and undissociated SH groups; see Fig. 1). The constants they calculated are given in Table 1. Elson and Edsall (1962) obtained similar values by measuring the pH-dependence of the intensity of the SH band in the Raman spectrum of cysteine; the SH group has a strong band at 2580 cm^{-1} which disappears when the group ionizes. The ionization constants of cysteine were also calculated by Grafius and Neilands (1955) on the basis of titration of S-methylcysteine and of cysteinebetaine. Jung et al. (1972) studied the ionization of the sulfhydryl and amino groups of cysteine and glutathione by NMR spectroscopy. They found, in agreement with Benesch and Benesch, that the pK of the SH group of glutathione was 9.2, and the macroscopic pK values of cysteine were 8.25 and 10.25. Rabenstein (1973) has determined the microscopic constants for glutathione by observation of the change of chemical shift of the NMR of carbon-bonded protons.

TABLE 1. Ionization Constants of Thiols*

Thiol	pK_A	pK_B	pK_C	pK_D	pK_{SH}
L-Cysteine	8.53 (8.65)	8.86 (8.75)	10.36 (10.05)	10.03 (9.95)	—
L-Cysteine ethyl ester	7.45	6.77	8.41	9.09	—
L-Cysteinylglycine	7.87	7.14	8.75	9.48	—
β-Mercaptoethylamine	8.35	—	—	—	—
β-Mercaptoethanol	—	—	—	—	(9.5)
β-Mercaptopropionic acid	—	—	—	—	10.3†
Mercaptoacetic acid	—	—	—	—	10.28‡
Glutathione	—	—	—	—	9.2 (8.7)

*The ionization constants of cysteine, cysteine ethyl ester, cysteinylglycine, β-mercaptoethylamine, and glutathione are taken from the data of Benesch and Benesch (1955), with the data of Grafius and Neilands (1955) for cysteine and the data of Calvin (1954) for glutathione and β-mercaptoethanol in parentheses. The ionization constants of 13 different thiols, including cysteine, were also determined by Danehy and Noel (1960). The values they obtained were generally close to those shown in the table. Jencks and Salvesen (1971) measured the pK values of mercaptoethanol and mercaptoacetate in water at 25°C and ionic strength 1.0 as 9.61 and 10.25, respectively.

†From the data of Cecil and McPhee (1959).

‡From the data of Lindley (1960).

Dixon and Tipton (1973) have pointed out that the group and molecular constants of Benesch and Benesch (1955) imply that the dissociation of the SH group of cystenine is almost the same as if 68% of the group had a pK of 8.36 (pK_2) and 32% of it had an independent pK of 10.53 (pK_3).

From the data presented in Table 1 it follows that the SH and NH_3^+ groups are acids of almost equal strength (in agreement with the proposal of Edsall); the acidic properties of the SH group are only a trace stronger than those of the NH_3^+ group of cysteine. The pK of the SH group in cysteine is depressed under the influence of the positively charged amino group. The pK values of the SH groups of cysteine ethyl ester and of cysteinylglycine are still lower. This is apparently due to the greater inductive effect of the carbonyl group in these compounds and by absence (in the ester) or weakening

TABLE 2. The pK Values of Some Aminothiols in Water – Methanol (1:1) Mixtures at 18°C (Franzen, 1957)

Aminothiol	pK_{SH}	pK_{NR_2}
$(C_2H_5)_2N - CH_2 - CH_2 - SH$	7.80	10.75
$(C_2H_5)_2N - [CH_2]_3 - SH$	8.0	10.5
$(C_2H_5)_2N - [CH_2]_4 - SH$	10.10	10.10
$(C_2H_5)_2N - [CH_2]_6 - SH$	10.10	10.10

(in the dipeptide) of the electrostatic effect of the negatively charged carboxyl group. In glutathione, as a consequence of the large distance between the SH and NH_3^+ groups, their mutual influence is diminished, and the pK of the SH group is higher than in the previously listed compounds. The influence of the distance between the SH and NH_3^+ groups on their ionization is clear from Table 2.

Benesch and Benesch (1955) calculated that at physiological pH (7.4) about 16% of the SH groups of cysteine ethyl ester were ionized, 11% of the SH groups of cysteinylglycine, 6% of the SH groups of cysteine, and 1% of the SH groups of glutathione.

The pK values of the SH groups of proteins vary within wide limits under the influence of different factors. One of these factors is the spatial disposition of the SH groups in the protein glo-

TABLE 3. The pK Values of SH Groups in the Active Sites of Enzymes

Enzyme	pK of SH group	Reference
Phosphoenolpyruvate carboxykinase	7.3	Barns and Keech, 1968
Papain	8.02	Smith and Parker, 1958
Papain	8.5	Lucas and Williams, 1969
Muscle 3-phosphoglyceraldehyde dehydrogenase, apoenzyme	8.1	Behme and Cordes, 1967
Muscle 3-phosphoglyceraldehyde dehydrogenase, holoenzyme and apoenzyme	8.0-8.1	MacQuarrie and Bernhard, 1971
Propionylcoenzyme A carboxylase	8.2	Edwards and Keech, 1967
Streptococcal protease	8.4	Gerwin et al., 1966
Ficin	8.55	Hollaway et al., 1964, 1971

bule. The so-called "masked" or "buried" SH groups, placed in a hydrophobic environment within the globule and having no contact with the solvent are characterized, apparently, by higher pK values than are groups placed on the surface of the globule. As can be seen in Table 3 the pK values of the majority of the SH groups in the active sites of enzymes vary within the limits 7-9. It is highly probable that these groups are on the surface of the enzyme macromolecules or in clefts on the surface. On the other hand the reactive SH groups of hemoglobin remain un-ionized at pH values below 9.5-10 (Snow, 1962; Guidotti, 1967).

According to Donovan (1964), some of the SH groups of aldolase have a pK of 10.5, while others (presumably completely buried within the globule) ionize only after the conformational change of the protein that takes place at a pH of about 11. On denaturation of aldolase by urea, which evidently "unmasks" the SH groups, their pK falls to 8.7.

One of the important factors that influence the value of the ionization constants of SH groups in protein molecules is the electrostatic effect of neighboring charged groups. From data on the ionization of low-molecular thiols one can conclude that the presence of a positively charged group in the immediate vicinity of an SH group lowers its pK to about 8.35 (the pK of β-mercaptoethylamine); in the absence of charged groups nearby the pK of an SH group may be about 9.5 (as in β-mercaptoethanol); in the presence of a close negatively charged group the pK of the SH group of a protein may rise to 10.3 (as in β-mercaptopropionic and mercaptoacetic acids).

The heat of ionization of SH groups is about 7 kcal/mole; this corresponds with a fall of 0.2 in the pK when the temperature is raised from 25 to 37°C (Benesch and Benesch, 1955; Cecil and McPhee, 1959).

2. Hydrogen Bond Formation

Hydrogen bonds in which a bivalent sulfur atom participates are markedly weaker than the analogous bonds in which oxygen participates, because of the comparatively low electronegativity of sulfur (2.6 electron volts for sulfur and 3.5 for oxygen). The low boiling point of H_2S in comparison with H_2O and of thioalcohols and thiophenols in comparison with their oxygen analogs is often cited as an illustration of the weakness of absence of intermolecular hydrogen bonds formed by sulfur. Nevertheless many authors have succeeded in demonstrating by infrared spectroscopy the existence of hydrogen bonds in which thiophenol takes part. They showed that thiophenol can form intermolecular hydrogen bonds with pyridine, aniline, and dibenzylamine, and also form dimers and polymers linked by hydrogen bonds (Gordy and Stanford, 1940; Calvin, 1954; Josien et al., 1957). Hopkins and Hunter (1942) obtained evidence for the formation of S ··· H − N hydrogen bonds in nonaqueous solutions of thioamides; they used the cryoscopic method for observing association of the molecules in solution, and found that replacement of the hydrogen atom in the −NH−CS− group prevents association of the thioamide molecules.

In contrast with thiophenol, aminothiols of the type $R_2N-[CH_2]_n-SH$ do not form, according to Plant et al. (1955), intermolecular hydrogen bonds that could be observed by IR spectroscopy. In such compounds the SH groups could clearly act only as proton donors in hydrogen bonds. The data of Plant et al. do not, however, exclude the possibility of the formation of hydrogen bonds in which sulfur is the proton acceptor. Donohue (1969) analyzed data on the crystal structure of a number of thio compounds (intermolecular distances and angles) and came to the conclusion that weak hydrogen bonds of the type N−H···S existed in thiourea, thioacetamide, dithiouracil, and some other substances (see also Hordvik, 1963).

Hydrogen bonds of the N−H···S type with N−S distances of 3.35-3.43 Å have been found in crystals of thiourea by neutron diffraction (Elcombe and Taylor, 1968), and O−H···S bonds in crystals of barium thiosulfate monohydrate (Manojlović-Muir, 1969).

Contradictions have arisen in the literature on the question of intramolecular hydrogen bond formation by SH groups. Cecil (1950, 1951) proposed the existence of such a bond between the SH and amino groups of glutathione on the basis of the fact that urea affects the kinetics of its reaction with $AgNO_3$. According to Pasynskii and Chernyak (1952) the oxidation of the SH group of cysteine by oxygen of the air is accelerated in the presence of high concentrations of urea.

Benesch and Benesch (1953) found that, although sodium sulfide and ethyl mercaptan give identical colorations with nitroprusside in water and in urea solutions, cysteine and its peptides, including glutathione, give a markedly more intense coloration in the presence of urea. They explained the increase in coloration in urea by breakage of hydrogen bonds of the type $S^{-}\cdots H-N$; such bonds lower the electron density at the sulfur atom and hence its reactivity. The data of Benesch and Benesch were confirmed by Poglazov et al. (1958). Boyer (1959) pointed out, however, that the reactivity of the SH groups of peptides, in particular glutathione, can be decreased also as a result of the formation of hydrogen bonds that do not involve sulfur (for example, bonds between carboxyl and amide groups); such bonds can give the molecule a conformation in which interaction of the SH group with the comparatively large nitroprusside molecule is sterically hindered. One should note that, according to Lindley (1962), urea at a concentration of 4 M has no influence on the rate of reaction of glutathione with chloroacetamide at pH 8.0. These data compel caution in interpreting the experiments of Benesch and Benesch with nitroprusside. Studies of the UV and Raman spectra of cysteine gave no indication of intramolecular hydrogen bonds between the SH and NH_2 groups (Edsall, 1965). Nevertheless enthalpy measurements for the ionization of cysteine indicate interactions between these groups (Wallenfels and Streffer, 1966). But X-ray analysis of glutathione (Wright, 1958) and L-cysteine (Harding and Long, 1968) crystals showed no sign of hydrogen bonding between the sulfur and nitrogen atoms. In the crystal lattice of cysteine the main Van der Waals contacts of the sulfur atoms are one with another; each sulfur atom has in addition one close contact of 3.44 or 3.48 Å with an oxygen atom.

3. Acylation of SH Groups.

Properties of Thioesters

Thiols react readily with various acylating agents, such as keten, acetyl chloride, acetic and succinic anhydrides, ethylthiotrifluoroacetate, etc. For example,

$$R{-}SH + (CH_3CO)_2O \rightarrow R{-}S{-}\underset{\underset{O}{\|}}{C}{-}CH_3 + CH_3COOH$$

The acylation of SH groups by phenylisocyanate was used by Desnuelle and Rovery (1949) in studying the enzyme urease:

$$C_6H_5{-}N{=}C{=}O + HS{-}R \rightarrow C_6H_5{-}NH{-}\underset{\underset{O}{\|}}{C}{-}S{-}R$$

However, because this reagent, like other acylating compounds, is rather unspecific, it has not received much application in studying the SH groups of proteins. Ethylisocyanate may be used, however, for protecting the SH group of cysteine in peptide synthesis (Guttmann, 1966). Acylating agents are usually used in protein chemistry for blocking amino, imidazole, and phenolic groups. But the SH groups react at the same time, unless they have been previously protected with a more specific reagent.

The ability of the SH groups of low-molecular thiols (coenzyme A, 4'-phosphopantetheine, lipoic acid, glutathione) and of some enzymes (glyceraldehyde-3-phosphate dehydrogenase, ficin, papain, thiolase, etc.) to form thioesters and to transfer acyl groups is of great biological importance. The acylating power of the thioester of a carboxylic acid is greater than that of its oxygen ester. Calvin (1954) explained this difference as follows. The acylating power of oxygen esters is weakened as a consequence of the existence of resonance form I. In contrast with oxygen, bivalent sulfur has little tendency to donate its electrons to form a double bond with carbon; consequently the existence of the analogous resonance form for thioesters is improbable. The electronic structure of thioesters is closer to form II, which has a greater degree of localization of positive charge on the carbon atom of the carbonyl group, and this explains the high acylating power of these compounds. Baker and Harris (1960) also considered the contribution of resonance form III, characterized by an electron shift from oxygen to sulfur; but its contribution to the resonance stabilization of thioesters is apparently small (Bruice and Benkovic, 1966):

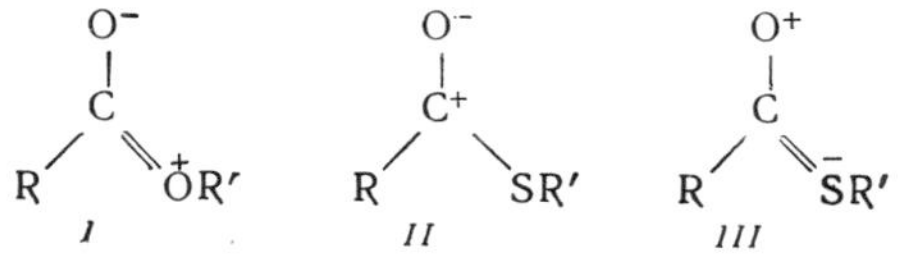

$$\underset{I}{R-C(-O^-)=\overset{+}{O}R'} \qquad \underset{II}{R-C^+(-O^-)-SR'} \qquad \underset{III}{R-C(-O^+)=\overset{-}{S}R'}$$

It should be noted that the C—S bond in thioesters is longer and weaker than the corresponding C—O bond in oxygen esters; this may result in lesser electronic repulsion of the attacking nucleophile and thus facilitate splitting of the ester linkage in nucleophilic displacement reactions at the carbonyl carbon atom (Lynen, 1970).

Thioesters are particularly reactive to nitrogen nucleophiles; they react with them some orders of magnitude faster than do O-esters. With oxygen nucleophiles the difference in reactivity between the two types of ester is not so marked (Connors and Bender, 1961; Bruice and Benkovic, 1966; Brubacher and Bender, 1966).

A characteristic of thioesters is the raised tendency of the proton on the α-carbon atom to dissociate. In their reactivity at the α-carbon, thioesters are reminiscent of ketones (Lynen, 1953). This property of thioesters is reviewed in the description of the reactions of acetyl-CoA (see Chapter 6, Section 1c).

Intramolecular transfer of an acyl group from a sulfur atom to a nitrogen atom (S→ N transfer) has been observed with some aminothiols. S-acyl derivatives of β-mercaptoethylamine rapidly undergo this reaction at pH values above 5; a substituted thiazolidine is formed as an intermediate (Wieland and Bokelman, 1952; Wieland and Horning, 1956; Martin and Hedrick, 1962):

$$\begin{array}{l} H_2C{-}CH_2 \\ H_2N \quad\ \ | \\ \qquad\ \ S \\ \qquad\ \ | \\ \qquad\ \ C{=}O \\ \qquad\ \ | \\ \qquad\ \ CH_3 \end{array} \rightarrow \left[\begin{array}{c} H_2C{-}CH_2 \\ |\qquad | \\ HN \quad S \\ \diagdown\ \diagup \\ C \\ \diagup\ \diagdown \\ CH_3 \quad OH \end{array}\right] \rightarrow \begin{array}{l} H_2C{-}CH_2 \\ |\qquad\ \ | \\ HN \quad SH \\ | \\ C{=}O \\ | \\ CH_3 \end{array}$$

An intramolecular transfer of an acyl residue from an SH group to the ε-amino group of a lysine residue has been observed in the active site of glyceraldehyde-3-phosphate dehydrogenase (Polgar, 1964, 1966a,b; Mathew et al., 1965, 1967; Park et al., 1966).

Intramolecular transfer of an acyl group from a sulfur atom to oxygen (S →O transfer) has also been described (Harding and Owen, 1954; Jencks et al., 1960; Martin and Hedrick, 1962):

$$\begin{array}{c} O \\ \| \\ CH_3{-}C \\ \diagdown \\ H{-}O \qquad\qquad S \\ \diagdown \qquad \diagup \\ [CH_2]_n \end{array} \rightleftarrows \begin{array}{c} O \\ \| \\ C{-}CH_3 \\ \diagup \\ O \qquad\qquad S{-}H \\ \diagdown \qquad \diagup \\ [CH_2]_n \end{array}$$

Hydroxylamine, hydrazine, semicarbazide, and other substituted hydrazines react readily with thioesters (much more rapidly than with O-esters), forming a free thiol and the acylated derivative of the hydroxylamine or hydrazine:

$$R{-}\underset{\underset{O}{\|}}{C}{-}SR' + NH_2OH \rightarrow R{-}\underset{\underset{O}{\|}}{C}{-}NHOH + R'SH$$

Hydroxylaminolysis of thioesters is much faster than alkaline or acid hydrolysis (Noda et al., 1953; Bruice, 1961; Bruice and Fedor, 1964). The conversion of thioesters into hydroxamic acids is used for their identification and quantitative estimation.

It is noteworthy that the thioesters of carboxylic acids are split by Hg^{2+} ions with the formation of the corresponding mercaptides (Sachs, 1921; Lynen et al., 1951; Stern, 1956), but are not split by p-mercuribenzoate at neutral pH (Sanner and Pihl, 1962). A thioester of N-ethylcarbamic acid ($R-S-CO-NH-C_2H_5$) is, however, split by p-mercuribenzoate and by mercuriacetate (Yanaihara et al., 1969). Under the influence of sodium borohydride thioesters (in contrast with O-esters) undergo reductive splitting with the formation of a thiol and an alcohol:

$$RCH_2\underset{\underset{O}{\|}}{C}-S-CoA \xrightarrow{NaBH_4} RCH_2CH_2OH + HS-CoA$$

4. Reaction with Cyanate

Cyanate reacts with the SH group of cysteine with the formation of S-carbamoylcysteine:

$$RS^- + HNCO + H_2O \rightleftarrows RS-\underset{\underset{O}{\|}}{C}-NH_2 + OH^-$$

In addition to SH groups, α- and ε-amino groups, carboxyl and imidazole groups, and also the phenolic groups of tyrosine all react with cyanate. The hydroxyl groups of serine and threonine do not as a rule react with cyanate; but the reactive hydroxyl group of the serine residue in the active center of chymotrypsin and of other proteases undergoes carbamoylation with formation of a urethane.

SH groups are carbamoylated much faster than any other functional groups of a protein; the difference in rate constants is from one to three orders of magnitude (Table 4). According to Stark et al. (1960), half of the SH group of cysteine or glutathione reacts in 3-4 min in 0.1 M KNCO at 25°C; in this time the amino groups are not appreciably carbamoylated.

The reaction of SH groups with cyanate is readily reversible; the S-carbamoyl group is stable at a pH below about 5-6, but is rapidly split at pH 8.0 and above (carbamoylamino groups are sta-

TABLE 4. Rate of Reaction of Various Functional Groups with KNCO (Stark, 1967)

Group being modified	pK_a	k_i^*, $M^{-1} \cdot min^{-1}$	Temperature, °C
α-NH_2 (glycine)	9.60	$2.1 \cdot 10^{-2}$	30
α-NH_2 (glycylglycine)	8.17	$1.4 \cdot 10^{-1}$	30
ε-NH_2 (ε-aminohexanoic acid)	10.75	$2.0 \cdot 10^{-3}$	30
SH (cysteine)	8.3	4.0	25
Imidazole	7.15	$1.8 \cdot 10^{-1}$	25

*The constants are given for the bimolecular reaction that follows the following equations: rate = $k_i[NCO^-][R\text{-}NH_3^+]$ (for amines), or rate = $k_i[NCO^-][R\text{-}SH]$ (for cysteine). At a pH above 5.5 cyanic acid is virtually completely ionized and the total concentration of cyanate can be expressed as $[NCO^-]$.

ble under these conditions). This allows cyanate to be used for reversible blocking of protein SH groups. Stark et al. (1960; Stark, 1964) studied the reaction of cyanate with the SH groups of β-lactoglobulin and reduced ribonuclease A. They called attention to the fact that solutions of 8 M urea, often used for denaturing proteins, can contain cyanate up to a concentration of 0.02 M, enough to modify SH and NH_2 groups. Urea may be freed from cyanate by recrystallization or by exposure to pH 2 before use.

The pH dependence of the reaction of cyanate with SH groups illustrates an important general principle. Many reactions of the SH group follow curves of the type shown in Fig. 2a (and c), i.e., their rates are proportional to the concentration of mercaptide ion, $[RS^-]$. The reaction with cyanate, however, like some other reactions, follows the kind of curve shown in Fig. 2b (and d). Paradoxically this curve follows [RSH], the concentration of the undissociated form, over the pH range in which the thiol group dissociates. But it may easily be seen that this bell-shaped curve expresses the product of $[RS^-]$ and the concentration of the protonated form of some group that is predominantly unprotonated at this pH. In the present case the reactive species appear to be HNCO and RS^- (the kinetics gives no evidence for this; exactly the same pH dependence would result if they were NCO^- and RSH, the major forms). At pH values well below the pK of HNCO the rate rises with increasing pH (Fig. 2b,d) because the very low con-

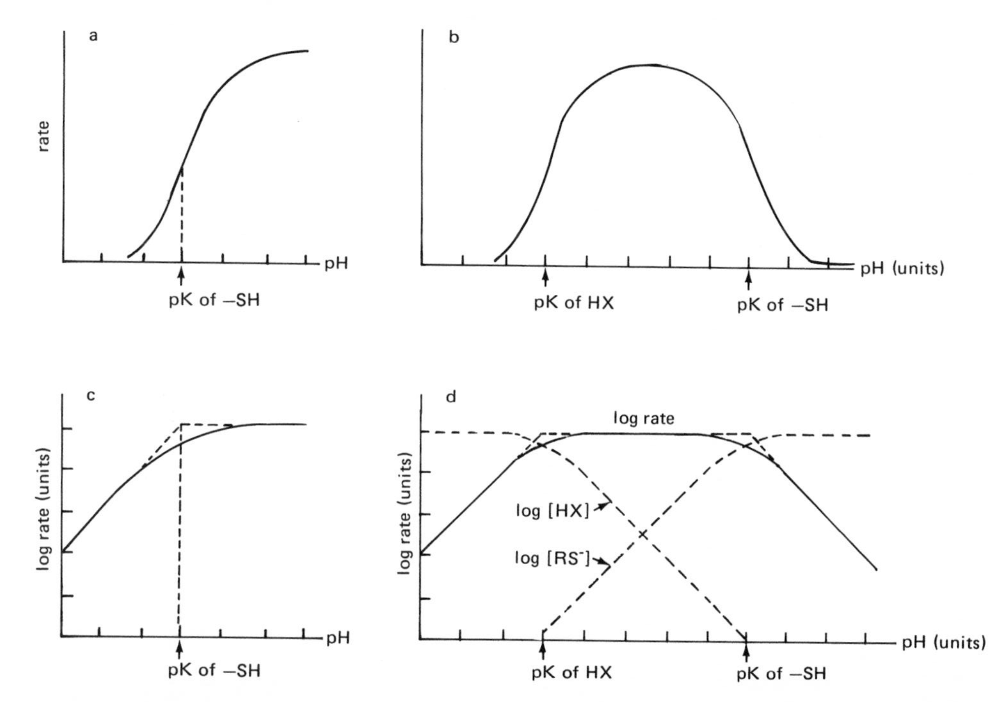

Fig. 2. Possible pH dependences of reactions that involve RS^- ions. a) Curve for rate $\propto [RS^-]$; b) curve for rate $\propto [RS^-] \times [HX]$; c, d) as a, b, but logarithms of rates are shown.

centration of RS^- is rising, whereas [HNCO] is almost constant; above the pK of HNCO and below that of RSH there is a plateau; because each unit that the pH rises increases $[RS^-]$ tenfold and diminishes [HNCO] by the same amount; above the pK of RSH further raising of the pH continues to diminish [HNCO], but without a compensating rise of $[RS^-]$. The same principles apply to the amino group.

5. Reaction with Carbonyl Compounds

Thiols readily combine with the reactive carbonyl groups of aldehydes and of some ketoacids. The combination takes place at neutral pH and room temperature; in more forcing conditions ketones react analogously (Schubert, 1936a; Ratner and Clarke, 1937; Calvin, 1954). As a result of the reaction of thiols with aldehydes and ketones hemimercaptals and hemimercaptols (thiohemiacetals and thiohemiketals) respectively are formed:

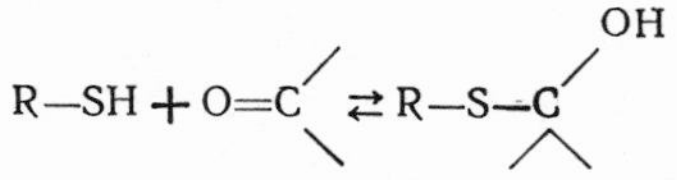

$$R{-}SH + O{=}C\langle \rightleftarrows R{-}S{-}C(OH)\langle$$

Combination with these of a second molecule of thiol leads to the formation of a full mercaptal (or mercaptol):

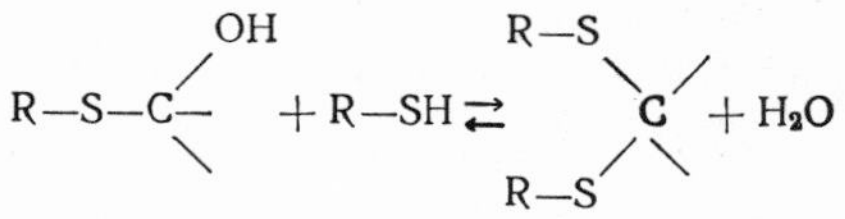

$$R{-}S{-}C(OH)\langle + R{-}SH \rightleftarrows (R{-}S)_2C\langle + H_2O$$

These reactions are analogous to the combination of alcohols with aldehydes with the formation of hemiacetals and acetals. Thiols, however, have much greater reactivity toward the carbonyl group; the products of reaction with thiols are more stable than the corresponding hemiacetals and acetals (Campaigne, 1961; Jencks, 1964). On interaction with aldehydes of 1,2-dithiols (e.g., BAL) cyclic mercaptals are formed:

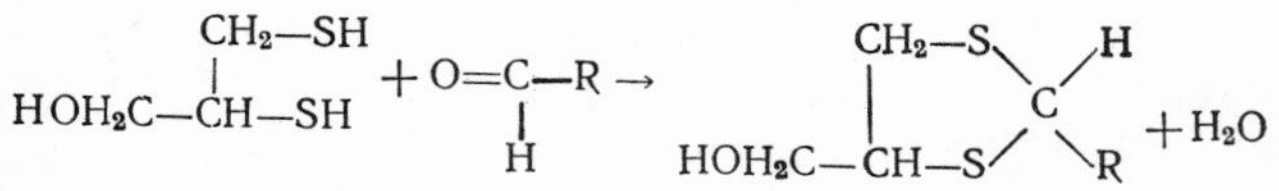

$$HOH_2C{-}CH(SH){-}CH_2{-}SH + O{=}CH{-}R \rightarrow HOH_2C{-}\underset{|}{CH}{-}S{-}C(H)(R){-}S{-}CH_2\ (\text{ring}) + H_2O$$

Lienhard and Jencks (1966) and Barnett and Jencks (1969) studied the equilibria and kinetics of reactions of thiols with acetaldehyde. They found that weakly acidic thiols (with which protein chemistry is mainly concerned, as they include most alkyl mercaptans) react mainly by a specific base catalyzed pathway, although a pathway showing general acid catalysis becomes significant at low pH.

The main pathway thus appears to be

$$R-S-H \rightleftharpoons R-S^- + H^+$$

$$R-S^- + \;>C{=}O + H_2O \rightleftharpoons R-S-\overset{|}{\underset{|}{C}}-OH + OH^-$$

and the attack of the mercaptide ion on the carbonyl compound was the rate determining step. For thiols of pK above 8, the rate constant of the reaction with respect to mercaptide ion was independent of the pK of the thiol.

The equilibrium constant for the reaction $R-SH + CH_3-CHO \rightleftharpoons R-S-CHOH-CH_3$ changes less than fivefold for a variation of 10^7-fold in the dissociation constant of R−SH (from ethanethiol of pK 10.25 to thioacetic acid of pK 3.2); evidently the affinities of RS^- for H^+ and for aldehyde plus proton vary exactly in step.

The ability of SH groups to react readily with carbonyl compounds explains the role of these groups in the mechanism of action of a number of enzymes whose substrates contain the carbonyl group (see Chapter 6, Section 1). Since the mercaptide ion reacts with only the unhydrated carbonyl group, we may postulate that some enzymes also catalyse the dehydration of aldehyde substrates. There are reports in the literature of inhibition by low-molecular thiols of the activity of enzymes whose substrates are aldehydes; this inhibition is presumably caused by combination of the SH groups of the added thiol with the substrate (Rothschild and Barron, 1954; De Barreiro, 1969).

The combination of thiols with the double bond of the carbonyl group is in some ways analogous with their combination with alkenes (see below). But it differs in that the reaction of thiols with the carbonyl group is readily reversible. The products of this reaction (hemimercaptals) dissociate in aqueous solution and hence often react with nitroprusside to give the color characteristic of free thiols; they are also oxidized by iodine to form a disulfide and the original carbonyl compound (Schubert, 1936a, 1951).

An interesting property of hemimercaptals is their ability to react readily with amino compounds:

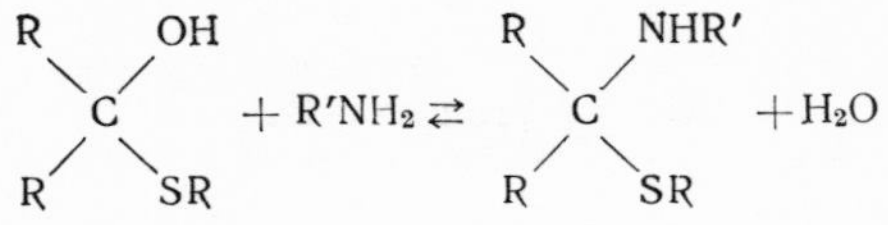

$$\underset{R}{\overset{R}{>}}C\underset{SR}{\overset{OH}{<}} + R'NH_2 \rightleftarrows \underset{R}{\overset{R}{>}}C\underset{SR}{\overset{NHR'}{<}} + H_2O$$

When 1-amino-2-thiols (for example cysteine) react with aldehydes, thiazolidines are formed (Schubert, 1935; 1936; Ratner and Clarke, 1937; Bloch and Clarke, 1938):

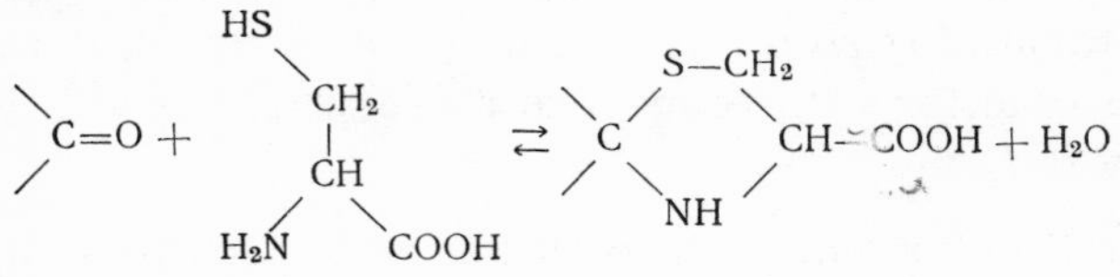

$$>C{=}O + HS{-}CH_2{-}CH(NH_2){-}COOH \rightleftarrows >C\underset{NH}{\overset{S-CH_2}{\big\langle}}CH{-}COOH + H_2O$$

Thiazolidines are easily hydrolyzed, and also react with Ag^+ and Hg^{2+} ions; they are oxidized markedly slower than are free thiols (Basford and Huennekens, 1955).

An interesting reaction is that of cysteine and other 1-amino-2-thiols with pyridoxal phosphate, the phosthetic group of many enzymes of amino acid metabolism and of phosphorylase. This interaction also leads to the formation of the thiazolidine ring (Heyl et al., 1948; Matsuo, 1957; Abbott and Martell, 1970). There has been disagreement in the literature on whether the formation of the thiazolidine goes through a hemimercaptal or a Schiff base (Buell and Hansen, 1960; Bergel and Harrap, 1961). From kinetic studies on the reaction of pyridoxal phosphate with L-cysteine and its analogs, Mackay (1962) concluded that in the first step of the reaction both hemimercaptal and Schiff base are formed; only the latter, however, cyclizes according to his data to form a thiazolidine:

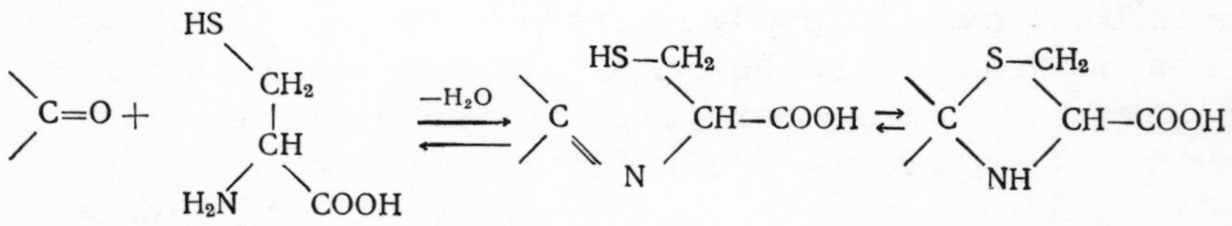

$$>C{=}O + HS{-}CH_2{-}CH(NH_2){-}COOH \underset{}{\overset{-H_2O}{\rightleftarrows}} >C{=}N{-}CH(CH_2{-}SH){-}COOH \rightleftarrows >C\underset{NH}{\overset{S-CH_2}{\big\langle}}CH{-}COOH$$

6. Alkylation and Arylation of SH Groups

a. Addition to an Activated Double Bond

Thiols have the ability to combine with polarized or easily polarizable double bonds in compounds such as acrylonitrile, acrylamide, ethyl acrylate, acrolein, maleic acid, N-ethyl maleimide, vinyl sulfones, and others. These reactions are irreversible, and usually go rapidly in alkaline media, which indicates that the mercaptide ions take part in them. The triplet bond of alk-3-ynoyl-CoA esters adds similarly to the SH group of acetoacetyl-CoA thiolase (Bloxham et al., 1973).

The interaction of the SH group of a number of thiols (thioglycolic acid, cysteine, glutathione, penicillamine, β-mercaptoisoleucine) with acrylonitrile was studied by Friedman et al. (1965). They found that the rate of reaction increased rapidly with increasing pH, and when the pH exceeded the pK it began to approach an asymptotic value. On the basis of the kinetic data the authors postulated the following mechanism for the cyanoethylation of SH groups:

$$RSH + H_2O \rightleftarrows RS^- + H_3O^+$$
$$RS^- + CH_2{=}CH{-}CN \rightleftarrows RS{-}CH_2{-}\bar{C}H{-}CN$$
$$RS{-}CH_2{-}\bar{C}H{-}CN + H_3O^+ \rightleftarrows RS{-}CH_2{-}CH_2{-}CN + H_2O$$

Friedman et al. established that SH groups that are attached to a tertiary carbon atom (as in penicillamine and β-mercaptoisoleucine) are markedly less reactive than those on a primary carbon atom. They also compared the reaction rates of acrylonitrile with β-alanine ($NH_2{-}CH_2{-}CH_2{-}COOH$) and β-mercaptopropionic acid ($SH{-}CH_2{-}CH_2{-}COOH$); the SH and NH_2 groups of these compounds have the same basicity (identical pK values) and the same steric environment; hence comparison of the reaction rates of these compounds permitted evaluation of the relative nucleophilicities of the SH and NH_2 groups. It was found that β-mercaptopropionic acid reacted with acrylonitrile 280 times faster than did β-alanine.

Weil and Seibles (1961; Seibles and Weil, 1967; see also Pummer and Hirs, 1964) proposed the use of acrylonitrile for blocking the SH groups formed on reduction of the disulfide bonds of proteins. It turned out, however, that acrylonitrile reacts not only with SH groups, but also with α- and ε-amino groups of proteins, particularly at pH 8-9 (Levin, 1965; Kalan et al., 1965). The reaction with amino groups can be minimized by decreasing the pH, lowering the excess of reagent and the time of incubation with the protein. Cavins and Friedman (1968) worked out conditions in which the SH groups of proteins can be selectively blocked by acrylonitrile in the presence of β-mercaptoethanol. These conditions are: pH 7.0, temperature 30°C, and molecular ratio of reagent to the total SH groups of the protein and of the β-mercaptoethanol of 1:1. Under these conditions the reaction of the SH groups of a protein with methyl acrylate is complete within 5 min, with acrylonitrile in 20 min, with acrylamide in 100 min; other amino acids, including lysine, are not affected.

N-ethylmaleimide (NEM) has been widely used for studying the SH groups of proteins:

$$RS^- + H^+ + \begin{array}{c} HC(=O)\text{–} \\ HC{=}CH \text{ ring: } HC{=}HC,\ C{=}O,\ N\text{–}C_2H_5 \end{array} \rightarrow RS\text{–}CH\text{–}CH_2 \text{ (succinimide ring with } C{=}O,\ N\text{–}C_2H_5)$$

The rate of reaction of NEM with cysteine and other thiols rises rapidly with increasing pH (Lee and Samuels, 1964; Schneider and Wenck, 1969) (see Table 5). Brubacher and Glick (1974) have tabulated its rate of reaction with several enzymes.

Under mildly alkaline conditions the adduct derived from NEM and cysteine (compound I) readily undergoes an intramolecular

TABLE 5. The Influence of pH on the Rate of Reaction of N-ethylmaleimide with Thiols (Schneider and Wenck, 1969)

Thiol	pK of the SH group	Second-order rate constant for the reaction at 25°C, $M^{-1} \cdot sec^{-1}$	
		pH 4.0	pH 5.0
Cysteine	8.5	1.73	15.5
N-acetylcysteine	9.7	0.41	2.44
4-Mercaptomethylimidazole	9.5	9.40	94.5

transamidation to form a thiazine derivative (compound II) (Smyth et al., 1960, 1964):

$$\underset{\text{I}}{HO_2C-\underset{NH_2}{\underset{|}{CH}}-CH_2-S-\overset{}{\underset{OC\diagdown N(C_2H_5)\diagup CO}{CH-CH_2}}} \longrightarrow \underset{\text{II}}{HO_2C-\underset{HN-C(=O)}{CH}\diagup CH_2-S\diagdown CH-CH_2-CO-NH-C_2H_5}$$

When compound I is treated with 1 M NaOH at room temperature, the solution gives a positive nitroprusside test for SH groups. Compound I is rather stable toward acid; only on prolonged acid hydrolysis (for 72 h at 110°C in 6 M HCl) does it form ethylamine and S-(1,2-dicarboxyethyl)-L-cysteine with an 88% yield.

According to Smyth et al. (1960, 1964) NEM reacts not only with SH groups, but also with the imidazole group of histidine and the α-amino groups of amino acids and peptides. The ratio of ethylamine to S-(1,2-dicarboxyethyl)cysteine in an acid hydrolysate of a protein treated with NEM can be used as a measure of the specificity of the reaction. This ratio would be equal to 1 if NEM reacted only with SH groups (provided that the excess NEM was removed from the protein before hydrolysis), and more than 1 if the NEM reacted also with other groups. Brewer and Riehm (1967) studied the reaction of NEM with the functional groups of ribonuclease and lysozyme; they found that NEM at a concentration of 0.1 M and pH 8.0 alkylated up to 80% of the lysine residues and some of the histidine residues in these proteins; ethylamine and N-(1,2-dicarboxylethyl)-L-lysine were found among the products of acid hydrolysis of the treated proteins. It is important to note that, according to the data of Brewer and Riehm, amino groups would be alkylated only to a minute extent, if at all, under the conditions in which NEM is usually used for modifying SH groups (a concentration of 1 mM and pH 6–7). Thus the choice of appropriate conditions for the reaction, in particular carrying it out without much excess of NEM and at neutral pH, can significantly raise the specificity of the reagent. Even if these precautions are observed however, one must not *a priori* exclude the possibility that NEM may react with some amino group that possesses increased reactivity as a result of its particular environment in the protein.

It is known, for example, that in carboxyhemoglobin NEM readily alkylates not only the SH group of the β-chain, but also the α-amino group of the α-chain (Guidotti and Konigsberg, 1964). Another derivative of maleimide, N-(N'-acetyl-4-[^{35}S] sulfamoylphenyl)-maleimide, selectively reacts at pH 7.3 with the active ε-amino group of glutamate dehydrogenase, in spite of the presence of free SH groups in the protein (Holbrook and Jeckel, 1969).

Several authors have synthesized colored derivatives of NEM in order to use them to label the SH groups of proteins:

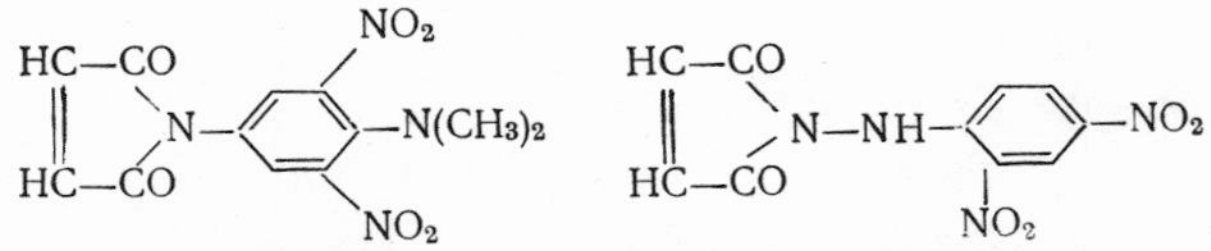

N-(4-dimethylamino-3,5-dinitrophenyl)-maleimide (Witter and Tuppy, 1960)

N-2,4-dinitroanilino-maleimide (Clark-Walker and Robinson, 1961)

N-(4-dimethylamino-3,5-dinitrophenyl)-maleimide was successfully used for labeling SH groups in glyceraldehyde-3-phosphate dehydrogenase (Gold and Segal, 1964), taka-amylase A (Narita and Akao, 1965), and stem bromelain (Murachi and Takahashi, 1970). The peptides that contain an SH group labeled with this reagent are colored yellow, and this helps in identifying and purifying them. Richards et al. (1966) synthesized the polymer polyvinyl (N-phenylenmaleimide) which they proposed for use for irreversible binding of thiols and removing them from solution. N,N'-Hexamethylene-bis-maleimide was used as a bifunctional reagent which selectively reacted with the SH groups of serum albumin and united its molecules into dimers (Zahn and Lumper, 1968). Another substance used as a bifunctional reagent was bis-(N-maleimidemethyl)ether (Hasselbach and Taugner, 1970; Freedman and Hardman, 1971; Simon et al., 1971):

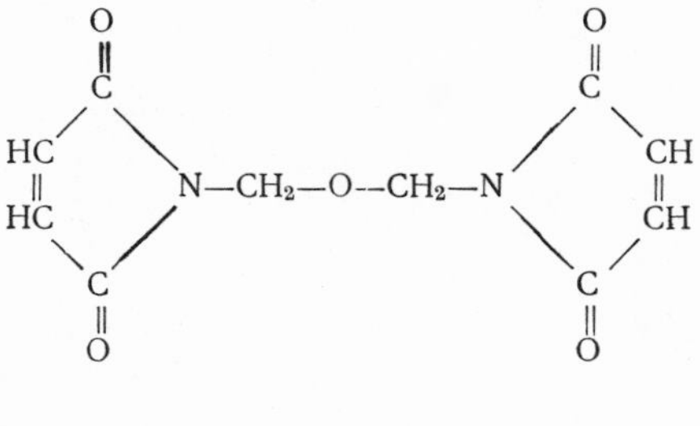

A compound that may prove to be a useful reagent is N-[p-(2-benzoxazolyl)phenyl]maleimide:

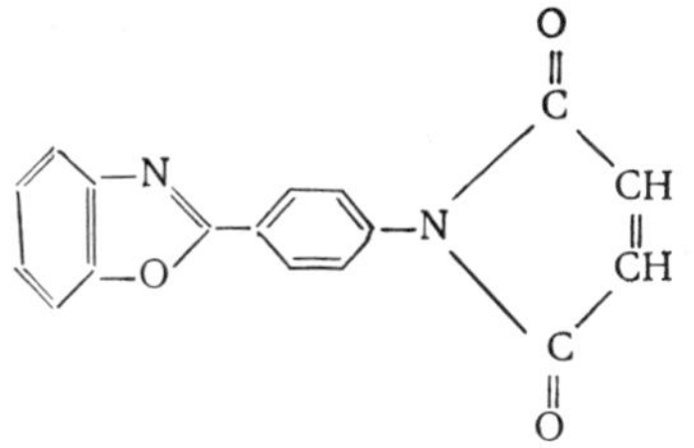

This compound has no appreciable fluorescence, but the product of its combination with the SH group of cysteine ethyl ester is strongly fluorescent, with emission maxima at 352 and 365 nm (Kanaoka et al., 1968). A new sulfhydryl reagent with long-lived fluorescence, N-(3-pyrene)-maleimide, was recently synthesized by Weltman et al. (1973).

Rohrbach et al. (1973) have found a useful reagent in 4,4'-bis-dimethylaminodiphenylcarbinol. Below pH 6 it exists largely as a resonance-stabilized carbonium (immonium) ion:

$$Me_2N-C_6H_4-CHOH-C_6H_4-NMe_2 \rightleftharpoons Me_2N^+=C_6H_4=CH-C_6H_4-NMe_2 + OH^-$$

It has intense absorption at 612 nm, the value depending on the nature of the medium, partly because this affects the position of the above equilibrium. In 4 M guanidine hydrochloride the extinction is 70,800 $M^{-1} \cdot cm^{-1}$, and the disappearance of this blue color as the compound reacts with sulfhydryl groups can be used for their assay:

$$Me_2N^+=C_6H_4=CH-C_6H_4-NMe_2 + R-SH \rightarrow Me_2N-C_6H_4-CH(SR)-C_6H_4-NMe_2$$

b. Reactions with Quinones

Thiols add to the double bond in the ring of a quinone forming the thioether of the hydroquinone or quinone. Snell and Weissberger (1939) studied the reaction between thioglycolate and p-benzoquinone; they found that this reaction leads to the formation of a substituted hydroquinone when excess thiol is used, but gives the oxidized product in the presence of excess benzoquinone. They consider that the first product of the reaction is the substituted hydroquinone, which is then oxidized by excess quinone; other authors have agreed (Schubert, 1947; Fieser and Turner, 1947). Nickerson et al (1963), however, came to a different conclusion when they studied the reaction of 2-substituted 1,4-naphthoquinones with glutathione. They showed that the SH group of glutathione combines with the carbon atom in position 2 or 3 of the quinone ring. The reaction went fastest when chlorine was the substituent on the carbon atom with which the thiol group combines, and more slowly when it was hydrogen; it did not go at all with CH_3 or OH groups as substituents. Thus the rate of formation and yield of thioether depended on the electron-withdrawing properties of the substituent on the carbon atom. From these findings it follows that the combination of glutathione is a nucleophilic substitution. According to Nickerson et al. the thioether of the quinone is formed without the preliminary accumulation of the thioether of the hydroquinone; the atom of hydrogen eliminated on nucleophilic attack of the RS^- ion on the carbon of the ring is abstracted by a second molecule of quinone:

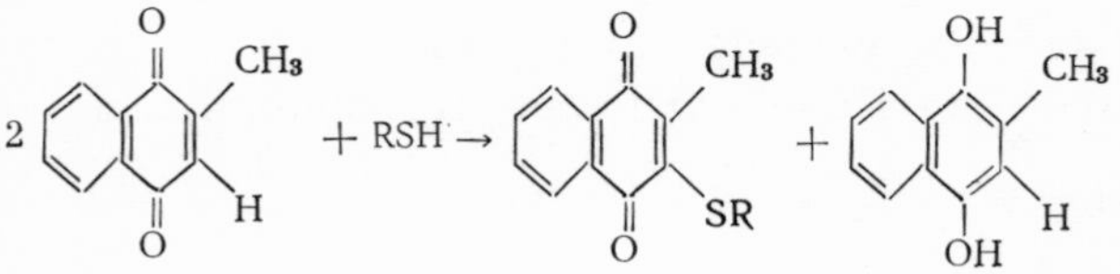

O, CH_3, 2, + RSH →, O, CH_3, +, OH, CH_3, H, O, SR, O, H, OH

The hydroquinone is then reoxidized by the oxygen of the air with formation of the original quinone and of hydrogen peroxide. Either mechanism is likely to start by addition of the mercaptide ion to the quinone as follows:

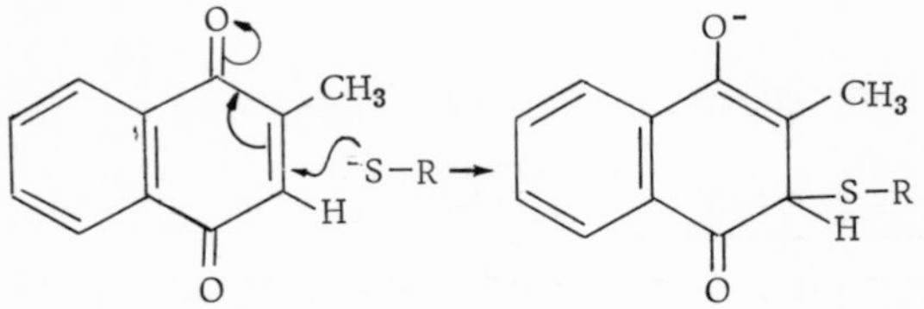

O, CH_3, ^-S-R →, H, O, O^-, CH_3, S−R, H, O

This first product can be protonated to form a tautomer of the hydroquinone thioether, but the findings of Nickerson et al. that the hydroquinone thioether does not reduce the original quinone under the reaction conditions suggest that this tautomer may be oxidized by the quinone faster than it can tautomerize to its more stable form.

Nakai and Hase (1968) found that when 2-methyl-1,4-naphthoquinone reacts with cysteine the first product is a thioether with an absorption maximum at 420–430 nm; formation of this requires half a mole of O_2 per mole of reacting quinone. In contrast with the thioethers formed by reaction with glutathione and thioglycolate, the product of the reaction of cysteine with 2-methyl-1,4-naphthoquinone is unstable and undergoes intramolecular dehydration with formation of a thiazine ring; this reaction was earlier described by Kuhn and Hammer (1951):

O, O, $-CH_3$ —Cysteine, $^1/_2O_2$→ O, O, $-CH_3$, $-S-CH_2$, CH, H_2N, COOH
↓ $-H_2O$
OH, CH_3, S, N, CH_2, C—COOH ⇆ O, CH_3, S, N, CH_2, CH—COOH

The compound formed is further converted into an insoluble polymer. Nakai and Hase also described the reaction of 2-methyl-1,4-naphthoquinone with the SH group of papain and of serum albumin.

One should bear in mind that quinones not only can alkylate thiols, but also can oxidize them to disulfides. Quinones also react with the amino groups of amino acids and proteins, but more slowly than with SH groups.

c. Reactions with Haloacids and Their Amides

The addition of SH groups to the double bond of N-ethylmaleimid, acrylonitrile, and other ethylenic compounds leads to alkylation of the SH groups. Compounds containing alkyl halide groups (iodoacetate, iodoacetamide, α-iodopropionate, bromoacetate, β-bromoethylamine, chloroacetate, chloroacetophenone, etc.) can also be used to alkylate SH groups.

The reaction of SH groups with halogen compounds amount to a nucleophilic attack of the RS^- ion on the methylene carbon with liberation of a halide ion, e.g.,

$$RS^- + \underset{\underset{I^{\delta-}}{|}}{\overset{\delta+}{CH_2}}{-}COO^- \rightarrow RS{-}CH_2{-}COO^- + I^-$$

This reaction is rapid only when a carbonyl group is adjacent to the methylene group, but the carbonyl group can belong to groupings as diverse as ketones and carboxylate ions. There is presumably conjugation of the entering and leaving nucleophiles with the carbonyl group in the transition state.

According to Barron (1951) the half-times of reaction of iodoacetate, bromoacetate, and chloroacetate with the SH group of cysteine at pH 7.4 and 28°C are 5.5, 7.7, and 117.5 min respectively; fluoroacetate does not alkylate SH groups at all. Iodoacetamide reacts faster than iodoacetate with the SH group of cysteine; with N-acetyl-cysteine the rate constants in aqueous solution at 25°C expressed with respect to the mercaptide ion are 33.0 and 4.35 $M^{-1} \cdot sec^{-1}$ respectively (MacQuarrie and Bernhard, 1971).

The rate of reaction of SH groups with iodoacetate and iodoacetamide increases with rising pH. Rapkine (1933), for example, found the relative rates for reaction of the SH group of cysteine with iodoacetate at pH 5.6, 7.02, and 8.36, to be 0.14, 1, and 2.1. According to Smythe (1936) the rate of reaction of β-mercaptoethylamine with iodoacetate at pH 7.1 is about 10 times higher than at pH 6.1. These data indicate that the reacting species is the mercaptide anion, RS^-, and not the undissociated SH group. Cecil and McPhee (1959) calculated that the rates of reaction of a number of thiols with iodoacetate at pH 7.0 were directly proportional to the degree of ionization of the SH groups. This means that the more "acidic" SH groups (with lower pK values) are alkylated faster than groups with high pK values, and that basic strength has little effect on the reactivity of a mercaptide ion. The factors that lower the pK of an SH group (the proximity of an amino group with its positive charge, for example) increase the rate of alkylation, and conversely the factors that raise the pK (such as proximity of a carboxyl group, or lengthening of the chain between SH and amino groups) diminish the reaction rate (Hagen, 1956).

Lindley (1960, 1962) and Gerwin (1967) studied the kinetics of alkylation of thiols with chloroacetamide and with chloroacetate.

Chloroacetamide reacted with glutathione at a rate proportional to the amount of RS^-, and with the SH group of streptococcal protease the rate followed a similar curve against pH. This did not exactly fit a theoretical dissociation curve, but Gerwin explained this by making the reasonable assumption that the pK of the thiol group changed with pH as the charge on the rest of the molecule (especially nearby amino or phenolic groups) changed. Thus the rate was again proportional to the concentration of RS^-. The rate of reaction of the protein was 50-100 times that of glutathione.

There is yet another important factor that influences the rate of reaction of haloacids with the SH groups of proteins, namely the electrostatic interaction of the ionized carboxyl group of the reagent with the charged groups of the protein (or prosthetic group) that are close to the SH group under attack. As an example we shall cite the data of Evans and Rabin (1968) on the kinetics of the reaction of iodoacetamide and iodoacetate with the SH group in the active center of liver alcohol dehydrogenase. The rate of reaction of this SH group with the neutral molecule of iodoacetamide is independent of pH within the range of 6 to 9.6; the rate at which this group reacts with iodoacetate at pH 6-7 is much higher than with iodoacetamide, but diminishes sharply as the pH is raised from 7 to 9. The authors explain this fact by the presence close to the SH group of a positively charged group of pK 7.9, which attracts the iodoacetate anion and thus accelerates the alkylation of the SH group.

According to Gerwin (1967), the curve of the rate of alkylation of the SH group of streptococcal protease against pH, which is sigmoid for chloroacetamide following the concentration of the RS^- species, has a bell-shaped form for chloroacetate, with a broad maximum over a plateau from about pH 6 to about pH 9. This is not observed for the alkylation of glutathione, so it appears to be connected with the particular ionic environment of the SH group. A possible explanation follows. The plateau and the fall in rate as the pH is raised from 8 to 10, i.e., over the range where the reactive RS^- form becomes predominant, suggest that some group largely unprotonated above pH 5 needs to be protonated for rapid reaction (see Chapter 1, Section 4). Thus over the pH range 6-8 the rate may be constant because increase in the reactive RS^- form (10-fold for each pH unit) is eactly balanced by decrease in protonation of the other group. The fall above pH 8 is explained by the fact that as the pK of the SH group is approached the RS^- form no longer increases so rapidly, and hence there is less to compensate for the decrease in protonation of the other group. Such another group could be a carboxyl whose ionized form prevented binding of the chloroacetate ion. It should be noted that chloroacetate reacts with the SH group of the protease much faster than chloroacetamide in acid solution; the reverse is true in alkaline solution. Gerwin (1967) explains this by an alteration in the charge of the microenvironment of the SH group from positive at low pH to negative at high pH.

The rate of alkylation of the catalytically active SH group of muscle glyceraldehyde-3-phosphate dehydrogenase depends on the presence of bound coenzyme; combination of NAD^+ with the protein greatly increases the rate of reaction of the SH group with iodoacetate but slows the reaction with iodoacetamide (Boross et al., 1969; Fenselau, 1970).

Haloacids and their amides react not only with SH groups, but also with the thioether group of methionine to form sulfonium salts (Toennies and Kolb, 1945; Gundlach et al., 1959a,b):

$$ICH_2COO^- + \underset{\substack{| \\ [CH_2]_2 \\ | \\ CHNH_2 \\ | \\ COOH}}{S}-CH_3 \rightarrow CH_3-\underset{\substack{| \\ [CH_2]_2 \\ | \\ CHNH_2 \\ | \\ COOH}}{\overset{+}{S}}-CH_2COO^- + I^-$$

The reaction with thioethers, unlike that of SH groups, depends little on pH and can be carried out in acid medium (down to pH 1.7–2) (Gundlach et al., 1959a; Vithayathil and Richards, 1960; Stark and Stein, 1964; Link and Stark, 1968). This is not surprising since the reactive, neutral form of the thioether exists over a wide pH range; an acid pH may help to unfold a protein and expose methionine residues. According to Stark and Stein (1964), the alkylation of the thioether grouping of methionine by iodoacetate proceeds 2–3 times faster than with iodoacetamide. The rate constants for the alkylation by iodoacetate of free methionine, N-acetylmethionine, and the methionine residues of ribonuclease A in 8 M urea were 7.8, 12.4, and 15.5 $M^{-1} \cdot h^{-1}$ respectively at 40°C and pH 3.5.

The carboxymethylsulfonium salt of methionine is completely destroyed on heating in 6 M HCl (110°C, 20 h), with the formation of S-carboxymethylcysteine (about 50%), methionine (about 20%), homoserine (about 5%) and its lactone (about 6%), and homocystine (about 2%) (Gundlach et al., 1959a). Naider and Bohak (1972) found that the sulfonium salt reacts with thiols to regenerate an intact methionine residue:

$$\begin{array}{c} R-\underset{\underset{O}{\|}}{C}-CH_2-\overset{+}{S}-CH_3 \\ \quad\quad\quad | \\ \quad\quad\quad [CH_2]_2 \\ \quad\quad\quad | \\ -NH-CH-CO- \end{array} \xrightarrow{R'-S^-} \begin{array}{c} S-CH_3 \\ | \\ [CH_2]_2 \\ | \\ -NH-CH-CO- \end{array}$$

The high yield obtained shows that of the three carbon atoms that are linked to sulfur the one activated by a neighboring carbonyl group is far the most electrophilic with respect to the attacking thiolate anion. The thiolysis is rapid at 25°C and pH 7-9 in a thiol concentration of 5-100 mM. The products of alkylation of other functional groups of proteins are not expected to be susceptible to thiolysis under these conditions.

Haloacids also react with the unprotonated forms of imidazole, amino, and phenolic groups of proteins (Michaelis and Schubert, 1934; Schubert, 1936b, 1951; Korman and Clarke, 1956; Barnard and Stein, 1959; Gurd, 1967). According to Cotner and Clagett (1973) the extent and rate of reaction of tyrosine residues in proteins with iodoacetamide increases sharply as the pH is raised from 7 to 10. The O-carboxamidomethyltyrosine formed is completely hydrolyzed with regeneration of tyrosine residues on incubation for 48 h in 6 M HCl at 105°C. Takahashi et al. (1967) described a reaction of iodoacetate with the γ-carboxyl group of a glutamic acid residue in the active site of ribonuclease T_1; this reaction occurs at pH 5.5 and leads to the formation of the γ-carboxymethyl ester of the glutamic acid:

$$I\,CH_2COO^- + \begin{array}{c} CH_2-COO^- \\ | \\ CH_2 \\ | \\ -NH-CH-CO- \end{array} \rightarrow \begin{array}{c} CH_2COOCH_2COO^- \\ | \\ CH_2 \\ | \\ -NH-CH-CO- \end{array} + I^-$$

The reactions of haloacids with the groups just listed needs harsher conditions, e.g., a higher concentration of reagent or a higher temperature, than the reaction with SH groups. Thus Goddard and Michaelis (1935) noted that if the reaction is carried out under mild enough conditions (at room temperature and a pH of about 9) it is mainly the SH groups of a protein that undergo alkylation. Anfinsen and colleagues (Anfinsen, 1958; Sela et al., 1959) confirmed these findings and showed that iodoacetate does not alkylate the amino groups of polylysine or the imidazole ring of histidine to any marked degree if it is incubated with them for two hours at room temperature and pH 8.5. When they incubated reduced pancreatic ribonuclease with iodo[^{14}C]acetate under these conditions, only its SH groups were alkylated. On the other hand Stein and Moore (1962) showed that iodoacetate (but not iodoacetamide!) at pH 5.5-6.0* and 40°C alkylated the imidazole group of one of the histidine residues in the molecule of native pancreatic ribonuclease, and that this reaction was accompanied by the loss of enzymic activity; on carrying out the reaction at pH 8.5-10.0 and 40°C it was mainly the ε-amino groups of lysine that underwent alkylation. In addition to histidine and lysine residues, one residue of methionine is also accessible to slow carboxymethylation in native ribonuclease (Fruchter and Crestfield, 1965; Goren and Barnard, 1970).

Weinryb (1968) observed alkylation of methionine residues when horse radish peroxidase was treated with iodoacetate at acid pH, and of histidine and possibly lysine residues at pH 7 and above. Similarly Wallis (1972) carboxymethylated the four methionine residues of ox growth hormone, and Colman (1968; see also Edwards et al., 1974) one methionine residue of isocitrate dehydrogenase. Thus in studies of the interaction of haloacids and their amides with proteins it is necessary to consider the possibility that not only SH groups, but also other functional groups, may be alkylated.† Appropriate choice of reaction conditions, in particular the use of only a small excess of reagent and a relatively short

---

*The alkylation of the imidazole of histidine usually goes faster at more alkaline pH. The unusual pH optimum for the alkylation of imidazole in ribonuclease is explained by the proximity of two imidazole groups in the tertiary structure of the protein. One of these needs to be protonated to attract the negatively charged iodoacetate, orientate it, and thus accelerate the alkylation of the other (unprotonated)group (Stein, 1964).

†Lee and Westheimer (1966) observed an unusual enzyme inactivation by iodoacetate in a study of acetoacetate decarboxylase. They found that inactivation occurred only in the light and in the presence of air. The first step is, in their opinion, the photochemical dissociation of iodoacetate into a free radical and an iodine atom:

$$ICH_2COO^- \xrightarrow{h\nu} I\cdot + \cdot CH_2COO^-$$

The iodine then reacts with the SH group of the enzyme:

$$E{-}SH + I\cdot \rightarrow ES\cdot + H^+ + I^-$$

The E–S• radical that is formed is then oxidized by oxygen from the air to form the sulfonate group of cysteic acid.

incubation, allows alkylation to be limited largely or exclusively to SH groups, since they usually react with alkylating reagents much faster than any other groups in a protein molecule. Among the haloacids most specific for SH groups are, apparently, α-iodopropionic acid and its amide. The latter alkylates the SH group of cysteine 40 times more slowly, but amino, imidazole, and thioether, groups 80 times more slowly than iodoacetamide (i.e., its specificity is increased twofold) (Wallenfels and Eisele, 1968).

Bradbury and Smyth (1973) showed that S-carboxymethylcysteine can cyclize if it is N-terminal, since the side-chain carboxyl group condenses with the amino group to form a fairly stable 6-membered ring. This complication is avoided if 3-bromopropionate is used. Although the activating effect of an adjacent carboxyl group is then largely lost, it reacts with SH groups at an adequate rate; cysteine exhibits a half-life of 10 min at pH 8 and 30°C in 0.2 M 3-bromopropionate.

d. Reactions with Ethylene Oxide and Ethylene Imine

Danehy and Noel (1960) studied the kinetics of the reactions of twelve different thiols, including cysteine, with ethylene oxide. They found that the rate constants increased with pH over a range characteristic for each thiol. With the single exception of thiophenol the rate was proportional to the concentration of mercaptide ion, so they concluded that this was the reactive species, as follows:

$$R{-}S{-}H \rightleftharpoons R{-}S^- + H^+$$

$$R{-}S^- + \underset{\diagdown O \diagup}{CH_2{-}CH_2} \rightarrow R{-}S{-}CH_2{-}CH_2{-}O^-$$

$$R{-}S{-}CH_2{-}CH_2{-}O^- + H^+ \rightarrow R{-}S{-}CH_2{-}CH_2{-}OH$$

The values of the second-order rate constants for the various $R{-}S^-$ ions afforded a measure of the relative nucleophilicity of these ions with respect to ethylene oxide. In general, the higher the pK of the thiol, the greater the intrinsic reactivity of the corresponding mercaptide ion.

The reaction of the SH groups of peptides and proteins with ethylene imine (Raftery and Cole, 1963, 1966) has attracted the attention of investigators in recent years. This reaction leads to the selective blocking of SH groups and tothe conversion of cysteine residues into residues of S-(β-aminoethyl)-cysteine:

$$\begin{matrix}-CO \\ \quad\diagdown \\ \quad CH-CH_2-S^- \\ \quad\diagup \\ -NH\end{matrix} + H^+ + \underset{NH}{CH_2-CH_2} \rightarrow \begin{matrix}-CO \\ \quad\diagdown \\ \quad CH-CH_2-S-CH_2-CH_2 \\ \quad\diagup \qquad\qquad\qquad\quad | \\ -NH \qquad\qquad\qquad\quad NH_2\end{matrix}$$

The peptide bonds formed by these residues (which resemble the side chain of lysine) can be split by trypsin. Hence the aminoethylation of the SH groups of proteins opens up further points at which polypeptide chains can be specifically broken (Levin, 1965; Guest and Yanofsky, 1966; Slobin and Singer, 1968). Aminoethylation of SH groups can also be achieved with β-bromoethylamine (Lindley, 1956), but it reacts much more slowly than ethylene imine. One should bear in mind that ethylene imine can modify methionine residues in proteins under some conditions (Schroeder et al., 1967).

Cysteine can also be converted into a basic amino acid by trimethylaminoethylation with (2-bromoethyl)trimethylammonium bromide at pH 9.1; the product of the reaction, $Me_3N^+-CH_2-CH_2-S-CH_2-CH(NH_2)-COOH$, is called 4-thialaminine (Itano and Robinson, 1972). Thialaminine peptide bonds are resistant to trypsin.

e. Reactions with O-Methylisourea and Methyl-p nitrobenzenesulfonate

O-Methylisourea is usually used for guanidination of the amino groups of proteins:

$$H_2N-\overset{\displaystyle O-CH_3}{\overset{|}{C}}=NH_2^+ + R-NH_2 \rightarrow H_2N-\overset{\displaystyle NH-R}{\overset{|}{C}}=NH_2^+ + CH_3OH$$

Banks and Shafer (1970) investigated the reaction of O-methylisourea with cysteine in D_2O at 28°C by NMR spectroscopy; they observed that methylation of the SH group occurred in addition to guanidination of the amino group. Study of the pH dependence of the rate of S-methylation showed that the reaction apparently took place by nucleophilic attack of the mercaptide ion on the methyl group of the O-methylisouronium cation:

$$H_2N-\overset{\displaystyle O-CH_3}{\overset{|}{C}}=NH_2^+ + RS^- \rightarrow RSCH_3 + H_2N-\overset{\displaystyle O}{\overset{\|}{C}}-NH_2$$

The second-order rate constant for the reaction of the RS^--form of β-mercaptoethanol with the O-methylisouronium cation at 28°C in water was $1.1 \times 10^{-2}\ M^{-1} \cdot min^{-1}$. On incubation of O-methylisourea for 24 h at pH 8.9 with arginine, glutamine, aspartic acid, glutamic acid, methionine, threonine, or serine, no reaction other than guanidination of the amino group was observed. At pH 8.9 the methylation of the SH group of cysteine was faster than guanidination of the amino groups of amino acids and of glycyl–glycine. Banks and Shafer (1970) consider that methylation of SH groups can occur under the conditions normally used for guanidinating proteins (long incubation of the protein in a concentrated, alkaline solution of O-methylisourea). Banks and Shafer (1972) have successfully used O-methylisourea for the selective methylation of the SH groups of papain.

The use of methyl-p-nitrobenzenesulfonate has recently been suggested for the selective methylation of SH groups in proteins. Heinrikson (1970, 1971) found that cysteine residues in proteins and peptides may be quantitatively converted into the S-methyl derivatives by reaction with a moderate excess of this reagent at pH 8.6 and 37°C; the reaction is performed in solutions containing acetonitrile with or without denaturing agents such as guanidine salts or urea. The pH dependence of the second-order rate constant for the methylation of reduced glutathione indicates, as might be expected, that the mercaptide ion is the reactive species. Amino-acid analysis of various S-methylated proteins and peptides revealed no modifications other than the quantitative conversion of cysteine into S-methylcysteine. According to Heinrikson, the advantages of methyl-p-nitrobenzenesulfonate over methyl iodide, the methylating agent most commonly used to date, derive from its selectivity and the ease with which it may be used. Rochat et al. (1970) found extensive losses of S-methylcysteine and of methionine in experiments with methyl iodide; these losses seemed to be due to the formation of S-dimethylsulfonium compounds which would then decompose during the subsequent acid hydrolysis. No such complications have been met with methyl-p-nitrobenzenesulfonate. Nevertheless Jacobson and Stark (1973) methylated the SH group of aspartate transcarbamylase with methyl iodide; the group had a half life of about 15 min at pH 8.5 in 0.04 M reagent in denaturing conditions. The methylation of SH groups may be particularly valuable in work on the roles of these groups in proteins because of the small size and neutral character of the substituent.

f. Arylation

The SH groups of low molecular thiols and proteins are arylated on reaction with fluorodinitrobenzene (FDNB), chlorodinitrobenzene (CDNB), and 2,4,6-trinitrobenzene-1-sulfonic acid. These compounds are usually thought of as reagents for the amino groups of proteins, but as a rule they react with SH groups faster. Thus Wallenfels and Streffer (1966; Wallenfels and Eisele, 1970) found with free amino acids at 40°C in 0.1 M KCl at pH 8.4 saturated with FDNB that the pseudo-first-order rate constants in sec^{-1} were: SH group, 100×10^{-3}; phenolic group, 1.09×10^{-3}; α- and ε-amino groups, 0.60×10^{-3}; and imidazole, 0.06×10^{-3}. As an example of the reaction of SH groups in a protein we shall cite the data of Gold (1968), who studied the reaction of the functional groups of phosphorylase b with FDNB and CDNB at pH 8.0. He found that four SH groups of this protein react with FDNB and CDNB considerably faster than the remaining SH, amino, and phenolic groups.

Zahn and Trauman (1954) established that below pH 5.2 FDNB reacts only with the SH group of cysteine and does not react with its NH_2-group or with the NH_2-group of glutamic acid; they noted that the use of a weakly acidic medium allowed selective dinitrophenylation of the SH groups alone in proteins. In a more alkaline medium there can be a migration of a dinitrophenyl residue from the SH group of cysteine to its NH_2 group (Wallenfels and Streffer, 1966); Wallenfels and Eisele, 1970).

In studies of the reaction of FDNB with proteins, it was found that FDNB selectively reacts with two SH groups of muscle aldolase at pH 6.0 (Cremona et al., 1965) and with one SH group of fructose-1,6-diphosphatase at pH 7.5 (Pontremoli et al., 1965). S-Dinitrophenyl groups can be easily removed by treating the protein with β-mercaptoethanol at pH 8 and 22°C (Shaltiel, 1967):

$$\text{Protein}-S-C_6H_3(NO_2)_2 + RSH \rightarrow \text{Protein}-SH + RS-C_6H_3(NO_2)_2$$

An interesting possibility is the β-elimination of S-dinitrophenyl groups in a protein in alkaline medium with the formation of dehydroalanine residues:

$$\begin{matrix}-CO \\ \quad\diagdown \\ \quad\quad CH-CH_2-\mathbf{SH} \\ \quad\diagup \\ -NH\end{matrix} + F-C_6H_3(NO_2)-NO_2 \xrightarrow[22^\circ]{pH\ 5.5} \begin{matrix}-CO \\ \quad\diagdown \\ \quad\quad CH-CH_2-S-C_6H_3(NO_2)-NO_2 \\ \quad\diagup \\ -NH\end{matrix} + HF$$

$$\xrightarrow[40^\circ]{pH\ 10.5} HS-C_6H_3(NO_2)-NO_2 + \begin{matrix}-CO \\ \quad\diagdown \\ \quad\quad C{=}CH_2 \\ \quad\diagup \\ -NH\end{matrix}$$

With the aid of this reaction an investigation was made of the role of the two SH groups in aldolase whose selective blocking by FDNB leads to a 2- to 3-fold increase of the enzyme's activity. It turned out that these two groups *per se* are not essential for activity, since their removal with the formation of dehydroalanine residues gave an enzyme with normal activity (Cremona et al., 1965).

The reaction of SH groups with trinitrobenzenesulfonic acid (TNBS) was studied by Kotaki et al. (1964) and also by Freedman and Radda (1968). These latter authors showed that the SH group of N-acetylcysteine reacts with TNBS much faster than the amino groups of the majority of amino acids and peptides. S-trinitrophenyl groups have a lower extinction coefficient at 340 nm than N-trinitrophenyl groups; they are unstable in alkaline medium but completely stable in acid; according to the data of Kotaki et al. they withstand heating in 6 M HCl to 110°C. Fields (1971) found rate constants of about 4000 M^{-1} min^{-1} for the reaction of TNBS with SH groups, and he also observed S $\rightarrow$ N transfer of the trinitrophenyl group.

Birkett et al. (1970) proposed 7-chloro-4-nitrobenz-2-oxa-1,3-diazole (NBD-chloride) as a reagent for protein SH groups:

$$O_2N-\text{[benzoxadiazole]}-Cl + RS^- \longrightarrow O_2N-\text{[benzoxadiazole]}-S-R + Cl^-$$

The rate of reaction of NBD-chloride with the SH group of N-acetylcysteine is 30 times greater at pH 8 than at pH 5.5; the reaction product has an absorption maximum at 425 nm, and a fluorescence (emission) maximum at 545 nm, and can be easily distinguished by its spectral properties from the product of reaction of the amino group. At pH 7.0 NBD-chloride reacts, apparently, only with SH groups.

g. Mixed Reactions

Olomucki and Diopon (1972, Diopon and Olomucki, 1972) have used ethyl chloroacetimidate and 4-chloro-3,5-dinitrophenacyl-bromide as bifunctional reagents for proteins. These substances form cross-links between sulfhydryl, amino, and imidazole groups.

7. Reaction with Metal Ions and with Organic Mercury Compounds

When SH groups react with metal ions, mercaptides are formed:

$$R{-}SH + Me^{+} \rightleftarrows R{-}SMe + H^{+}$$

In contrast with the reactions of alkylation considered above, mercaptidation reactions are reversible, but their equilibria greatly favor formation of the weakly dissociating mercaptides. Univalent cations of mercury, silver, copper, and gold, and bivalent cations of mercury, lead, copper, cadmium, and zinc, and also compounds of tervalent arsenic and antimony have a particularly high affinities for SH groups. The old name for thiols, "mercaptans," shows that their characteristic ability to react with mercury compounds was known at the dawn of the development of organic chemistry.*

When bivalent metal ions (Hg^{2+}, Cd^{2+}, etc.) combine with dithiols, weakly dissociating compounds of the following types are formed:

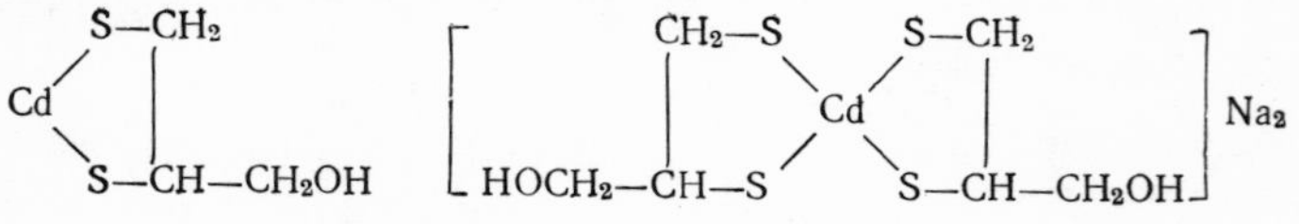

```
    S—CH2                     CH2—S       S—CH2
   /    |                     |     \   /     |
Cd      |                     |      Cd       |          Na2
   \    |                     |     /   \     |
    S—CH—CH2OH         [ HOCH2—CH—S       S—CH—CH2OH ]
```

* The term "mercaptan" comes from the Latin words "mercurio aptum," which mean fitted for mercury."

The logarithms of the stability constants of the complexes of Zn^{2+} ions with thiols and with EDTA are: mercaptoethanol, 5.9; mercaptoethylamine, 9.9; cysteine, 9.9; dithiothreitol, 10.3; dimercaptopropanol (BAL), 13.5; EDTA, 16.4 (Sillén and Martell, 1964; Cornell and Crivaro, 1972).

Silver and mercury ions in low (particularly equimolar) concentrations selectively react with the SH groups of proteins, as their affinity for these groups greatly exceeds their affinity for other functional groups of the protein. According to the findings of Benesch and coworkers (Benesch and Benesch, 1948, 1950; Benesch et al., 1955) and of Turpaev (1954) the presence of excess of any of the known natural amino acids has no effect on the titration of cysteine and of glutathione by Ag^+ ions in ammonium or tris buffers. One should nevertheless bear in mind that heavy metal ions, in particular mercury, if used in high concentrations, can combine, although less strongly, with other functional groups of proteins: imidazole, carboxyl, amino, disulfide, and others (Haarman, 1943; Gurd and Murray, 1954; Smith et al., 1955b; Gurd and Wilcox, 1956; Perkins, 1961; Resnik, 1964; Leslie, 1967).

The specificity of the reaction of heavy metal ions with SH groups is increased in the presence of some anions and chelating agents which compete with the functional groups of the proteins for binding the metal ions. Thus, according to Kolthoff et al. (1954), when serum albumin is titrated with Hg^{2+} ions in acetate, phosphate or borate buffers from pH 4.0 to 9.2, a secondary combination of the metal ions is observed with other functional groups of the proteins besides SH groups. But addition to the medium of ammonium or chloride ions completely suppresses the unspecific binding of mercury ions, and guarantees their selective interaction with the SH groups of the protein (see also, Poglazov and Baev, 1961).

When mercuric chloride interacts with simple thiols or proteins, two kinds of compounds arise: $RS-Hg-Cl$ and $(RS)_2Hg$. Hughes (1950) observed the formation of dimers from mercaptalbumin when it reacts with $HgCl_2$:

$$
\begin{array}{c}
\text{Protein}-\text{SH} + \text{HgCl}_2 \rightleftarrows \text{Protein}-\text{S}-\text{Hg}-\text{Cl} \\
+ \\
\text{Protein}-\text{SH} \\
\downarrow\uparrow \\
(\text{Protein}-\text{S})_2\text{Hg}
\end{array}
$$

Whether the Hg^{2+} ion combines with one or two SH groups in a protein molecule depends on the spatial relationship of the SH groups. Consequently for the quantitative determination of SH groups in proteins it is preferable to use unifunctional organic mercury compounds of the type R–Hg–X (p-chloromercuribenzoate, p-chloromercuriphenylsulfonate, phenylmercury acetate, methylmercury iodide, etc.) which can react with only a single SH group, for example,

$$^{-}OOC-C_6H_4-Hg^{+} + HS-R \rightleftarrows {}^{-}OOC-C_6H_4-Hg-SR + H^{+}$$

p-Mercuribenzoate ion

Organic mercury compounds are among the most specific reagents for protein SH groups. The dissociation constants of their mercaptides are less than those of the mercaptides of silver, but somewhat higher than the dissociation constants of the mercaptides formed by the Hg^{2+} ion. Thus the dissociation constant of the mercaptide of cysteine of the type $R-S-Hg-CH_3$ is $10^{-15.7}$, and the dissociation constant of its mercaptide with Hg^{2+} is $10^{-20.3}$ (Simpson, 1961). Consequently Hg^{2+}, but not Ag^{+}, can displace organomercurial reagents from SH groups. The dissociation constants of mercaptides and of the complexes formed by various ligands with the CH_3Hg^{+} ion are given below (Simpson, 1961):

Ligand	pK
Cysteine (SH group)	15.7
Histidine (NH_2 group)	8.8
Imidazole	7.3
Pyridine	4.8
Ammonia	8.4
OH^-	9.5
Phenolate	6.5
Acetate	3.6
Cl^-	5.45

From these data it is clear that the affinity of the CH_3Hg^{+} ion for the SH group is much greater than for other ligands; nevertheless organomercurial reagents can combine not only with SH groups, but with other groups in proteins, although much less strongly (Klapper, 1970; Duke et al., 1971; Autor and Fridovich, 1970).

Recently chloromercurinitrophenols were proposed as reagents for SH groups (McMurray and Trentham, 1969):

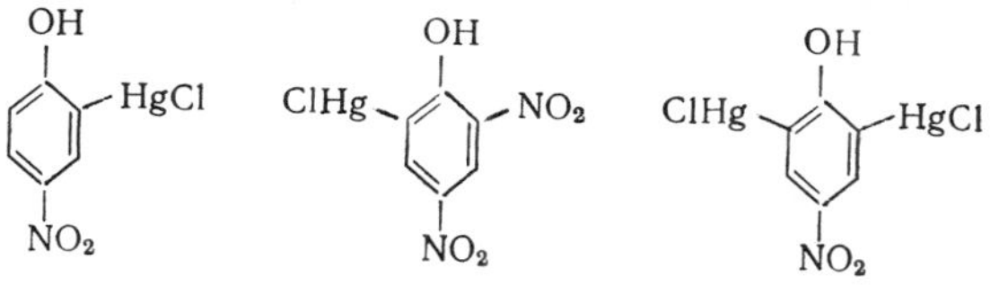

The extinction of these compounds depends on the degree of ionization of the phenolic hydroxyl which changes on combination of the thiol (in particular on whether its combination is accompanied by displacement of more weakly held ligands, such as, for example, EDTA) and also under the influence of the local environment in the protein. Thanks to these properties the molecule of mercurinitrophenol combined with a cysteine residue can act as a "reporter" or indicator group which "signals" conformational changes in the protein (applied to creatine kinase by Quiocho and Thomson, 1973).

The intramolecular transfer for organomercurial compounds from some SH groups of proteins to others has been described in a number of works (the reaction of "transmercaptidation"). Such a transfer was first observed by Szabolcsi et al. (1960), and confirmed by Smith and Schachman (1971), when they investigated preparations of glyceraldehyde-3-phosphate dehydrogenase in which some of the SH groups had been blocked with p-mercuribenzoate; on incubation of such preparations a "disproportionation" of the enzyme molecules took place, i.e., formation of reactivated molecules and of denatured molecules all of whose SH groups were blocked. Gruber et al. (1962) found that the partial blocking of the SH groups of lactate dehydrogenase with the use of eight equivalents of p-mercuriphenylsulfonate (p-MPS) led to inhibition of the enzymic activity by 97%; but after 47 h the activity was restored to 30% and the extinction at 255 nm (due to the absorption of mercaptides) did not fall. The authors explained this "self-reactivation" of the enzyme by transfer of p-MPS from SH groups that were essential for the catalysis to SH groups that bound p-MPS more strongly but had no direct relationship with activity. The reaction of "transmercaptidation" was also observed in studies of the SH groups of lipoamide dehydrogenase (Casola and Massey, 1966), and of the complex of actin with heavy meromyosin (Nakata and Yagi, 1969).

8. Reactions with Arsenic Compounds

Monothiols react with compounds of tervalent arsenic to form easily hydrolyzable mono- and dithioarsenites:

$$R{-}SH + O{=}As{-}R' \rightleftarrows R'{-}As\begin{matrix} OH \\ SR \end{matrix} + RSH \rightleftarrows R'{-}As\begin{matrix} SR \\ SR \end{matrix} + H_2O$$

Dithiols react with arsenoxides or with arsenite to form cyclic dithioarsenites, which are markedly more stable than the mono- or dithioarsenites that arise on reaction with monothiols. Data on the rate of hydrolysis of thioarsenites obtained by reaction of lewisite with mono- and dithiols are given below. The rate of hydrolysis was measured by the time of decoloration of porphyrindin, which goes over to the leuco-form on reaction with the liberated SH group (Stocken and Thompson, 1946):

Thiol	Time of decoloration, min
Cysteine	0.25
Glutathione	2.5
2-Mercaptoethanol	3.5
2,3-Dimercaptopropanol	> 180
1,3-Dimercaptopropanol	> 180

Five membered rings are particularly stable, and they arise by interaction of the arsenic compounds with 1,2-dithiols ("vicinal" dithiols):

$$\begin{matrix} CH_2{-}SH \\ | \\ CH{-}SH \\ | \\ R \end{matrix} + R'{-}As{=}O \underset{k_2}{\overset{k_1}{\rightleftarrows}} \begin{matrix} CH_2{-}S \\ | \\ CH{-}S \\ | \\ R \end{matrix}\!\!>\!As{-}R' + H_2O$$

From Table 6 it is clear that the dissociation constant of the complex of arsenite with 2,3-dimercaptopropanol is diminished almost twofold in comparison with the dissociation constant of the complex with 1,3-dimercaptopropanol. It is interesting that 7-membered rings formed by the interaction of arsenite with the erythro- and threo-isomers of 1,4-dimercapto-2,3-dihydroxybutane are more stable than the 6-membered ring formed on reaction with 1,3-dimercaptopropanol.

TABLE 6. Dissociation Constants of the Complexes of Dithiols with Arsenite and Rate Constants for the Formation and Breakdown of These Complexes (Zahlen and Cleland, 1968)

Dithiol	Dissociation constant, μM	k_1, $M^{-1} \cdot min^{-1}$	k_2, min^{-1}
1,4-Dimercapto-2,3-dihydroxy-butane (threo isomer)*	0.33	28 000	0.0092
1,4-Dimercapto-2,3-dihydroxy-butane (erythro isomer)†	0.37	27 000	0.010
1,3-Dimercaptopropanol	0.49	69 500	0.034
2,3-Dimercaptopropanol	0.26	18 000	0.0046

* Dithiothreitol.
† Dithioerythritol.

The high affinity of vicinal dithiols for arsenite and arsenoxides is the basis of the successful application of BAL (2,3-dimercaptopropanol) and unithiol (sodium 2,3-dimercaptopropanesulfonate) for treating poisoning by arsenic compounds. The reaction shown also explains the strong inhibitory action of arsenite and of arsenoxides on enzymes whose active sites contain dithiol groupings. This action was first discovered by Peters (1936, 1963; Peters et al., 1946) on the pyruvate oxidase system, which, as turned out later, contains the covalently bound dithiol cofactor, lipoic acid. The use of arsenite and arsenoxides has also allowed demonstration of the presence of dithiol groupings (pairs of SH groups that belong to cysteine residues and are close in space) in the active site of lipoamide dehydrogenase (see Chapter 6, Section 1b) and of some aldehyde dehydrogenases (Jacoby, 1958; Nirenberg and Jacoby, 1960). High sensitivity of an enzyme to inhibition by low concentrations of arsenite or arsenoxides (about 10–100 μM) is a sign of the presence of the dithiol grouping. Enzymes that are inhibited by preparations of arsenic are completely or partially reactivated on addition of 2,3-dimercaptopropanol; monothiols have little effect in such cases.

Peters (1963; Lotspeich and Peters, 1951) noticed a difference in action of two types of tervalent arsenic compounds: the singly (R–As=O) and doubly (R–As(Cl)–R) substituted. Compounds of the first type are more effective in blocking dithiol groupings, and compounds of the second type cannot form cyclic structures and react with SH groups singly. It should be emphasized that apparently neither type of arsenic compound can react with any group of proteins other than the SH groups.

The inhibition of some enzymes (succinic semialdehyde dehydrogenase, aldehyde dehydrogenase, thiol transacylase, glutamine synthetase, luciferase, and acetyl-CoA carboxylase) by arsenite becomes possible or is greatly increased in the presence of mono- or dithiols. Thus, for example, the maximum inhibition of glutamine synthetase is obtained on adding to it equimolar quantities of arsenite and of 2,3-dimercaptopropanol. Possibly the role of the thiol is reduction of a disulfide group in the protein to form a pair of SH groups that then react with arsenite. A different explanation was proposed by Wu (1965) who considered that a complex of arsenite and dithiol reacts with an SH group of glutamine synthetase:

$$\text{Protein—SH} + \begin{array}{c} \text{AsO}^- \\ \text{S} \quad \text{S} \\ | \qquad | \\ H_2C\text{——}CH\text{—}CH_2OH \end{array} \rightleftarrows \begin{array}{c} \text{Protein—S—As} \\ \text{S} \quad \text{S} \\ | \qquad | \\ H_2C\text{——}CH\text{—}CH_2OH \end{array} + OH^-$$

The inhibition evoked by this complex is removed on addition of excess dithiol or cysteine. In distinction with glutamine synthetase and the other enzymes listed above, the "classical" dithiol containing enzymes, such as pyruvate oxidase, are usually inhibited at very low concentrations of arsenite in the absence of any thiols (pyruvate oxidase is inhibited by 50% at an arsenite concentration of 17 μM), and the inhibition is not removed by cysteine (Peters et al., 1946; Stocken and Thompson, 1946).

9. Oxidation of SH Groups

The SH groups of low molecular thiols and proteins undergo oxidation (dehydrogenation) under mild conditions with the formation of inter- or intramolecular disulfide bonds under the action of iodine, ferricyanide [$Fe(CN)_6^{3-}$], tetrathionate ($Na_2S_4O_6$), porphyrindin, o-iodosobenzoate, and hydrogen peroxide:

$$2R\text{—SH} \xrightarrow{\text{Oxidizing agent}} R\text{—S—S—}R + 2H^+ + 2e^-$$

This reaction has been used for the quantitative determination of SH groups. Of the reagents just listed, o-iodosobenzoate has the greatest specificity, and it reacts with SH groups at pH 7.0 according to the following equation:

$$C_6H_4(COO^-)-IO + 2R-SH \rightarrow R-S-S-R + C_6H_4(COO^-)-I + H_2O$$

The drawback common to all oxidizing agents used as reagents for SH groups is their ability to oxidize not only SH but also various other groups in proteins, in particular tyrosine, tryptophan, and methionine residues. On treatment with iodine, iodination of the rings of tyrosine and histidine can also occur (Fraenkel-Conrat, 1950; Wolff and Covelli, 1969). In determining SH groups quantitatively with oxidizing agents it is necessary to take account of the possibility of "overoxidation" of the groups, i.e., transition of sulfur to a state more oxidized than in a disulfide (formation of sulfinates or sulfonates) as a result of which the stoichiometry of the reaction changes (Chinard and Hellerman, 1954; Hird and Yater, 1961; Little and O'Brien, 1967a,b).

The rate and character of the oxidation of SH groups depends on the relative oxidation–reduction potentials of the SH group and oxidizing agent (Table 7), the concentration of reagents, the pH, and the temperature. An important factor is the spatial positioning of the SH groups in the protein. In those cases where contact between SH groups within the molecule of the protein is hindered, or when the protein contains only a single SH group, intramolecular S–S bonds are not formed, but sometimes, under favorable steric conditions, intermolecular S–S bonds arise, and the protein aggregates (Little and O'Brien, 1967a). "Lone," or sterically restricted, SH groups can, however, also be oxidized in another way, forming sulfenic, sulfinic, or sulfonic (cysteic) acids, according to the strength of the oxidizing agent:

$$RSH \longrightarrow RSOH \rightarrow RSO_2H \rightarrow RSO_3H$$
$$RSH \rightarrow RSI \rightarrow RSOH$$

When lone SH groups are oxidized by iodine at neutral pH, either sulfenyl iodides or sulfenic acids can arise; in either case two

TABLE 7. Oxidation–Reduction Potentials of Thiol–Disulfide Systems*

System RSH/RSSR	E_h at pH 7, volts	Reference
Cysteine	–0.33	Ghosh et al., 1932
	–0.22	Fruton, Clarke, 1934
	–0.21	Cleland, 1964
	–0.39	Borsook et al., 1937
	–0.14	Ryklan, Schmidt, 1944
	–0.33	Tanaka et al., 1955
	–0.22	Jocelyn, 1967
Glutathione	–0.35	Ghosh, Ganguli, 1935
	–0.23	Fruton, Clarke, 1934
	–0.24	Rost, Rapoport, 1964
	–0.25	Scott et al., 1963
Thioglycolic acid	–0.34	Ghosh et al., 1932
	–0.34	Kolthoff et al., 1955b
	–0.23	Fruton, Clarke, 1934
Dithiothreitol	–0.33	Cleland, 1964
Lipoic acid	–0.294	Sanadi et al., 1959
	–0.325	Ke, 1957
Insulin	–0.38	Weitzman, 1965
Thioredoxin	–0.24	Gonzalez et al., 1970

*The table was composed by Clark (1960) and supplemented by us. Direct measurement of the potentials of thiol–disulfide pairs is made difficult by the interaction of sulfur with a metallic electrode; other methods give, as shown in the table, contradictory results.

equivalents of iodine are consumed per SH group oxidized:

$$RSH + I_2 \rightarrow RSI + H^+ + I^-$$

$$RSH + I_2 \xrightarrow{+H_2O} RSOH + 2H^+ + 2I^-$$

Hughes and Straessle (1950) first obtained evidence that the single SH group of human serum albumin is oxidized by iodine to a product that differs from a disulfide; they found that 2.2 equivalents of iodine were used per SH group of the albumin, although only one is needed for oxidation to a disulfide. Evidence in favor of formation of sulfenyl iodides was obtained in the oxidations by iodine of the SH groups of tobacco mosaic virus protein (Fraenkel-Conrat, 1955, 1959), glyceraldehyde-3-phosphate dehydrogenase

(Parker and Allison, 1969), dihydrofolate reductase (Kaufman, 1966), ovalbumin and lactoglobulin (Cunningham and Neunke, 1959, 1960, 1961; Cunningham, 1964).* Sulfenyl iodides can react with thiols (e.g., mercaptoethanol, cysteine) and also with thiourea or thiouracil, forming mixed disulfides:

$$RSI + R'SH \rightleftarrows R{-}S{-}S{-}R' + I^- + H^+$$

$$RSI + H_2N{-}\underset{\underset{S}{\|}}{C}{-}NH_2 \rightleftarrows R{-}S{-}S{-}\underset{\underset{NH}{\|}}{C}{-}NH_2 + I^- + H^+$$

Some protein sulfenyl iodides can react with excess iodine to form rather unstable, yellow colored, sulfenyl periodides (Jirousek and Pritchard, 1971a,b):

$$RSI + I_2 \rightleftharpoons RSI_3$$

These in turn can react with free SH groups, or with tyrosine, forming sulfenyl iodides or iodotyrosine respectively.

Field and White (1973) reported the preparation and properties of the first simple sulfenyl iodide to be isolated as a pure crystalline solid, namely, $I{-}S{-}CMe_2{-}CH(NH{-}Cbz){-}CO{-}NH{-}C_6H_4{-}Cl$. It remains unchanged under ambient conditions for more than ten weeks; it is quite reactive, however, and decomposes in solution. Studies on this and related model compounds may be helpful in understanding the reasons for the stability of protein sulfenyl iodides. Aromatic sulfenyl halides with electron-withdrawing groups can be stable and are available as reagents (p. 115).

In several cases when SH groups are oxidized by iodine, the products are sulfenic acids instead of sulfenyl iodides. Trundle and Cunningham (1969) came to this conclusion in studying the reaction of radioactive iodine with creatine kinase. These authors showed that iodine oxidizes six SH groups of the enzyme and that each SH group consumes two equivalents of iodine. The oxidized protein after gel filtration through Sephadex contained hardly any iodine, which ruled out sulfenyl iodides as final products of the reaction; possibly, however, they were formed at an intermediate stage of the reaction. Oxidation of two reactive SH groups in creatine kinase to the sulfenic level presumably also occurs as the final result of reaction with 1-dimethylaminonaphthalene-5-sulfonyl

* It has been proposed that protein sulfenyl iodides or periodides play a part in the process of iodination of tyrosine in thyroid proteins (Cunningham, 1964; Fawcett, 1968; Jirousek and Pritchard, 1971b).

chloride at pH 6.1 and 4°C; the sulfenic acid may be formed by breakdown of a labile thiolsulfonate intermediate (Brown and Cunningham, 1970):

$$RSH + Ar-SO_2-Cl \rightarrow RS-SO_2-Ar + H^+ + Cl^-$$

$$RS-SO_2-Ar + H_2O \rightarrow RS-OH + Ar-SO_2H$$

It is known that aliphatic sulfenic acids and sulfenyl iodides are extremely unstable and are easily split during attempts to isolate them. If sulfenic acids are really formed on oxidation of SH groups in some proteins (which has not yet been rigorously demonstrated) then they must be stabilized by their micro-environment in the protein. A number of authors (Parker and Allison, 1969; Little and O'Brien, 1969) have obtained indirect evidence for the formation of sulfenic acids on oxidation of the catalytically active SH group of glyceraldehyde-3-phosphate dehydrogenase by an equivalent quantity of o-iodosobenzoate or hydrogen peroxide at 0°C. The derivative is devoid of dehydrogenase activity but can hydrolyze acyl phosphates (Allison and Connors, 1970). Reaction with arsenite can aid in elucidating the oxidation products formed from SH groups in proteins; arsenite reduces sulfenic acids to thiols, but is apparently unable to reduce disulfides (Gutmann, 1908; Parker and Allison, 1969; Hartman, 1970). When sulfenic acids or sulfenyl halides react with excess thiol, regeneration of the SH groups occur; the reaction apparently goes via a mixed disulfide:

$$RSOH + R'SH \rightleftharpoons RS-SR' + H_2O$$

$$RS-SR' + R'SH \rightleftharpoons RSH + R'S-SR'$$

Since the fourth residue in peptide sequence from the reactive one is another cysteine, it is not surprising that the presumed sulfenic derivative is easily converted into an internal disulfide.

When tetrathionate reacts with the catalytically active SH groups of glyceraldehyde-3-phosphate dehydrogenase, papain, ficin, and streptococcal protease, the sulfenyl–thiosulfate derivatives of the proteins are formed (Pihl and Lange, 1962; Sanner and Pihl, 1963; Liu, 1967) 1967; Gleisner and Liener, 1973):

$$Enzyme{-}SH + S_4O_6^{2-} \rightarrow Enzyme{-}S{-}S_2O_3^- + S_2O_3^{2-} + H^+$$

The sulfenyl–thiosulfate derivative of the dehydrogenase is stable at 0°C; on warming or addition of urea it is destroyed as a result of intramolecular reaction with the SH group that is situated nearby with the formation of an intramolecular disulfide and liberating thiosulfate (Parker and Allison, 1969) (cf. p. 72):

$$\text{Enzyme}\begin{smallmatrix}\diagup S-S_2O_3^- \\ \diagdown SH\end{smallmatrix} \rightarrow \text{Enzyme}\begin{smallmatrix}\diagup S \\ | \\ \diagdown S\end{smallmatrix} + S_2O_3^{2-} + H^+$$

The oxidation of SH groups by tetranitromethane is interesting. Sokolovsky et al. (1969) found that the main product of oxidation of reduced glutathione was the disulfide, but that among the products of the reaction there was also a glutathione derivative that resembled a sulfinic acid in a number of its properties. On the basis of the stoichiometry of the reaction Sokolovsky and coworkers postulated the formation of a sulfenyl nitrite or nitro compound as an intermediate:

$$RSH + C(NO_2)_4 \rightarrow RSNO_2 + C(NO_2)_3^- + H^+$$

This could react with a second molecule of the thiol to form a disulfide and liberate a nitrite ion:

$$RSNO_2 + RSH \rightarrow RS{-}SR + NO_2^- + H^+$$

or it could undergo hydrolysis to form a sulfenic acid, which would be spontaneously oxidized to a sulfinic acid:

$$RSNO_2 + H_2O \rightarrow RSOH + NO_2^- + H^+$$
$$RSOH + {}^1/_2O_2 \rightarrow RSO_2H$$

The relative yields of the reaction products (disulfide and sulfinate) depended on the molar excess of the tetranitromethane and on the nature of the thiol (and, in the case of a protein, presumably on the environment of the SH group).

SH groups are oxidized not only when oxidizing agents are added, but also "spontaneously," by oxygen from the air (the so-called "auto-oxidation"). The possibility of auto-oxidation must always be kept in mind in work with enzyme systems and proteins *in vitro*. The oxidation of SH groups by molecular oxygen takes place at an appreciable rate in the presence of catalytic quantities of metal ions that are able to form complexes with the thiols being oxidized (in particular, iron and copper ions) (Lamfrom and Nielson, 1957; Franzen, 1957; Tarbell, 1961; Cavallini et al., 1968). Complexing any traces of metal ions with the aid of chelating agents markedly increases the stability of thiols and of those enzymes whose activity depends on the presence of free SH groups (Torchinskii, 1958).*

The speed of oxidation of SH groups is greatly influenced by the nature of neighboring groups and the distance between the SH groups in the molecule. This was clearly demonstrated by Barron et al. (1947; Barron, 1951) and also by Overberger and Ferraro (1962). From their findings it appears that the rate of oxidation of dithiols is diminished on increasing the distance between the SH groups in the molecule, and also under the influence of neighboring electronegative groups such as carboxyl groups (i.e., groups that raise the pK of the SH group; Tables 8 and 9). In general the SH groups with higher pK values (e.g., the SH group of glutathione) are oxidized more slowly than SH groups with lower pK values (e.g., the SH group of cysteine). This fact indicates that the mercaptide ions is oxidized more easily than the undissociated SH group. The following findings also support this supposition:

1. The pH optimum for the oxidation of cysteine and glutathione is in the alkaline region (Lyman and Barron, 1937; Little and O'Brien, 1967b, 1969).
2. The rate of oxidation of dithiols increases with increasing pH (Barron et al., 1947; Philipson, 1962).
3. The rate of oxidation of aminothiols is directly proportional to the concentration of RS^- ions (Franzen, 1957).

In the opinion of Benesch and Benesch (1955) the oxidation of cysteine takes place in two stages:

$$RS^- \rightarrow RS\cdot + e$$
$$RS\cdot + RS\cdot \rightarrow RS{-}SR$$

At low pH the first step is rate determining. Because of this the rate of oxidation increases as the pH is raised and as the concentration of mercaptide ion therefore increases. At a high enough concentration of mercaptide ions, however, the overall rate of oxidation is determined by the second step. This goes faster when the free radicals are electrically neutral. The charge of the radicals formed by cysteine in the $^-OOC-CH(CH_2SH)-NH_3^+$ form is

* The addition of various thiols, such as β-mercaptoethanol, 2,3-dimercaptopropanol, and, especially, dithiothreitol, is used successfully for "protection" of the SH groups of enzymes from oxidation.

TABLE 8. The Oxidation of Dithiols by Atmospheric Oxygen at pH 7.0 and 22.4°C (Barron et al., 1947)

Dithiol	Half-time of oxidation, min: in the presence of $CuCl_2$	Half-time of oxidation, min: in the absence of $CuCl_2$
$H_2C(SH)-CH(SH)-CH_2(OH)$	10.5	196
$H_2C(SH)-CH_2-CH_2(SH)$	350	500
$H_2C(SH)-[CH_2]_4-CH_2(SH)$	500	500
$H_2C(SH)-CH(SH)-C(OH){=}O$	55.0	500
$H_2C(SH)-CH(SH)-C(OCH_3){=}O$	10.4	80
$H_2C(SH)-CH(SH)-CH_2-CH_2(OH)$	22.0	240
$H_2C(SH)-CH(SH)-CH_2-OOC-CH_2-CH_3$	23.6	500
$H_2C(SH)-CH(SH)-CH_2-OOC[CH_2]_2CH_3$	123	500

TABLE 9. Relative Rates of Oxidation of Dithiols in Dimethylformamide (Overberger and Ferraro, 1962)

Thiol	Relative rate	Thiol	Relative rate
$CH_3-CH(SH)-(CH_2)_3-CH(SH)-CH_3$	1	$CH_3-CH(SH)-CH(SH)-CH_3$	61.2
$HO-CH_2-CH_2-SH$	5.2	$-[CH_2-CH(SH)-CH_2-CH(SH)-CH_2]-_{n/2}$	280
$CH_3-CH(SH)-[CH_2]_2-CH(SH)-CH_3$	11.9		

zero. The radicals formed from the $^{-}OOC-CH(CH_2SH)-NH_2$ form, which predominates at high pH, carry a negative charge. This slows down the recombination of radicals and lowers the overall rate of oxidation. The hypothesis of Benesch and Benesch explains the diminution in the rate of oxidation of cysteine that is observed, according to many authors, at pH values above 7.4-9.0.

Heaton et al. (1956) studied the oxidation in aqueous solution at pH 8.5 of a series of peptides of formula Cys$-(Gly)_n-$Cys, where n was varied from 0 to 4. The formation of the intramolecular disulfide link was maximal when n = 4.

10. The Thiol – Disulfide Exchange Reaction and Its Significance

Among oxidants for the SH group, disulfides have a unique place, since their reaction with thiols is highly specific. This reaction has received the name of the thiol–disulfide exchange:

$$R'S^- + R''S{-}SR'' \underset{k_2}{\overset{k_1}{\rightleftharpoons}} R'S{-}SR'' + R''S^-$$

$$R'S^- + R'S{-}SR'' \underset{k_4}{\overset{k_3}{\rightleftharpoons}} R'S{-}SR' + R''S^-$$

As can be seen from the equations, the reaction consists of two stages of nucleophilic displacement with formation in the intermediate stage of a mixed disulfide. The concentration of this form in the reaction mixture at equilibrium is quite appreciable (Eldjarn and Pihl, 1957; Jocelyn, 1967). The equilibrium and rate constants for the first and second stages of the reaction between reduced glutathione (R'SH) and cystine (R''$-$S$-$S$-$R'') at 37°C are presented below (Jocelyn, 1967):

Equilibrium constants*	pH 7.4	pH6.1
K_1	3.9	5.8
K_2	3.2 (3.7)	3.8
K_3	1.2 (0.79)	1.5
Rate constant, $M^{-1}\cdot min^{-1}$		
k_1	610	80
k_4	450	20

* $K_1 = K_2 \cdot K_3$; $K_2 = [R''SSR'][R''SH]/[R''SSR''][R'SH]$; $K_3 = [R'SSR'][R''SH]/[R'SSR''][R'SH]$. In parentheses are given the data of Gorin and Doughty (1968), obtained at pH 6.6 and 25°C.

From these data it is clear that the exchange reaction proceeds at an appreciable rate under physiological conditions of pH and temperature; on lowering the pH the rate of reaction decreases.

Kinetic investigations carried out with low-molecular thiols have shown that the thiol—disulfide exchange takes place with participation of the mercaptide ion; the rate of the reaction depends on the nature of the disulfide and also on the nucleophilicity and the concentration of the RS^- ions that carry out nucleophilic substitution at a sulfur atom of disulfide (Fredga, 1937; Bersin and Steudel, 1938; Fava and Iliceto, 1953; Kolthoff et al., 1955a; Eldjarn and Pihl, 1957b; Fava et al., 1957; Lamfrom and Nielsen, 1958; Foss, 1961; Smith et al., 1964; Lumper and Zahn, 1965; Jocelyn, 1967). Lamfrom and Nielsen (1958) found, for example, that the rate constant of reaction of the $R{-}S^-$ form of thioglycolate with cystine is 20,000 times larger than for the SH form. The disulfide exchange reaction that occurs in neutral or alkaline medium and is catalyzed by thiols goes by an analogous mechanism (see Chapter 2, Section 8):

$$R'S{-}SR' + R''S{-}SR'' \overset{RS^-}{\rightleftarrows} 2R'S{-}SR''$$

If an isolated SH group is present in a protein or if there are intermolecular interactions of proteins and peptides, the exchange reaction can stop at the stage of formation of a mixed disulfide. Pihl and coworkers (Pihl and Lange, 1962; Sanner and Pihl, 1963) observed formation of a mixed disulfide when glyceraldehyde-3-phosphate dehydrogenase or papain was treated with excess of the monosulfoxide of cystamine (diaminodiethyldisulfide):

$$\text{Enzyme}{-}SH + \underset{\substack{\| \\ O}}{RS}{-}SR \rightleftarrows \text{Enzyme}{-}S{-}SR + RSOH$$

$$2RSOH \rightarrow RSO_2H + RSH$$

Mixed disulfides of hemoglobin, serum albumin, and reduced lysozyme, formed by reaction with cystine, have been described (Taylor et al., 1963; Isles and Jocelyn, 1963; Kanarek et al., 1965), and also mixed nitrophenyl—disulfide derivatives of a number of enzymes, which arise when the enzymes are treated with Ellman's reagent, 5,5'-dithiobis(2-nitrobenzoic acid) (see, for example, Torchinskii and

Sinitsyna, 1970):

$$\text{Enzyme}-S^- + O_2N-C_6H_3(COO^-)-S-S-C_6H_3(COO^-)-NO_2 \rightleftarrows$$
$$\text{Enzyme}-S-S-C_6H_3(COO^-)-NO_2 + {}^-S-C_6H_3(COO^-)-NO_2$$

Ellman's reagent is a particularly interesting disulfide. The nitrothiophenolate ion is an especially good leaving group, so this disulfide is very reactive. The electron density is removed from the sulfur atom so that the product is unreactive:

$$S^--C_6H_3(COO^-)-NO_2 \longleftrightarrow S{=}C_6H_3(COO^-){=}NO_2^-$$

This displaces the equilibrium of the exchange reaction so that only a small excess of reagent is needed to achieve stoichiometric release of nitrothiophenolate. Brocklenhurst and Little (1973) have tabulated the equilibrium constant for the reaction with papain at various pH values. The same quinonoid resonance moves the absorption maximum of the ion released up to 412 nm. These two features together make the reaction useful for the quantitative determination of SH groups (Chapter 3, Section 1c). The charged carboxyl groups give the reagent solubility.

Other disulfides with particularly good leaving groups are similarly useful. These include pyridine disulfides, reactive in acid solution (p. 116), and compounds of the type R-S-S-CO-OMe (Brois et al., 1970) which have been used to modify papain (G. Lowe and M. R. Bendall, unpublished work).

Boross (1969) observed an interesting series of secondary thiol–disulfide exchange reactions during a study of the interaction of Ellman's reagent with pig muscle glyceraldehyde-3-phosphate dehydrogenase. This enzymes consists of four subunits (each of mol. wt. about 36,000) each of which contains four SH groups; one of these groups, the most reactive, fulfills a catalytic role in the active site of the enzyme (see Chapter 6, Section 1). On addition to the enzyme of one equivalent of Ellman's reagent (calculated per subunit) one equivalent of the nitrothiophenolate anion is rapidly liberated, with complete inactivation of the enzyme:

$$\text{Protein}-S^- + RS-SR \rightleftarrows \text{Protein}-S^1-SR + RS^- \qquad (a)$$

After a little time, however, a second equivalent of nitrothiophenolate is released, and this is explained by the entry of a second nearby SH group into the reaction:

$$\text{Protein}\underset{\displaystyle |\atop S^-}{—}S^1—SR \rightleftarrows \text{Protein}\left|\begin{matrix} / S^1 \\ \backslash S^2 \end{matrix}\right. + RS^- \qquad \text{(b)}$$

The reaction does not stop here; on longer incubation 50% of the original enzymic activity is restored while some of the protein is completely denatured and precipitates. Boross explained this phenomenon as follows: formation of the internal disulfide in reaction (b) destabilizes the protein molecule; the remaining two SH groups which were previously masked now come to the surface and enter into thiol—disulfide exchange, with the result that in some molecules all the SH groups are oxidized and in others all are reduced:

$$\text{Protein}\left|\begin{matrix} S^1 \\ HS^4 \quad S^3H \quad S^2 \end{matrix}\right. + \text{Protein}\left|\begin{matrix} S^1 \\ HS^4 \quad S^3H \quad S^2 \end{matrix}\right. \rightleftarrows \text{Protein}\left|\begin{matrix} S \\ S—S \quad S \end{matrix}\right. + \text{Protein}\begin{matrix} SH \\ HS \quad SH \quad SH \end{matrix} \qquad \text{(c)}$$

Some proteins exist *in vivo* in the form of mixed disulfides with low-molecular thiols. Thus King (1961) observed that a fraction of human serum albumin contained a mixed disulfide of the protein with cysteine and, to a lesser extent, with glutathione; this explains the long-known fact that serum albumin contains a nonintegral number (2/3) of SH groups per molecule (see also Andersson, 1966; Modig, 1968). A mixed disulfide with glutathione has also been found in horse muscle acylphosphatase (Ramponi et al., 1971).

Ferdinand et al. (1965) found that the inactive form (zymogen) of streptococcal protease contains a mixed disulfide of the type protein —S—S—R, where R—S— is a residue of a volatile mercaptan. Reduction of this disulfide by thiol—disulfide exchange converts the zymogen into the active enzyme (Bustin et al., 1970). The plant protease papain can contain a mixed disulfide formed by the SH group of the protein with cysteine (Klein and Kirsch, 1969a,b); the active form of papain arises only on reduction of this disulfide and liberation of the SH group of the protein (see Chapter 6, Section 1).

The thiol—disulfide exchange reaction plays an important role in many other biochemical processes. First of all we should note its significance in the renaturation and perhaps in the biosynthesis of proteins that contain S—S bonds (see Chapter 7, Section 2).

The aggregation and polymerization of some proteins also apparently proceeds by means of this reaction. Huggins et al. (1951) showed that the formation of gels by concentrated solutions of serum albumin and γ-globulin in the presence of urea is a result of a chain of thiol–disulfide exchange reactions. Finally, it is possible that the regulation of the activity of some intracellular enzymes occurs by means of this reaction. Natural disulfides, those like cystamine for example, can react reversibly with the SH groups of enzymes to form mixed disulfides. Such a regulation mechanism was postulated for the activity of fructose-1,6-diphosphatase, which is activated at neutral pH by Ellman's reagent, cystamine, and other disulfides (Pontremoli et al., 1967; Little et al., 1969) and also for phosphofructokinase; blocking one SH group of phosphofructokinase with Ellman's reagent leads to loss of the allosteric properties of the enzyme and to a more than five-fold increase in its activity at the substrate concentrations found *in vivo* (Forest and Kemp, 1968).

A special type of enzymes that catalyze thiol–disulfide exchange, the thiol–disulfide transhydrogenases (EC 1.8.4), has been found in cells. One such enzyme, isolated from yeast in a homogeneous state, catalyzes the thiol–disulfide exchange between glutathione and a number of low-molecular disulfides, particularly L-cystine, but is completely inactive toward the S–S bonds of proteins (Nagai and Black, 1968). On the contrary, a nucleoprotein isolated from sea urchin eggs catalyzes thiol–disulfide exchange between protein molecules (Sakai, 1967). Glutathione-insulin transhydrogenase (EC 1.8.4.2) from rat liver catalyzes the exchange between dihydrolipoate, glutathione, or cysteine on the one side and insulin on the other (Spolter and Vogel, 1968). An enzyme has also been found in rat liver that catalyzes the exchange reaction between glutathione and the mixed disulfide of L-cysteine and glutathione (Eriksson and Mannervik, 1970):

$$\text{G—SH} + \text{R—S—S—G} \rightleftarrows \text{R—SH} + \text{G—S—S—G}$$

A transhydrogenase that catalyzes the exchange reaction between glutathione and asymmetric disulfides, including the disulfide of glutathione with coenzyme A, has been discovered in ox kidney and in rat tissues (Change and Wilken, 1966). The "activating" enzyme from rat liver that accelerates thiol–disulfide interchange on reoxidation of reduced proteins also belongs to the group of transhydrogenases (De Lorenzo et al., 1966a,b) (see Chapter 7, Section 2).

Various hypotheses to explain the protective action of aminothiols in irradiation pay particular attention to mixed disulfides. Eldjarn and Pihl (1956, 1957a,b, 1960; Eldjarn et al., 1956; Pihl and Sanner, 1963a,b) supposed that the formation of mixed disulfides with the SH groups of proteins and enzymes lay at the basis of the protective action of cystamine and other disulfides and of amino thiols. When a free radical attacks a mixed disulfide, one of the sulfur atoms is reduced and the other is oxidized; the degree of damage to the enzyme is therefore cut in half. Other authors suppose, however, that the protective action of amino thiols is based on their removal of the free radicals that are formed during radiolysis (Bacq and Alexander, 1961; Gorin, 1965; Graevskii, 1969; see also Winstead, 1967; Brown, 1967).

The formation of mixed disulfides can be used for protecting SH groups during various chemical modifications of proteins; the protective groups are easily removed on addition of thiols. Thus Kilmartin and Rossi-Bernardi (1969) protected the reactive SH groups of hemoglobin by reaction with cystamine before modifying certain of its amino groups with cyanate. Protein mixed disulfides with 5-thio-2-nitrobenzoate can be used as intermediates in the synthesis of other S-substituted derivatives of enzymes. Because the nitrothiophenolate anion is a better leaving group than the alkyl mercaptide ion, nucleophilic attack on the disulfide displaces it preferentially. Thus Vanaman and Stark (1970) achieved essentially quantitative formation of the S-cyano (R−S−CN) and S-sulfo ($R{-}S{-}SO_3^-$) derivatives of the catalytic subunit of aspartate transcarbamylase by treating the mixed disulfide formed from the protein and Ellman's reagent with cyanide and sulfite respectively.

11. Reaction with Sulfenyl Halides

Thiols react in acid media with sulfenyl halides to form mixed (asymmetric) disulfides:

$$R'SH + Cl{-}SR'' \rightarrow R'S{-}SR'' + HCl$$

Parker and Kharasch (1960) used this reaction for the preparation of a number of mixed disulfides. Fontana et al. (1968a) found that 2-nitro-, 4-nitro-, and 2,4-dinitrophenylsulfenyl chlorides selectively reacted in acidic solvents with the residues of cysteine and tryptophan in proteins; the reaction with tryptophan leads to formation of a thioether:

Indole (NH) with CH₂—CH(CO—)(NH—) + RSCl → indole (NH, SR at C-2) with CH₂—CH(CO—)(NH—) + HCl

The mixed disulfides formed in the reaction of nitrophenylsulfenyl halides with SH groups are stable in acid medium, but are destroyed in alkali, liberating nitrothiophenol. They are easily reduced by excess β-mercaptoethanol, thioglycolic acid, and sodium borohydride. The use of nitrophenylsulfonyl chlorides for modifying SH groups in proteins is restricted by the necessity of carrying out the reaction in 50% acetic acid and by the instability of the reaction product (disulfide) in alkaline media.

Azobenzene-2-sulfenyl bromide (I) has a number of advantages compared with nitrophenylsulfenyl chlorides; it is highly soluble in water, apparently because of its salt-like structure (II) (Fontana et al., 1968b):

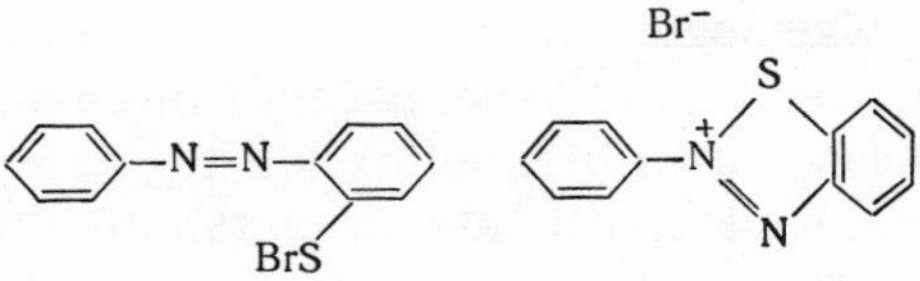

Br⁻

S

—N=N— —N⁺

BrS N

I *II*

Azobenzenesulfonyl bromide does not interact with the indole ring of tryptophan or with any other groups in proteins apart from the SH groups; the reaction can be carried out in aqueous buffer solution at pH 5; the product of the reaction (a mixed disulfide) is stable in acid solution but is easily destroyed in even weakly alkaline solution, and this restricts the use of the reagent in studying the role of SH groups in proteins.

12. Concluding Remarks

The data presented in this chapter illustrate the high reactivity of SH groups and the unique diversity of the chemical reactions in which these groups participate. In most of the reactions of the SH group it takes part in the form of the mercaptide ion $R-S^-$. Accordingly the reactivity of SH groups depends on their pK and on the nucleophilicity of the $R-S^-$ ions. In proteins both these factors are determined by a great extent by the interactions of the SH group with the neighboring functional groups in the molecule.

In the table below reagents for SH groups are listed and the degree of their specificity is assessed.

Reagents for SH Groups in Proteins

Character of reaction	Reagent	Specificity
Mercaptide formation	p-Mercuribenzoate	Highly specific
	p-Mercuriphenylsulfonate	"
	Salyrganic acid (Mersalyl)	"
	(O-carboxymethylsalicyl-[3-hydroxymercuri-2-methoxy-propyl]amide)	"
	Phenylmercury acetate	"
	Methylmercury iodide	"
	Methylmercury bromide	"
	2-Chloromercuri-4-nitrophenol	"
	Mercuric chloride	"
	Silver nitrate	
	Arsenoxides of the R−As=O type	Particularly highly specific for dithiols
	Arsenite	"
Alkylation	Iodoacetate, bromoacetate, chloroacetate and their amides	Relatively specific under mild conditions
	N-(4-iodophenyl)iodoacetamide	"
	4-Bromoacetamido-2-nitrophenol	"
	2,2'-Dicarboxy-4'-iodoacet-amidodiazobenzene	"
	α-Iodopropionate	"
	β-Bromoethylamine	"
	Methyl iodide	"
	Ethyleneimine	"
	ω-Chloroacetophenone	"
	Methyl-p-nitrobenzenesulfonate	"
Addition to a double bond	N-Ethylmaleimide	Relatively specific (at concentrations not above 1 mM)
	N-(4-dimethylamino-3,5-di-nitrophenyl)maleimide	"
	N,N'-hexamethylene-bis-maleimide	"
	Bis(N-maleimidemethyl)ether	"
	Acrylonitrile	"

Reagents for SH Groups in Proteins (cont'd)

Character of reaction	Reagent	Specificity
Arylation	Fluoro-2,4-dinitrobenzene	Relatively specific (at a pH not above 5.5)
	7-Chloro-4-nitrobenz-2-oxa-1,3-diazole	Relatively specific
Oxidation	o-Iodosobenzoate	Highly specific (at pH 7.0)
	Tetrathionate	"
	Tetranitromethane	Relatively specific
	Iodine	Poorly specific
	Ferricyanide	"
	Porphyrindin	"
	Hydrogen peroxide	"
Thiol-disulfide exchange	5,5'-Dithio-bis-(2-nitrobenzoic acid)	Highly specific
	2,2'- and 4,4'-Dithiodipyridine	"
	2,2'-Dithio-bis-(5-nitropyridine)	"
	6,6'-Dithiodinicotinic acid	"
	β-Hydroxyethyl-2,4-dinitrophenyl disulfide	"
	Cystine	"
	Oxidized glutathione	"
	Cystamine	"
	Cystamine monosulfoxide	"
Sulfenylation	Azobenzene-2-sulfenyl bromide	Highly specific
	p-Nitrophenylsulfenyl chloride	Relatively specific

The most specific reagents are the disulfides, arsenite and the arsenoxides, and p-mercuribenzoate and the other organomercurial compounds. The remaining reagents for SH groups, including oxidants and alkylating agents, are only relatively specific; under some conditions they can react not only with SH but also with other functional groups of proteins. The question can therefore arise: Why are the less specific reagents needed when reagents exist that have a strict specificity for SH groups?

In practice both kinds are necessary. The advantage of the alkylating and oxidizing reagents over the organomercurials consists in the fact that they distinguish better between the various types of SH groups in a protein and allow selective blocking of the more accessible or more reactive SH groups in a molecule, leaving untouched the less accessible or less reactive. Thus, for example, in alcohol dehydrogenase p-mercuribenzoate blocks all 24 of the SH groups of the molecule of this enzyme; in contrast with p-mercuribenzoate, iodoacetate selectively alkylates only the two essential SH groups, which are evidently in the active site or close to it (Li and Vallee, 1963, 1965).

In pig heart aspartate aminotransferase, on the other hand, only two SH groups are accessible to iodoacetate, and these are unnecessary for enzymic activity (Polyanovskii and Telegdi, 1965). Two of the five SH groups of the aminotransferase (calculated per subunit) can be alkylated by iodoacetate without drop in activity; the subsequent blocking of one of the three remaining SH groups by p-mercuribenzoate leads to inhibition of the enzyme by 95% (Torchinskii and Sinitsyna, 1970).

Thus to obtain the full characterization of the SH groups of a particular enzyme, and to identify the essential groups, it is useful to study their interaction with thiol reagents of various types. Those SH groups that cannot react with p-mercuribenzoate are, as a rule, also incapable of reacting with alkylating, disulfide, and oxidizing reagents. On the other hand those SH groups that are inaccessible to alkylation of oxidation can often be blocked by p-mercuribenzoate. For identifying the SH groups that are situated in the active sites of enzymes, in addition to the reagents listed above (particularly the alkylating ones) one can use pseudosubstrates and substrate-like inhibitors (see Chapter 5).

Chapter 2

Chemical Properties of S–S Groups and Methods for Their Scission

1. The Geometry and Optical Properties of the Cystine Molecule

Valuable information on the spatial structure of cystine has been obtained by x-ray analysis of crystals of N,N'-diglycyl-L-cystine dihydrate. The geometry of the molecule of this disulfide is shown in Fig. 3, which shows that the dihedral angle between the two C–S bonds of the disulfide is 101° and that the length of the sulfur-sulfur bond is 2.04 Å. Similar data were obtained on X-ray crystallographic analysis of L-cystine hydrochloride (Steinrauf et al., 1958).

The bond between two bivalent sulfur atoms is stronger than the bond between two oxygen atoms; the dissociation energies of the S–S bond (in alkyldisulfides) and of the O–O bond (in dibutyl peroxide) are about 70 and 39 kcal/mole respectively (Franklin and Lumpkin, 1952; Parker and Kharasch, 1959).

Rotation about the S–S bond is greatly hindered; various authors calculate the energy barrier for such rotation to be from 3 to 14 kcal/mole (Scott et al., 1950, 1952; Fehér and Schulze-Rettmer, 1958; Bergson and Schotte, 1958; Bergson, 1962). This barrier is a consequence of the mutual electrostatic repulsion between the nonbonding p-electrons of the sulfur atoms. The repulsion is maximal when the dihedral angles is 0° or 180°, and minimal when it is 90° (Parker and Kharasch, 1959).

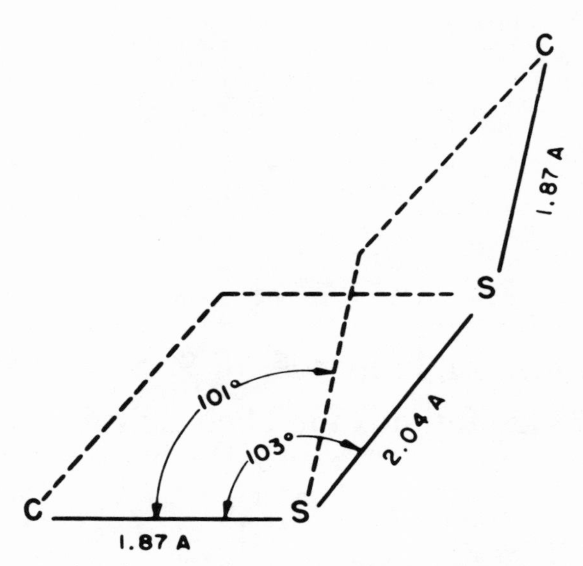

Fig. 3. The geometry of the molecule of N,N'-diglycyl-L-cystine in its dihydrate (Yakel and Hughes, 1954).

Dissymmetry (chirality) is intrinsic to the disulfide bond (Fig. 4), and this is the cause of the high optical rotation of cystine, which is greater by an order of magnitude than the rotations of the other amino acids (the specific rotations of L-cystine and of L-cysteine for the sodium D-line in 88% formic acid are −285° and +8° respectively; Turner et al., 1959).

The absorbance maximum of the disulfide groups of cystine lies at about 250 nm. In this region negative circular dichroism is observed for L-cystine and for oxidized glutathione (Fig. 5). In S—S-containing peptides and proteins the corresponding band of circular dichroism is at 250-270 nm; this band can be negative (e.g., in insulin) or positive (e.g., in oxytocin) (Beychok, 1965; Beychok and Breslow, 1968; Urry et al., 1968). The difference in sign apparently depends on the particular nature of the asymmetric environment of the S—S group or on its inherent chirality. Several

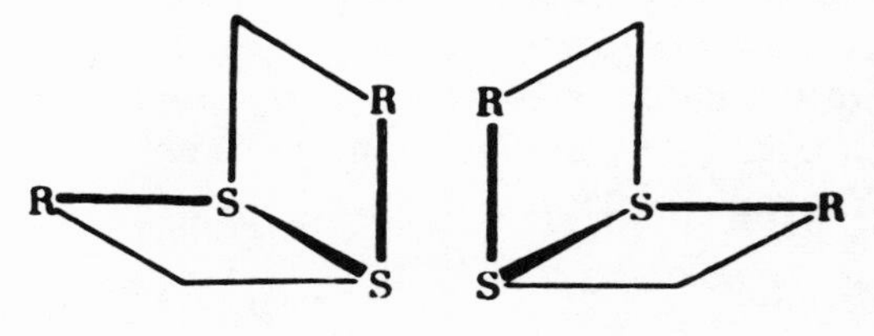

Fig. 4. The dissymmetric disulfide group (the forms with right- and left-handed screws; Strem et al., 1961).

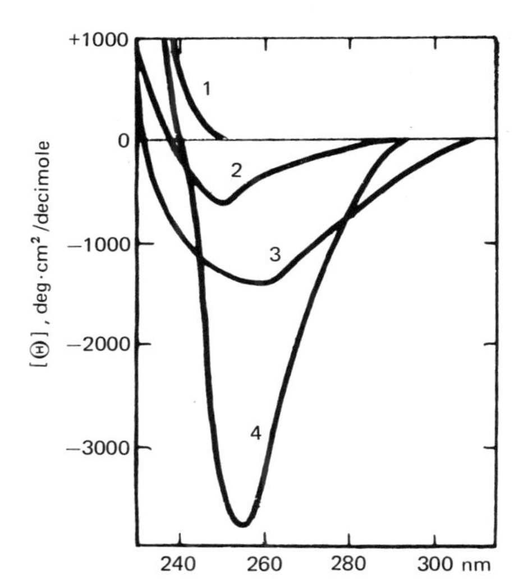

Fig. 5. Circular dichroism spectra of cystine and related compounds (Beychok, 1965). 1) Cysteinesulfinic acid in water; 2) cystinesulfoxide in 0.1 N H_2SO_4; 3) oxidized glutathione in water; 4) cystine in water (0.01%).

authors have established a correlation between the chirality of the disulfide group in model compounds and the sign of the two Cotton effects it generates in the 250-300 nm and 230-250 nm regions (Carmack and Neubert, 1967; Dodson and Nelson, 1968; Claeson, 1968; Ludescher and Schwyzer, 1971). The determination of the chirality of the S−S groups of peptides and proteins, however, from the sign of their Cotton effects, is hindered by the contributions of the peptide group and aromatic residue chromophores, and also by the influence of the asymmetrical environment in the protein.

On x-ray analysis of ribonuclease Wyckoff et al. (1970) found that three of the S−S groups in its molecule had left-handed chirality and the fourth apparently right-handed. In lysozyme the two conformations of the S−S group were also observed, the mirror images of each other; the dihedral angle about the S−S bond in the lysozyme molecule is $100 \pm 5°$ (Blake et al., 1967). In chymotrypsin three cystine residues are right-handed and two left-handed; the dihedral angles around the S-S bonds vary from 87° to 126° (Birktoft and Blow, 1972).

The contribution of S−S groups to the optical rotation of proteins was studied by Würz and Haurowitz (1961), Tanford et al. (1967), and Coleman and Blout (1968). The analysis made by Coleman and Blout of the dispersion curve of N,N'-diacetyl-L-cystine bis-(methylamide) showed the presence of two negative disulfide

Cotton effects at 262 and 199 nm, the rotatory strengths of which were -1.64×10^{-40} erg·cm^3 and -33.7×10^{-40} erg·cm^3 respectively. The molar rotation of the disulfide group at 589 nm is 436°, and at 210 nm (the trough of the Cotton effect) it is about −24,000°. According to the calculations of Coleman and Blout the contribution of disulfide Cotton effects to the optical rotation at 210 nm can exceed 10% in the case of a polypeptide containing two S—S links per 100 residues, 25 to 40 of which are in an α-helix.

2. Reaction with Metal Ions

Silver and mercury ions accelerate the hydrolytic breakage of S—S bonds in alkaline medium. Cecil and McPhee (Cecil, 1950; Cecil and McPhee, 1957, 1959) studied the kinetics of the reaction of disulfides with silver salts. They showed that the initial rate of hydrolysis was proportional to the concentration of Ag^+ ions and on this basis they came to the conclusion that the role of the metal ions is not solely the displacement of the equilibrium of the hydrolytic splitting of the S—S bond; in their opinion the metal ions combine with the S—S bond with formation of a complex which is then hydrolyzed by nucleophilic attack by hydroxide ion (see also Klotz and Campbell, 1962):

$$2RS{-}SR + 2Me^+ \rightleftarrows 2\begin{bmatrix} RS{-}SR \\ | \\ Me \end{bmatrix}^+$$

$$2\begin{bmatrix} RS{-}SR \\ | \\ Me \end{bmatrix}^+ + 2OH^- \rightleftarrows 2RSMe + 2RSOH$$

$$2RSOH \rightleftarrows RSO_2H + RSH$$

$$RSH + Me^+ \rightleftarrows RSMe + H^+$$

$$\overline{2RS{-}SR + 3Me^+ + 2OH^- \rightleftarrows 3RSMe + RSO_2H + H^+}$$

The final products of the reaction are a mercaptide and a sulfinic acid; each mole of disulfide forms 1.5 moles of mercaptide.

Brown and Edwards (1969) found that $HgCl_2$ and $Hg(NO_3)_2$ added in high concentrations reacted rapidly in acid medium with cyclic and linear disulfides, in particular with α-lipoic acid, splitting the S—S bond; they proposed the following reaction mechanism:

$$RS{-}SR + Hg(NO_3)_2 \rightarrow \begin{matrix} R{-}S\cdots S{-}R \\ \vdots \quad\; \vdots \\ O_2NO\cdots HgNO_3 \end{matrix} \rightarrow RS{-}NO_3 + RS{-}HgNO_3$$

An important question concerns the degree of resistance of the S—S bonds of proteins to ions of heavy metals that are used for titrating the SH groups of the proteins. The data in the literature lead to the conclusion that the S—S bonds of native proteins do not, as a rule, react with metal ions (at least in the absence of great excess) at room temperature and pH 4-8 (Hughes, 1954, Edelhoch et al., 1953; Gurd and Wilcox, 1956; Forbes and Hamlin, 1968, 1969). In harsher conditions (high concentrations of metal ions, strongly alkaline media, or neutral media at raised temperatures), however, metal ions and organomercurial compounds can cause splitting of the S—S bonds of proteins. Thus, according to Cunningham et al. (1957), p-mercuribenzoate catalyses the splitting of S—S bonds in cystine, insulin, and ribonuclease at pH 7.0 if the incubation is carried out at 83°C for two hours.

In contrast with cystine, the sulfoxide of cystine reacts rapidly with $HgCl_2$ and $AgNO_3$ at pH 2-5 and with organomercurial compounds at pH 4-7 and 20°C with the formation of a mercaptide and of sulfinic acid (Maclaren et al., 1965; Savige and Maclaren, 1966):

$$R-\underset{\underset{O}{\|}}{S}-S-R + CH_3HgI + H_2O \rightarrow R-S-HgCH_3 + R-SO_2H + HI$$

As Maclaren and coworkers showed, the sulfoxide of cystine is formed during the oxidation of certain proteins by dilute performic acid under mild conditions. The formation of sulfoxides can thus be the cause of raised results in titrations of protein SH groups.

3. Reaction with Sulfite

Clarke (1932) first showed that cysteine and cysteine-S-sulfonate are formed by reaction of sulfite with cystine according to the following equation:

$$RS-SR + SO_3^{2-} \rightleftarrows RS^- + RS-SO_3^-$$

Cecil and McPhee (1955, 1959; McPhee, 1956) studied the kinetics of the reaction of a number of low-molecular disulfides with sulfite; they found that the rate of the reaction with the HSO_3^- ion was extremely small in comparison with the rate of the reaction with the SO_3^{2-} ion. The rate of the reaction depends largely on the charge on the groups in the disulfide molecule that are close to the S—S bond; negatively charged groups, e.g., carboxyl groups, markedly inhibit the reaction, while positively charged groups raise the rate of the reaction many times. The presence of branched hydrocarbon chains close to the S–S bond greatly slows the reaction and occasionally prevents it completely (in, for example, the disulfide of penicillamine; Rosenthal and Oster, 1954). The preferential attack by SO_3^{2-} on one of the sulfur atoms in asymmetric disulfides can be explained by the influence of steric factors (Van Rensburg and Swanepoel, 1967).

Fava and Iliceto (1958) studied the isotope exchange reaction between alkylthiosulfonates and $^{35}SO_3^{2-}$:

$$RS{-}SO_3^- + \overset{*}{S}O_3^{2-} \rightleftarrows RS{-}\overset{*}{S}O_3^- + SO_3^{2-}$$

According to their data the exchange reaction goes readily only when the replacing agent can approach the S–S bond at an angle of 180°. The larger the radical R, the slower the replacement reaction proceeds. The relative rates of reaction when R is methyl, ethyl, isopropyl, and tertiary butyl are 100, 50, 0.7, and 0.0006 respectively.

These data indicate the nature of the factors that can determine the reactivity to sulfite of the S–S bonds in proteins. Among these factors, therefore, are the electrostatic influence of neighboring polar groups and favorable steric conditions for nucleophilic attack by SO_3^{2-} on the S–S bond. When proteins are denatured the reactivity of the S–S bonds usually increases greatly. According to Cecil and coworkers (Cecil and Loening, 1960; Cecil and Wake, 1962) there are marked differences in the reactivity of interchain and intrachain S–S bonds in proteins. The interchain bonds (e.g., the bonds between the A and B chains in insulin) usually react readily with sulfite at pH 6.5-7.0 in the absence of denaturing agents. The splitting by sulfite of intrachain bonds (for example in lysozyme, ribonuclease, serum albumin, and the bond between residues 6 and 11 in the A chain of insulin) occurs only in the presence of urea, guanidinium salts or mercaptide-forming reagents. The resistance of the intrachain S–S bonds to sulfite was explained by Cecil by the fact that they form a kind of ring structure stabilized by noncovalent interactions. X-ray structural analysis of insulin showed, in complete agreement with the chemical findings, that the interchain S–S bonds are accessible to solvent (although to different degrees) but that the intrachain bond between residues 6 and 11 of the A chain is completely masked and forms part of the nonpolar core of the molecule (Blundell et al., 1971).

Mercaptide-forming reagents block the SH groups that arise on sulfitolysis and thus shift the equilibrium of the reaction to the right. A convenient analytical method for determining the S–S content of proteins is based on this (see Chapter 3, Section 2). Leach (1960) found that when S–S bonds of proteins are broken by sulfite in the presence of $HgCl_2$ and urea, one Hg^{2+} ion combines with two SH groups; he proposed that the product of the reaction can have structure I or II; in his view structure II, in which the mercury ion unites two sulfur atoms that previously formed an S–S bond, was the more probable:

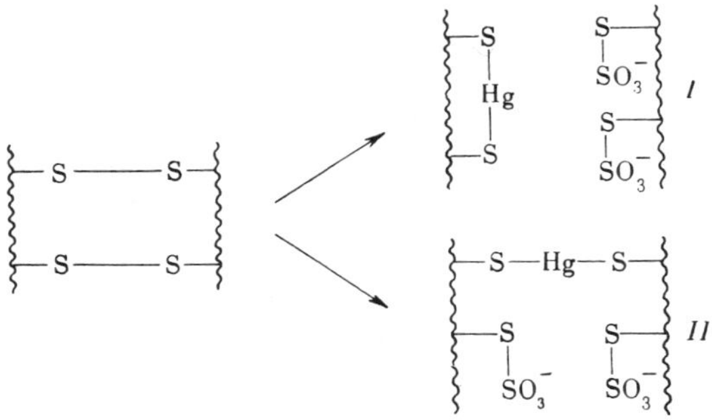

Milligan and Swan (1962) expressed the opinion that product II arises as a result of a series of reactions including exchange between S–S and thiosulfonate groups. A possible reaction sequence is shown below (Crewther et al., 1965):

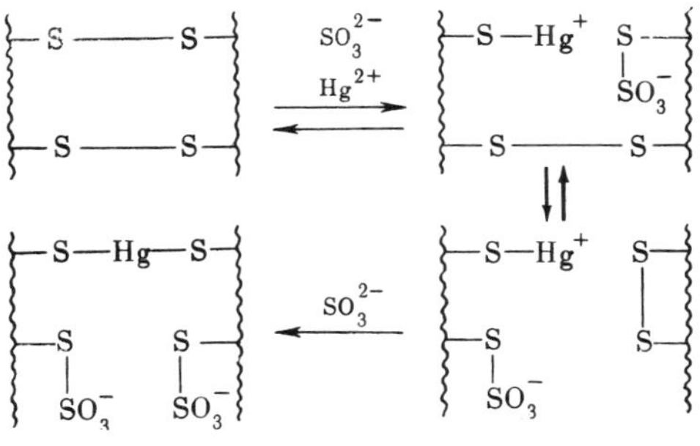

The SH groups that arise on splitting of S–S groups by sulfite are often unstable and can be oxidized by atmospheric oxygen. To obtain a stable final product in the form of S–sulfoprotein (which is essential in order to separate polypeptide chains that were joined by S—S bridges, and in other structural investigations) the reaction with sulfite may be carried out in the presence of mild oxidizing agents: Cu^{2+} (Kolthoff and Stricks, 1951a,b; Swan, 1957b; Leach et al., 1963), o-iodosobenzoate or tetrathionate (Bailey and Cole, 1959; Cole, 1967). The role of these oxidizing agents is apparently that they reconvert the SH groups into disulfides which can react with sulfite; the cycle of oxidation—reduction reactions continues until every S—S bond in the protein is converted into two thiosulfonate groups (residues of S-sulfocysteine):

$$RS{-}SR + SO_3^{2-} \rightarrow RS{-}SO_3^- + RS^-$$

$$\text{(}RS^- \xrightarrow{\text{Oxidizing agent}} RS{-}SR\text{)}$$

The ease of such a conversion varies between different proteins. Thus, for example, it occurs in ribonuclease only in the presence of 8 M urea and o-iodosobenzoate. In chymotrypsin complete conversion of the S—S bonds into thiosulfonate groups proceeds in the absence of urea and of o-iodosobenzoate; the oxygen of the air is a sufficiently effective oxidizing agent in this case.

In insoluble proteins, e.g., keratin, the SH groups that arise on sulfitolysis are, for steric reasons, not always able to form new intramolecular S—S bonds. Milligan and Swan (1962) proposed that oxidative sulfitolysis in the presence of tetrathionate and copper ions may proceed without the intermediate formation of such bonds; in the presence of tetrathionate, for example, they thought that the following series of reactions took place:

$$RS{-}SR + SO_3^{2-} \rightleftarrows RS^- + RS{-}SO_3^-$$

$$RS^- + {}^-O_3S{-}S{-}S{-}SO_3^- \rightleftarrows RS{-}S{-}SO_3^- + S_2O_3^{2-}$$

$$RS{-}S{-}SO_3^- + SO_3^{2-} \rightleftarrows RS{-}SO_3^- + S_2O_3^{2-}$$

Amino acid analysis of the S-sulfoproteins showed that sulfite reacted selectively with the S—S groups of proteins and did not touch other amino acid residues with the exception of tryptophan, the loss of which was 5-7% (Bailey and Cole, 1959). The S-sulfonate groups are stable at neutral pH, but are rapidly destroyed upon acid hydrolysis of the protein (Lindley, 1959), and also on treatment with excess thiol (Footner and Smiles, 1925; Swan, 1957b):

$$RS{-}SO_3^- + R'S^- \rightleftarrows RS{-}SR' + SO_3^{2-}$$

$$RS{-}SR' + R'S^- \rightleftarrows R'S{-}SR' + RS^-$$

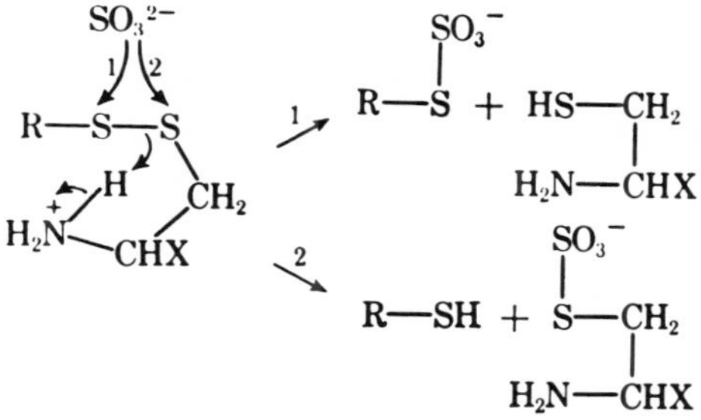

$$R{-}S{-}S{-}CH_2{-}CHX{-}\overset{+}{N}H_2{-}H + SO_3^{2-} \xrightarrow{1} R{-}S{-}SO_3^- + HS{-}CH_2{-}CHX{-}NH_2$$

$$\xrightarrow{2} R{-}SH + {}^{-}O_3S{-}S{-}CH_2{-}CHX{-}NH_2$$

Fig. 6. The reaction of sulfite with the mixed disulfide of a protein with cysteine or β-mercaptoethylamine (Chan, 1968). R = protein; X = COOH group (in cysteine) or H (in β-mercaptoethylamine).

Chan (1968) proposed an improved method of preparing S-sulfo derivatives of proteins. According to his recipe, the protein (2 mg/ml) is treated with Na_2SO_3 (0.05 M) in 0.1 M tris-HCl buffer (pH 8.4) in the presence of atmospheric oxygen, 8 M urea and catalytic quantities of cysteine (0.2 mM). The cysteine may be replaced in this system by β-mercaptoethylamine, but not by β-mercaptoethanol or dithiothreitol, which indicates the importance of the amino group. To explain the effect of the amino group, Chan proposed the reaction scheme shown in Fig. 6. This figure shows the reaction to occur by means of nucleophilic attack of the SO_3^{2-} ion on one of the sulfur atoms of the disulfide; this attack is facilitated by the fact that the atom of sulfur that is being displaced can acquire a proton from the nearby protonated amino group which thus fulfills the role of a general acid catalyst. The preparation of S-sulfo-derivatives of proteins by the method of Chan avoids side reactions that can occur when tetrathionate or copper ions are used. The catalytic activity of S-sulfo-aldolase prepared by this method is completely restored on addition of excess β-mercaptoethanol.

A different approach to the complete conversion of disulfide groups into their S-sulfo derivatives is that of Vanaman and Stark (1970; see also Stark and Crawford, 1972). They first reduced the S–S bonds to SH groups and then used Ellman's reagent to form the mixed disulfide. The asymmetrical nature of this mixed disulfide (Chapter 1, Section 10) allowed quantitative formation of the S-sulfo derivative of the protein when the disulfide was treated with sulfite.

4. Reaction with Cyanide

The reaction of disulfides with cyanide is in many respects analogous with their reaction with sulfite; it also takes place by nucleophilic displacement of one of the sulfur atoms of the disulfide:

$$RS{-}SR + CN^- \rightleftarrows RS^- + RS{-}CN$$

Gawron et al. (1961, 1964; Gawron, 1966) studied the dependence of the reaction on pH, temperature and cyanide concentration. They found that the bimolecular rate constants of the reactions of cystine and its derivatives with the cyanide anion are much smaller than the corresponding rate constants for reaction with the sulfite ion. These rates are greatly increased on protonation of the amino group of cystine. The cause of this increase may not be simply a direct electrostatic effect, but may be partly intramolecular catalysis by a hydrogen ion:

$$CN^- \; S(R){-}S{\frown}N^+H_3 \rightarrow NC{-}S(R) + HS{\frown}NH_2$$

When cyanide reacts with cystine the products are cysteine together with β-thiocyanoalanine, which readily cyclizes to form 2-aminothiazoline-4-carboxylic acid or its tautomer 2-iminothiazolidine-4-carboxylic acid (Schöberl et al., 1951):

$$NCS{-}CH_2{-}CH(NH_2){-}COOH \rightarrow H_2C{-}CH{-}COOH\ (\text{ring: } S{-}C(NH_2){=}N) \rightarrow H_2C{-}CH{-}COOH\ (\text{ring: } S{-}C({=}NH){-}NH)$$

When the S—S bond of oxidized glutathione is split by cyanide at a pH below 8, an iminothiazolidine is formed (Catsimpoolas and Wood, 1966):

$$\begin{array}{l} CH_2{-}CH{-}CONH{-}CH_2{-}COOH \\ SCN \;\; NH{-}CO{-}[CH_2]_2{-}CH(NH_2){-}COOH \end{array} \rightarrow \begin{array}{l} H_2C{-}CH{-}CONH{-}CH_2{-}COOH \\ S \;\; N{-}CO{-}[CH_2]_2{-}CH(NH_2){-}COOH \\ \;\; C{=}NH \end{array}$$

A more alkaline medium was unfavorable for cyclization, since it led to elimination of thiocyanate ion. The peptide bond whose nitrogen belongs to the iminothiazolidine is easily hydrolyzed. Thus a method is provided for the specific splitting of peptides on the NH-side of cystine residues, and Catsimpoolas and Wood (1966) applied this scission to oxytocin and to pancreatic ribonuclease. Jacobson et al. (1973) found optimum scission in guanidine hydrochloride at pH 9. They produced the residues of thiocyanoalanine either by consecutive treatment with Ellman's reagent and cyanide, or by treatment with 2-nitro-5-thiocyanatobenzoic acid. This was made by treating Ellman's reagent first with cyanide and then with cyanogen bromide (Patchornik and Degani, 1971).

At pH 8 and above residues of β-thiocyanoalanine in proteins are destroyed with elimination of thiocyanate ion and formation of dehydroalanine (α-aminoacrylate) residues (Catsimpoolas and Wood, 1964, 1966):

$$NCS{-}CH_2{-}\underset{\diagdown CO{-}}{\overset{\diagup NH{-}}{CH}} \xrightarrow{OH^-} SCN^- + CH_2{=}\underset{\diagdown CO{-}}{\overset{\diagup NH{-}}{C}}$$

The latter can react with the SH group of cysteine to form lanthionine, as Cuthbertson and Phillips (1945) first suggested:

$$CH_2{=}\underset{\diagdown CO{-}}{\overset{\diagup NH{-}}{C}} + HS{-}CH_2{-}\underset{\diagdown CO{-}}{\overset{\diagup NH{-}}{CH}} \rightarrow \underset{{-}OC\diagup}{\overset{{-}HN\diagdown}{CH}}{-}CH_2{-}S{-}CH_2{-}\underset{\diagdown CO{-}}{\overset{\diagup NH{-}}{CH}}$$

Swan (1961) showed that thiocyanate ion is not eliminated on incubation of S-cyanokerateine at pH 8.2 in the absence of cysteine; on dialysis against a solution of cysteine thiocyanate is eliminated and lanthionine is formed. On the basis of these experiments Swan suggested that lanthionine is formed as a result of direct attack of an ionized SH group (either preformed or arising by splitting of an S–S bond in the protein) on a residue of β-thiocyanoalanine:

$$\underset{{-}OC\diagup}{\overset{{-}HN\diagdown}{CH}}{-}CH_2{-}S^- + NCS{-}CH_2{-}\underset{\diagdown CO{-}}{\overset{\diagup NH{-}}{CH}} \rightarrow \underset{{-}CO\diagup}{\overset{{-}NH\diagdown}{CH}}{-}CH_2{-}S{-}CH_2{-}\underset{\diagdown CO{-}}{\overset{\diagup NH{-}}{CH}} + SCN^-$$

Evidently the pH and other experimental conditions determine which of these pathways of lanthionine formation predominate in proteins treated with cyanide.

5. Reaction with Monothiophosphoric Acid

Neumann et al. (1964, 1967) used monothiophosphoric acid* for splitting the disulfide bonds of proteins. The oxidation–reduction potential of this acid is lower than that of cysteine; consequently it is incapable of the simple reduction of disulfides. According to Neumann and Smith (1967), the splitting of the S–S bond of cystine by the monothiophosphate anion is a nucleophilic heterolytic splitting, which goes according to the following equation:

$$RS{-}SR + SPO_3^{3-} \rightleftarrows RS{-}SPO_3^{2-} + RS^-$$

The reaction of monothiophosphate with cystine is reversible over the pH range 4.5-9.5; the rate of the reaction depends on pH, being maximal at about pH 9.7. The equilibrium constant of the reaction with cystine is 2.2×10^{-1} M^{-1} at pH 8.0 and 9×10^{-2} M^{-1} at pH 6.8.

When monothiophosphate reacts with proteins in the presence of oxygen, two thiophosphate groups combine per S–S bond split:

$$RS{-}SR + 2HPSO_3^{2-} + {}^{1}/{}_{2}O_2 \rightarrow 2RS{-}SPO_3^{2-} + H_2O$$

According to Neumann et al., the reaction of ribonuclease or lysozyme with monothiophosphate in the presence of 8 M urea leads to the splitting of all four S–S bonds and the formation of an inactive derivative of the enzyme which contains eight S-thiophosphate groups, one for each half-cystine residue. In the absence of urea monothiophosphate selectively splits two of the S–S bonds of ribonuclease with complete retention of the catalytic activity of the enzyme. Thiophosphate groups can easily be split off from the protein with β-mercaptoethanol. Berezin et al. (1972) used sodium monothiophosphate for splitting the S–S bonds of α-chymotrypsin. They treated the protein with a 100-fold excess of reagent at pH 7.8 for 16 h, and found that under these conditions two of the S–S

* Monothiophosphoric acid (phosphorothioate) is the derivative of orthophosphoric acid in which one of the oxygen atoms is replaced by a sulfur atom.

bonds of the molecule (out of five) were split with retention of catalytic activity.

6. Scission in Alkaline Medium

When disulfides are heated in strong alkali the S–S bonds are split and one of the reaction products is a thiol. Three different mechanisms have been put forward for the reaction:

a. A direct attack by hydroxide ion on the S–S bond (nucleophilic displacement) with formation of a mercaptide ion and a sulfenic acid (Schöberl, 1933; Schöberl and Rambacher, 1939):

$$RS{-}SR + OH^- \rightleftarrows RS^- + RSOH$$

The sulfenic acid is unstable and can undergo various further changes, such as

$$\begin{gathered} RCH_2SOH \rightarrow RCHO + H_2S \\ 2RCH_2SOH \rightarrow RCH_2SH + RCOOH + H_2S \\ 2RCH_2SOH \rightarrow RCH_2SH + RCH_2SO_2H \\ RCH_2CH_2SOH \rightarrow RCH{=}CH_2 + H_2O + S \end{gathered}$$

b. A primary splitting of the C–S bond, evoked by removal of the proton from the carbon atom that occupies the β-position with respect to the sulfur atom ("β-elimination") (Tarbell and Harnish, 1951):

$$\begin{array}{c} | \\ CO \\ | \\ CH{-}CH_2S{-}SCH_2{-}CH \\ | \\ NH \\ | \end{array} \begin{array}{c} | \\ CO \\ | \\ \\ | \\ NH \\ | \end{array} + OH^- \xrightarrow{-H_2O} \left[\begin{array}{c} | \qquad\qquad\qquad | \\ CO \qquad\qquad\quad CO \\ | \qquad\qquad\qquad | \\ CH{-}CH_2S{-}SCH_2{-}C^- \\ | \qquad\qquad\qquad | \\ NH \qquad\qquad\quad NH \\ | \qquad\qquad\qquad | \end{array}\right]$$

$$\left[\begin{array}{c} | \qquad\qquad\qquad | \\ CO \qquad\qquad\quad CO \\ | \qquad\qquad\qquad | \\ CH{-}CH_2S{-}SCH_2{-}C^- \\ | \qquad\qquad\qquad | \\ NH \qquad\qquad\quad NH \\ | \qquad\qquad\qquad | \end{array}\right] \rightarrow \begin{array}{c} | \\ CO \\ | \\ CH_2{=}C \\ | \\ NH \\ | \end{array} + \begin{array}{c} | \\ CO \\ | \\ CH{-}CH_2S{-}S^- \\ | \\ NH \\ | \end{array}$$

$$\begin{array}{c} | \\ CO \\ | \\ CH{-}CH_2S{-}S^- \\ | \\ NH \\ | \end{array} \rightarrow S + \begin{array}{c} | \\ CO \\ | \\ CH{-}CH_2S^- \\ | \\ NH \\ | \end{array}$$

The disulfide anion (persulfide) that is formed as an intermediate can give inorganic sulfide by reaction with a thiol or with water:

$$RSS^- + R'SH \rightleftarrows RS{-}SR' + HS^-$$
$$RSS^- + H_2O \rightleftarrows RSOH + HS^-$$

c. A primary splitting of the C—S bond, evoked by removal of the proton from the carbon atom that occupies the α-position with respect to the sulfur atom ("α-elimination") (Rosenthal and Oster, 1954, 1961):

$$RCH_2S{-}SCH_2R + OH^- \rightarrow [R\bar{C}HS{-}SCH_2R] + H_2O$$
$$RCHO + H_2S \xleftarrow{H_2O} RCH{=}S + RCH_2S^-$$

There are many papers on investigations of the mechanism of the splitting of disulfides in alkaline media, and their results are often somewhat contradictory. According to Wronski (1963) the stoichiometry and kinetics of the hydrolysis of cystine in 1 M NaOH at 60° agree with mechanism a. Zahn and Golsch (1963) came out in favor of either mechanism a or c starting from the fact of the instability of the S—S bond of α,α'-diphenylcystine. Anderson and Berg (1969) found that the splitting of the S—S bond of oxidized glutathione at pH 9.2-11.0 in the presence of p-mercuribenzoate followed mechanism a and led to the formation of the mercaptide of glutathione and the sulfinic acid. Donovan and White (1971) studied the alkaline cleavage of the disulfide bonds of ovomucoid and of low-molecular disulfides. They concluded that a thiol was formed in the rate-determining step of the cleavage; the observed changes in UV absorption spectra agreed with liberation of mercaptide ion and gave no evidence of the formation of persulfide ion. The data appear to favor mechanism a, but the evidence is far from conclusive. Many experimental data, on the other hand, agree well with mechanism b (Danehy, 1966; Gawron and Odstrchel, 1967). The data of Swan (Swan, 1957a; Stapleton and Swan, 1960), in particular, speak convincingly in favor of this mechanism. They argue from the extreme slowness of the splitting of the S—S bond in α,α'-dimethylcystine, in which the hydrogen atom that is necessary for ionization is replaced by a methyl group; α,α'-dimethylcystine is hardly hydrolyzed at all on boiling in 0.25 M NaOH for 3 h; under these conditions cystine is 89% split in 1 h. Schneider and Westley (1969) obtained evidence for the formation of persulfide anions, RS−S⁻ (predicted by scheme b) on incubating glutathione and insulin in 0.5 M NaOH. They noted that cystine residues in peptides and proteins form persulfide much more easily than does free cystine, and they explained this by the fall in electron density on the "β-carbon" of cystine (β in relation to the S−S bond) on formation of the peptide bond. Modifications of the structure of cystine that led to a fall in electron density at this atom (acetylation of the amino group, esterification of the carboxyl group, formation of a Schiff base with pyridoxal phosphate) facilitated the formation of persulfide.

The data of Donovan and White (1971) on the kinetics of splitting of S−S bonds in ovomucoid and in a number of low-molecular disulfides are given in Table 10. It is clear from this table that the S−S bonds of ovomucoid are split significantly more rapidly than those of aliphatic disulfides. This can perhaps be explained, on the basis of the β-elimination mechanism, by the electron-attracting properties of the peptide groups. Interestingly papain can apparently catalyze the alkaline scission of Ellman's reagent (Brocklehurst et al., 1972). Apparently its mixed disulfide with the reagent is more reactive to hydroxide ion than the reagent itself.

A number of supplementary factors can influence the splitting of the disulfide bonds of proteins in alkali; these must include the accessibility of the bonds to the solvent, the nature of the neighboring groups and the strain on the bonds.

When some disulfide-containing proteins are treated with alkali, residues of lanthionine or lysinoalanine [N^6-(DL-2-amino-2-carboxyethyl)-L-lysine] are formed (Horn et al., 1941, 1942; Cuthbertson and Phillips, 1945; Ziegler, 1964; Bohak, 1964; Crewther et al., 1967; Donovan and White, 1971). The formation of lanthionine can be explained by addition of the SH group of a cysteine residue to the double bond of dehydroalanine that has been formed by the hydrolysis of cystine by mechanism b:

$$\begin{array}{c} | \\ NH \\ | \\ CH{-}CH_2{-}SH \\ | \\ CO \\ | \end{array} + \begin{array}{c} | \\ NH \\ | \\ CH_2{=}C \\ | \\ CO \\ | \end{array} \rightarrow \begin{array}{ccc} | & & | \\ NH & & NH \\ | & & | \\ CH{-}CH_2{-}S{-}CH_2{-}CH \\ | & & | \\ CO & & CO \\ | & & | \end{array}$$

Lysinoalanine is evidently formed by the combination of the ε-

TABLE 10. Second-Order Rate Constants and Activation Energies for the Cleavage of Disulfide Bonds (Donovan and White, 1971)

Compound	pH	Ionic strength	k ($M^{-1} \cdot sec^{-1}$) at 40°C	Activation energy, kcal/mole
Ovomucoid	12.8	0.13	5.8×10^{-2}	18.8 ± 0.5
Ovomucoid	12.1	0.03	3.1×10^{-2}	19.9 ± 1.1
2,2'-Dithiodiethanol	13.6	0.54	1.6×10^{-4}	18.8 ± 0.6
3,3'-Dithiodipropionic acid	13.8	1.00	1.9×10^{-5}	19.4 ± 0.8
5,5'-Dithiobis (2-nitrobenzoate)	11.6	0.005	1.3	14.9 ± 0.3

amino group of a lysine residue with dehydroalanine (Bohak, 1964):

$$\begin{array}{l} | \\ NH \\ | \\ CH-[CH_2]_4-NH_2 \\ | \\ CO \\ | \end{array} + \begin{array}{r} | \\ NH \\ | \\ CH_2=C \\ | \\ CO \\ | \end{array} \rightarrow \begin{array}{l} | \\ NH \\ | \\ CH-[CH_2]_4-NH-CH_2- \\ | \\ CO \\ | \end{array}\begin{array}{l} | \\ NH \\ | \\ CH \\ | \\ CO \\ | \end{array}$$

The formation of lysinoalanine residues in proteins occurs only at a pH above 12. Under milder conditions (pH 9-11) the hydrolytic cleavage of S−S groups by mechanism a can occur (Andersson, 1970). The hydrolysis proceeds, however, to a significant extent only if the SH groups that are formed are removed from the equilibrium mixture by thiol−disulfide exchange or blocking with a thiol reagent.

7. Reduction with Thiols

The reduction of the S−S bonds of low-molecular disulfides and of proteins with thiols proceeds according to the following equations:

$$RS{-}SR + R'S^- \rightleftarrows RS{-}SR' + RS^-$$
$$RS{-}SR' + R'S^- \rightleftarrows R'S{-}SR' + RS^-$$

The mechanism of this reaction was discussed in Chapter 1, Section 10, where the reverse reaction of the oxidation of the SH groups of proteins by disulfides was described; the reaction takes place in two stages with the formation of a mixed disulfide at the intermediate step.

Complete reduction of disulfides is obtained only in the presence of excess thiol. Experiments with low-molecular disulfides have shown that the presence close to the S−S bond of negatively charged groups which repel RS^- ions, or of branched hydrocarbon chains, which can hinder their approach, greatly slows the reaction (Foss et al., 1957; Hird, 1962; Weber et al., 1970). Weber et al. (1970) investigated the thiol−disulfide exchange reaction between reduced glutathione and a series of S-(alkylmercapto)cysteines of the type Cys−S−S−R, where R is an alkyl group. They found that the rate of reaction was inversely proportional to the electron-donating inductive effect of the alkyl substituent. This effect leads to an increase of electron density on the S−S group and consequently hinders the attack of RS^- ions on it.

β-Mercaptoethanol, β-mercaptoethylamine, and, more recently, dithiothreitol have been used to split the S—S bonds of proteins. Previously β-mercaptoacetic (thioglycolic) acid was often used, but White (1960) showed that this reagent has a serious disadvantage. When it is stored in the pure (anhydrous) state, thioester bonds are slowly formed by intermolecular condensation. Hence preparations of thioglycolic acid that have not been very recently distilled not only reduce disulfide bonds, but in neutral solution they may also acylate the many types of nucleophilic groups that occur in proteins.

The reactivity of the S—S bonds varies from protein to protein; it depends on the accessibility of the S—S bond to the thiol, the nature of the neighboring groups, and the strength of the noncovalent interactions in the vicinity of the bond; usually interchain S—S bonds are split more easily than intrachain ones. In most proteins some or all of the S—S bonds resist reduction in the absence of denaturing agents such as urea, guanidinium salts, or sodium dodecyl sulfate. An exception is deoxyribonuclease, whose S—S bonds are split with unusual ease by β-mercaptoethanol at pH 7.2 and room temperature in the absence of urea (Price et al., 1969). As an example of the differences in reactivity between different S—S groups of the same protein, one can cite data on pepsin, on papain, and on trypsinogen. If native pepsin, which contains three S—S bonds, is treated with β-mercaptoethanol at pH 5-6, two of them are reduced; reduction of the third occurs only in the presence of 8 M urea or 3-4 M guanidine hydrochloride (Blumenfeld and Perlmann, 1961; Nakagawa and Perlmann, 1971). In papain only one of its three S—S bonds is reduced by 0.32 M β-mercaptoethanol in 8 M urea, and complete reduction of the S—S bonds can be achieved only in 6 M guanidine hydrochloride (Shapira and Arnon, 1969). In trypsinogen, the S—S bond between residues 179 and 203 is readily reduced by 0.5 mM dithioerythritol (or 0.1 M $NaBH_4$) in the absence of denaturing agents, the bond between residues 122 and 189 requires 10 mM dithioerythritol for its reduction, and the remaining four S—S bonds cannot be reduced in the native protein (Sondack and Light, 1971); from these data Sondack and Light divided the S—S groups into three types: fully exposed, partly buried, and buried (nonreactive in the native state).

The protein after reduction is usually separated from the excess thiol and the number of SH groups formed is determined by amperometric or spectrophotometric titration (see Chapter 3).

Anfinsen and Haber (1961) recommended the following procedure for the reduction of the S–S bonds or ribonuclease: 350 mg of the protein is dissolved in 10 ml of a freshly prepared 8 M solution of recrystallized urea; the pH is brought to 8.6 with 5% methylamine, after which β-mercaptoethanol is added (1 μl per mg of protein). The mixture is incubated for 4.5 h at room temperature under nitrogen, and the pH is then brought to 3.5 with acetic acid. The protein is separated from reagents by passage through a column of Sephadex G-25 previously equilibrated with 0.1 M acetic acid.

The SH groups that arise upon reduction of S–S bonds may be oxidized by atmospheric oxygen, or they may participate in thiol–disulfide exchange reactions; either of these can lead to aggregation of the protein. To avoid such complications and to obtain a stable preparation of the reduced protein the SH groups are usually blocked by treatment with iodoacetate, iodoacetamide, or ethyleneimine (see, for example, Humbel et al., 1968).

In recent years two new and efficaceous reagents have been used with great success for the reduction of the disulfide bonds of proteins. These were introduced by Cleland (1964) and are dithiothreitol and dithioerythritol, which are respectively the threo- and erythro-isomers of 2,3-dihydroxy-1,4-dimercaptobutane. These reagents react with S–S groups according to the following equations:

$$RS{-}SR + HS{-}CH_2[CHOH]_2CH_2{-}SH \rightleftarrows RSH + RS{-}SCH_2[CHOH]_2CH_2SH$$

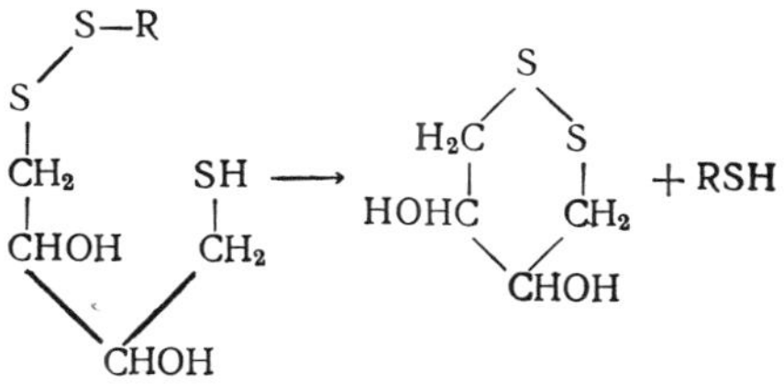

```
   S—R
  /
 S                            S
 |                           / \
 CH2        SH  ——→     H2C     S
 |          |            |      |     + RSH
 CHOH       CH2        HOHC     CH2
    \      /               \   /
     CHOH                  CHOH
```

The equations show that the reaction consists of a thiol–disulfide exchange whose second step occurs intramolecularly and leads to the formation of a stable cyclic disulfide. Thanks to this, the equilibrium of the reaction is greatly displaced to the right; the equilibrium constant for the reduction of cystine by dithiothreitol is 1.3×10^4 M.

Bewley et al. (1968) found that excess dithiothreitol at pH 8.1 quantitatively reduced the S–S bonds of hypophyseal growth hormone, lysozyme, serum albumin, and some other proteins, all in the absence of denaturing agents. Gorin et al. (1968) compared the rates of reduction of the S—S bonds of lysozyme by β-mercaptoethanol, 3-mercaptopropionate, β-mercaptoethylamine, and dithiothreitol; dithiothreitol gave the fastest reduction.

8. Disulfide Exchange

The disulfide exchange reaction leads to the formation of mixed disulfides:

$$R'S{-}SR' + R''S{-}SR'' \rightleftarrows 2R'S{-}SR''$$

The reaction mechanism differs in alkaline and acid media (see the reviews of Cecil and McPhee, 1959; Lumper and Zahn, 1965). In neutral and alkaline media the reaction is catalyzed by thiols, which, in the form of RS^- ions, carry out nucleophilic attack on a sulfur atom of the disulfide, just as in thiol—disulfide exchange:

$$R'S^- + R''S{-}SR'' \rightleftarrows R'S{-}SR'' + R''S^-$$

$$R''S^- + R'S{-}SR' \rightleftarrows R''S{-}SR' + R'S^-$$

Catalytic quantities of thiols can arise by hydrolytic cleavage of disulfides (see Section 6). Reagents that block SH groups (p-mercuribenzoate, N-ethylmaleimide) inhibit the exchange reaction in neutral or alkaline media (Ryle and Sanger, 1955).

The disulfide exchange reaction in a strongly acidic medium was first observed by Sanger (1953). He noticed that acid hydrolysates of insulin contained a significantly larger number of different cystine peptides than could correspond with the structure of the protein, and he explained this by disulfide exchange. This hypothesis was then confirmed by experiments with model disulfides. Ryle and Sanger (1955) found that thiols inhibited disulfide exchange in acid media (in contrast with exchange in alkaline media), and they worked out conditions of hydrolysis in which such exchange was reduced to a minimum.

Benesch and Benesch (1958) studied the mechanism of disulfide exchange in acidic media; they proposed that the exchange takes place through a sulfenium cation, which is formed by attack of a proton on the S—S bond:

$$R'S{-}SR' + H^+ \rightleftarrows \left[\begin{array}{c} R{-}\underset{\displaystyle H}{\underset{|}{S}}{-}S{-}R \end{array}\right]^+ \rightleftarrows R'SH + R'S^+$$

The sulfenium cation carries out an electrophilic displacement on a sulfur atom of the disulfide:

$$R'S^+ + R''S{-}SR'' \rightleftarrows R'S{-}SR'' + R''S^+$$
$$R''S^+ + R'S{-}SR' \rightleftarrows R''S{-}SR' + R'S^+$$

The inhibitory action of thiols is explained by their reaction with sulfenium ions, whose concentration they diminish:

$$RSH + R'S^+ \rightleftarrows RS{-}SR' + H^+$$

In agreement with this, Benesch and Benesch found that a number of compounds (e.g., sulfenyl chlorides) that form RS^+ ions in acid media catalyze disulfide exchange. Other catalysts under these conditions are selenate (Na_2SeO_4), selenite (Na_2SeO_3), tellurate (Na_2TeO_4), and metavanadate ($NaVO_3$); the catalytic action of these compounds is apparently based on their removal of thiols from the medium (Lawrence, 1969).

Lindley and Haylett (1968) proposed that the active intermediate in disulfide exchange is the S-monoxide of the disulfide $(R{-}\underset{\underset{\displaystyle O}{\|}}{S}{-}S{-}R)$; the S-monoxide readily undergoes an exchange reaction catalyzed by halide ions; this can explain why the exchange is faster in HCl than in H_2SO_4.

Glazer and Smith (1961) studied disulfide exchange between bis(dinitrophenyl)cystine and 12 proteins in 9.6 M HCl, and they worked out a method of determining the total content of cysteine plus cystine in a protein from the quantity of mixed disulfide formed. They found that the rate of reaction varied greatly with the protein, and they suggested that this difference was caused by the influence of polar groups in the peptides that contain the S−S bonds. Thus, for example, the positively charged groups of lysine or arginine residues, if situated close to the S−S bond, can, in their opinion, inhibit disulfide exchange by repelling the attacking sulfenium ion.

In addition to the disulfide exchanges catalyzed by bases and by acids, a third type has been described. This occurs with participation of free radicals (Eager and Savigo, 1963; Lumper and Zahn, 1965):

$$R'S{-}SR' \rightleftarrows 2R'S\cdot$$
$$R'S\cdot + R''S{-}SR'' \rightleftarrows R''S{-}SR' + R''S\cdot$$
$$R''S\cdot + R'S{-}SR' \rightleftarrows R''S{-}SR' + R'S\cdot$$

The RS• radicals can arise by splitting of disulfides at high temperatures (above 100°C) or under intense ultraviolet radiation (photolysis, Section 13).

Catalysis by iodine (radical mechanism?) was used by Nelander and Sunner (1972) to extend the findings of Haraldson et al. (1960) on the equilibrium constants for exchange reactions of simple alkyl disulfides. These constants are close to the statistical value of 4 for a number of pairs, but are considerably more in favor of the mixed disulfide when one of the alkyl groups is t-butyl. Thus, for example, at equilibrium $[t\text{-Bu}-S-S-Et]^2/[Et_2S_2][t\text{-Bu}_2S_2]$ is about 25. The instability of $t\text{-Bu}_2S_2$ does not appear to be due to strain, since ΔH for the exchange is zero, but to an entropy term that arises because the two t-butyl groups cannot rotate independently of each other — a "cogwheel effect" since they are mutually enmeshed.

9. Reduction by Sodium Borohydride

In recent years sodium borohydride ($NaBH_4$) has been increasingly used for reductive scission of the S—S bonds of proteins. Crestfield et al. (1960) used it to reduce the S—S bonds of ribonuclease, β-lactoglobulin, and lysozyme; Brown (1960) used it at 40-50°C to reduce those of ribonuclease, pepsin, ovalbumin, and serum albumin.

Light et al. (1969; Sondack and Light, 1971) found that $NaBH_4$ selectively reduces one of the six S—S bonds of native trypsinogen and that after this the trypsinogen still possesses the ability to be converted into partly active trypsin. Lawrence and Laskowski (1967) showed that $NaBH_4$ selectively reduces one S—S bond in ox pancreatic trypsin inhibitor with preservation of the inhibitory action of the protein. The reduction was carried out in the cold at pH 8.4 by addition to the protein solution, through which a current of nitrogen was passing, of an equal volume of freshly prepared 0.2 M $NaBH_4$.

Lawrence and Laskowski pointed out the following advantages of the borohydride method: 1) excess $NaBH_4$ can easily be destroyed by lowering the pH of the solution to 3; 2) the extent of reduction of the S—S bonds can be followed by reaction with 5,5'-dithiobis(2-nitrobenzoic acid) (see Chapter 3, Section 1c); 3) the absence of thiols in the incubation medium avoids the complication of disulfide exchange which can occur on partial reduction of S–S bonds. Lowering the concentration of $NaBH_4$ (to 0.1 M) and carrying out the reaction in the cold in the presence of 0.02% EDTA diminishes the danger of the splitting of peptide bonds, which several authors have pointed out (Kimmel and Parcells, 1960; Crestfield et al., 1963; Andersson, 1969).

Krull and Friedman (1967) reduced the S–S bonds of serum albumin and lysozyme with sodium hydride in dimethylsulfoxide; according to their data the reaction occurs with minimal side reactions.

10. Electrolytic Reduction

Dohan and Woodward (1939) first applied electrolytic reduction at a mercury cathode to the splitting of an S–S bond, in oxidized glutathione. Markus (1960, 1964) used this method to split the S–S bonds of insulin at pH 8.5 and with a potential of 10 V. He followed the course of the reduction by amperometric titration of SH groups. Markus found that the two interchain S–S bridges in insulin are reduced markedly more rapidly than the intrachain S–S bond in chain A. The electrolytic reduction did not involve any other bonds in the protein molecule, and can be carried out successfully under mild conditions in the absence of denaturing agents.

Leach et al. (1965) described the electrolytic reduction of S–S bonds in proteins in the presence of a low concentration of β-mercaptoethanol (0.07 M). The electrolytic reduction, "catalyzed" by the thiol, apparently takes place in two stages: reduction of the S–S bond in the protein by the thiol and the subsequent electrolytic reduction of the disulfide form of the "catalyst." By this method it was possible to reduce practically all the S–S bonds of lysozyme and of bovine serum albumin (under conditions that excluded oxidation of SH groups by atmospheric oxygen during the analysis).

According to Leach and co-workers, only a partial reduction of the S–S bonds of insulin, ribonuclease, lysozyme, and serum albumin is obtained in the absence of β-mercaptoethanol. The degree of electrolytic reduction of the S–S bonds depends on the pH of the solution, the nature of the added thiol, and the potential at the cathode. By varying these conditions it is possible to achieve selective reduction of protein S–S bonds (see also Berezin, et al., 1968, and for the selective reduction of one S–S bond of insulin, Zahn et al., 1972).

Nagy and Straub (1969) described the electrolytic reduction of a number of proteins in dilute buffer solutions of pH close to 7; they found that the reduction of the S–S bonds is facilitated by addition of 30% ethanol.

11. Reduction by Tertiary Phosphines

Recently tertiary phosphines have been successfully used for the reduction of the S–S bonds in proteins. Humphrey and coworkers (Humphrey and Hawkins, 1964; Humphrey and Potter, 1965) investigated the ability of these compounds to reduce low molecular disulfides. They found that tributylphosphine quantitatively reduced aromatic and aliphatic disulfides on incubation of a twofold excess of the reagent with the disulfide for 60 min at room temperature; the reaction takes place, apparently, according to the following equation:

$$RS{-}SR + (C_4H_9)_3P + H_2O \rightarrow 2RSH + (C_4H_9)_3PO$$

Sweetman and Maclaren (1966) used tris(hydroxymethyl)-phosphine, tris(diethylaminomethyl)phosphine, and tri-n-butyl-phosphine for reducing the S–S bonds of wool keratin. The first two of these reagents gave side reactions, which were apparently connected with the formation of formaldehyde when they decomposed in aqueous solution. The tributylphosphine was the most effective and specific reagent; it reduced 94% of the S–S bonds of keratin on addition of 2 moles per mole of disulfide at 20°C and almost neutral pH. To obtain the same degree of reduction with benzyl mercaptan it was necessary to add not less than a 10 to 20 fold excess.

In the opinion of Sweetman and Maclaren, the reaction of S–S bonds with a trialkylphosphine goes as follows:

$$\begin{matrix} R{-}S{-}S{-}R \\ + PR'_3 + H_2O \end{matrix} \rightarrow \left[\begin{matrix} R{-}\bar{S}{-}\bar{S}{-}R \\ HO^- \;\; \overset{+}{P}R'_3 \end{matrix} + H^+ \right] \rightarrow 2RS^- + 2H^+ + OPR'_3$$

Levison et al. (1969) described the reduction of the S–S bonds of human γ-globulin with the water-soluble phosphines tris(hydroxymethyl)phosphine and tris(carboxyethyl)phosphine. They found that the products of reduction of γ-globulin by phosphines and by mercaptoethanol were similar in electrophoretic mobility and in a number of other properties. According to their data S—S bonds can be reduced by tertiary phosphines in the presence of iodoacetamide. Maclaren and Sweetman (1966) also noted that the reduction of S—S bonds by tributylphosphine and the alkylation of the SH groups produced can in some cases be carried out simultaneously.

12. Desulfurization with Raney Nickel

Desulfurization with Raney nickel is widely used in organic chemistry. Cooley and Wood (1951) first applied this method to removal of sulfur from proteins; they treated egg albumin and casein with Raney nickel and directed attention to the fact that cystine residues are desulfurized more easily than methionine residues. Ivanov et al. (1963, 1967) found that treatment of insulin with Raney nickel for 4 h at pH 8-9 led to the scission of two S—S bonds (one intrachain and one interchain) with the formation of four alanine residues. Perlstein et al. (1971) investigated the desulfurization of cystine, methionine, and a number of proteins, using an atmosphere of H_2. They found that the desulfurization of methionine was much slower than that of cystine, and depended little on pH over the range 5-8. Desulfurization of cystine, however, was faster at pH 7.0 than at 5.0 or 8.0. At pH 7.0 and 22°C, cystine was completely desulfurized in 12 min, whereas methionine was essentially unchanged even after 10 h incubation (in this time only about 4% of methionine disappeared). Thus under these conditions the reaction exhibits a high specificity toward cystine, which at pH 7 is quantitatively transformed into alanine. Perlstein et al. found that treatment of α-lactalbumin with Raney nickel in aqueous solution at pH 7.0 and 22°C for 48 h led to the splitting of two S—S bonds, without touching methionine residues at all. The S—S bonds of native lysozyme proved unavailable for reaction with Raney nickel. After they had been reduced by β-mercaptoethanol in 8 M urea, only four of the eight cysteine residues could be desulfurized with Raney nickel.

The selective conversion of cysteine and cystine residues into alanine residues can give valuable information on the role of SH and S—S groups in proteins, since no bulky substituent is thereby introduced which may alter conformation. There may, however, be some danger of adsorption or denaturation at the nickel surface.

13. Photolysis

Szendrő et al. (1933) described the scission of the S—S bond of cystine by ultraviolet irradiation; they irradiated solutions of L-cystine in the absence of air and observed the formation of hydrogen sulfide and of cysteine, the latter in a yield of about 5%. Much work has been devoted to the study of the mechanism of the photolysis of cystine, and its results agree best with a free radical decomposition involving the breaking of S—S bonds in some molecules and C—S bonds in others (Lyons, 1948; Forbes and Savige, 1962; Bogle et al., 1962; Eager and Savige, 1963; Asquith and Hirst, 1969, Dixon and Grant, 1970):

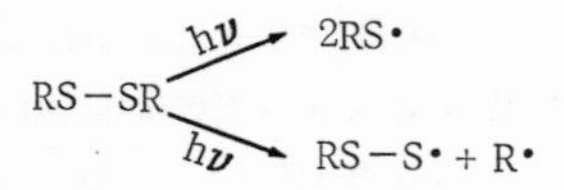

Asquith and Hirst (1969) also consider scission of C—N bonds probable.

The radicals formed undergo further conversions as a result of oxidation, reaction with solvent, or reaction with one another. Forbes and Savige (1962) irradiated alkaline solutions of cystine in air with light of wavelength 254 nm and identified the following products: alanine-sulfinic acid ($R-SO_2H$), cysteic acid ($R-SO_3H$), alanine-thiosulfonic acid ($R-SO_2SH$), S-sulfocysteine ($R-S-SO_3H$), hydrogen sulfide, and traces of serine, cysteine, alanine, lanthionine, and glycine. The yield of cysteine and alanine rose considerably when the irradiation was performed in the absence of air.

Asquith and Hirst (1969) generally confirmed the formation of these products, and observed in addition the appearance of pyruvic acid, NH_2OH and ammonia on irradiation of cystine at pH 1 or 10. At pH 1 bis-(2-amino-2-carboxyethyl)-trisulfide, which had been found in acid hydrolysates of wool and other proteins by Fletcher and Robson (1963), and bis-(2-amino-2-carboxyethyl)-tetrasulfide were also formed. According to their findings lanthionine is formed only in acidic media. They also reported that the quantum yield for the destruction of cystine is almost independent of pH; the rate of destruction was slightly higher at pH 10 than at pH 1. Their scheme for photolysis in air, which explains the formation of all the compounds noted, is shown in Fig. 7.

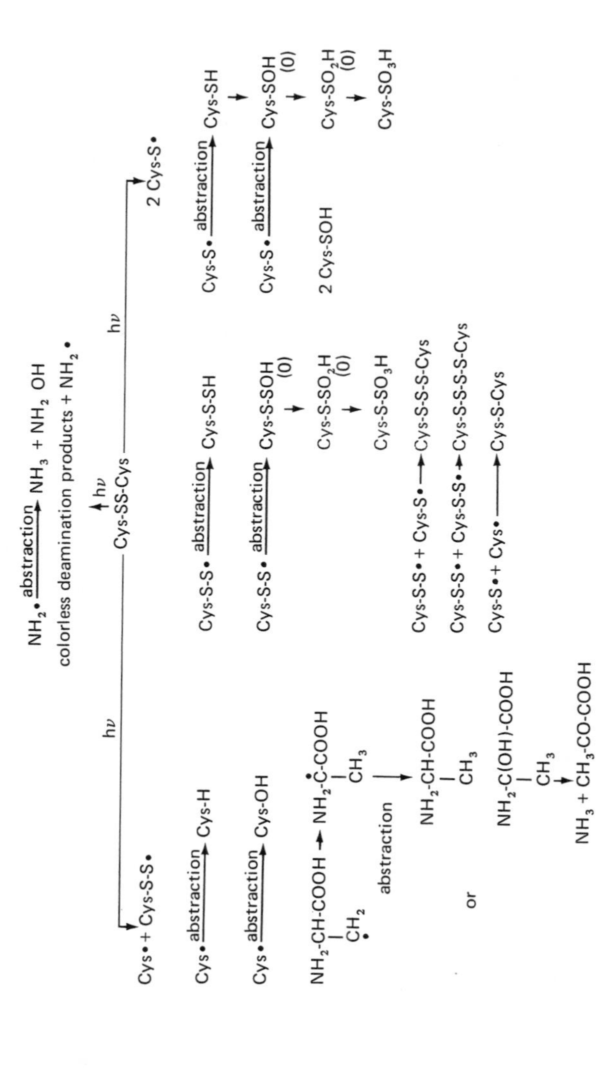

Fig. 7. The postulated reaction scheme for the photolysis of cystine (Asquith and Hirst, 1969).

The products formed on irradiating cystine under an atmosphere of nitrogen are far fewer than those formed in the presence of air (Asquith and Shah, 1971). The major decomposition products are cysteine, alanine, ammonia, sulfur, and hydrogen sulfide, and at pH 1 also lanthionine. The rate of photolytic degradation of cystine under a nitrogen atmosphere ($k = 9.82 \times 10^{-3}$ min^{-1}) is markedly faster than that in the presence of oxygen ($k = 5.50 \times 10^{-3}$ min^{-1}). On the basis of their quantitative determination of the products of photolysis Asquith and Shah considered that in a nitrogen atmosphere C—S fission probably predominates over S—S fission.

Swanepoel (1963) and Swanepoel and van Rensburg (1965) found that photolysis of cystine gives a very high yield of thiol if the irradiation with light of wavelength 254 nm is carried out in the presence of hypophosphorous acid or its salt. The photochemical process has chain character and includes a reaction of thiyl radicals (RS•) with the enol tautomer of hypophosphite to form a phosphoranyl radical:

$$RS-SR \xrightarrow{h\nu} 2RS^{\bullet}$$

$$RS^{\bullet} + PO_2H_3 \rightarrow RS\dot{P}O_2H_3$$

$$RS\dot{P}O_2H_3 + RS-SR \rightarrow [RS\overset{+}{P}O_2H_3]\bar{S}R + RS^{\bullet}$$

$$[RS\overset{+}{P}O_2H_3]\bar{S}R + H_2O \rightarrow H_3PO_3 + 2RSH$$

The splitting of the S—S bonds of proteins under the influence of ultraviolet radiation is of interest. Bersin (1933) observed activation of papain after a *short* ultraviolet irradiation, and explained it by splitting of an S—S bond and liberation of an SH group. Recently Dose and Risi (1972) confirmed Bersin's observations and concluded that the activation is due to scission of the mixed disulfide formed between the SH group of the active site of papain and free cysteine (see Chapter 6, Sec. 1). The activating effect of light on papain is exceptional; as a rule strong ultraviolet and other kinds of radiation lead to the inactivation of enzymes and of protein hormones, and indeed a longer irradiation inactivates papain. The cause of the inactivation is the destruction of the residues of cystine and of the aromatic amino acids, which are the most photosensitive residues in proteins (Luse and McLaren, 1963).

Several authors (Siebert et al., 1965; Dose, 1966; Risi et al., 1967) have noted that the quantum yields for the destruction of cystine residues in proteins are markedly greater than those for the destruction of free cystine and oxidized glutathione (in other words the photosensitivity of the S–S bond is greater in proteins than in free cystine). The photolytic destruction of the S–S bonds of at least some proteins is nonrandom, i.e., their cystine residues differ in photosensitivity. These differences apparently depend on energy transfer or chemical interactions (or both) between the cystine and other chromophoric groups, particularly tryptophan residues (Dose, 1968; Ghiron et al., 1971). Risi et al. (1967) found that the yields of cystine destruction are much greater in proteins that contain tryptophan. Since the usual fate of the radicals produced in the photochemical event is probably recombination, the chemical reactions observed may be the result of other reactions that compete only poorly with this recombination. It is therefore not surprising that the observed quantum yields vary as the environment affects the balance of this competition, even in the absence of chromophoric groups that can transfer energy to the S–S bond.

Several papers have been concerned with the connection between the photoinactivation of proteins and the scission of their S–S bonds. Augenstine and Ghiron (1961) studied the action of UV irradiation on trypsin and found that the inactivation of this enzyme was entirely due to the destruction of cystine residues. Siebert et al. (1965) found a direct correlation between the degree of inactivation of insulin by small doses of radiation and the selective scission of one of its three S–S bonds. In the case of papain, however, no simple correlation could be found between the UV-induced inactivation and the destruction of cystine (Dose and Risi, 1972). Risi et al. (1967) found that the most photosensitive S–S bonds of ribonuclease, trypsin, and lysozyme could be destroyed by irradiation without loss of activity. These bonds are accordingly not critical for maintaining the catalytically active conformation; the number of such bonds in ribonuclease was two or three out of the four that it contains.

Hexter and Westheimer (1971) photolyzed (at a wavelength of 254 nm) the derivatives of trypsin and chymotrypsin in which the active serine residue had been substituted with an ethyldiazomalonyl ($Et-O-CO-CN_2-CO-$) group. The main modified amino acid found after acid hydrolysis was S-carboxymethylcysteine, although its yield was small compared with that of glycolic acid. They inferred that the carbene produced by photolysis of the ethyldiazomalonyl group had occasionally attacked a disulfide bridge, although it had more often attacked water. It is also conceivable, however, that the photolysis was that of the disulfide bond, and that the nearby diazomalonyl group successfully competed with one of the R—S· radicals thus formed for reaction with the other of these radicals. If the explanation of Hexter and Westheimer is correct, and the fact that diazoacyl groups absorb about 30 times more strongly than disulfide groups at 254 nm supports it, then disulfides are not only photolyzed themselves, but also can react with the photolytic products of other chromophores.

14. Oxidation of S—S Groups. Identification of Cystine-Containing Peptides

The oxidation of disulfides can occur by one of the following pathways (Savige and Maclaren, 1966):

$$RSSR \xrightarrow{a} RSOSR \rightarrow RSOSOR \rightarrow RSOSO_2R \rightarrow RSO_2SO_2R \rightarrow 2RSO_3H$$

$$RSSR \xrightarrow{b} 2RSOH \rightarrow 2RSO_2H \rightarrow 2RSO_3H$$

$$RSSR \xrightarrow{c} RSSOH + ROH \rightarrow RSSO_2H \rightarrow RSSO_3H \rightarrow RSO_3H + H_2SO_4$$

When disulfides are oxidized by pathway a or b the S—S bond is split and the final product is 2 moles of sulfonic acid per mole of disulfide. Oxidation by pathway c proceeds by scission of the C—S bond and formation of equimolar quantities of sulfonic acid and sulfate. The main final product of the oxidation of cystine by performic acid, hydrogen peroxide, iodine, or bromine, is cysteic acid; alkaline solutions of permanganate or of hydrogen peroxide can apparently lead to splitting of the C—S bond and the formation of significant quantities of sulfate or cysteine-S-sulfonate ($RSSO_3H$) in addition to cysteic acid.

The majority of intermediate products of oxidation of disulfides are unstable, so it is difficult to observe them in proteins. Maclaren et al. (1965) obtained indirect evidence for the formation of cystine-S-monoxide, R—SO—S—R, when wool keratin or serum albumin is oxidized by dilute peracetic acid under mild conditions.

The S-monoxides (sulfoxides) and sulfones of disulfides, and sulfinic acids, react with thiols with the formation of mixed disulfides (Maclaren and Kirkpatrick, 1968):

$$R'{-}S{-}\overset{\overset{O}{\|}}{S}{-}R' + 2R''SH \rightarrow 2R'{-}S{-}S{-}R'' + H_2O$$

$$R'{-}S{-}\underset{\underset{O}{\|}}{\overset{\overset{O}{\|}}{S}}{-}R' + R''SH \rightarrow R'{-}S{-}S{-}R'' + R'SO_2H$$

$$R'{-}SO_2H + 3R''SH \rightarrow R'{-}S{-}S{-}R'' + R''{-}S{-}S{-}R'' + 2H_2O$$

The S-monoxide of cystine is split by Hg^{2+} and Ag^{+} ions and by organic mercurials considerably more easily than cystine itself (see Chapter 2, Section 2).

In contrast with the intermediate products of oxidation, cysteic acid is stable to acid hydrolysis of a protein and may easily be determined in the hydrolysate. Hence cystine is oxidized to cysteic acid before hydrolysis for the quantitative determination of the cystine residues of a protein (see Chapter 3, Section 2a). Oxidative scission of S—S bonds is also used for investigations of the primary structure of proteins. Performic acid is usually used for this purpose. One must bear in mind that this acid oxidizes not only the cystine and cysteine residues of proteins, but also their methionine and tryptophan; in very harsh conditions it can also oxidize tyrosine, serine and threonine residues.

The formation of the strongly acidic sulfonic groups on oxidative scission of S—S bonds greatly changes the electrophoretic mobility of cystine-containing peptides. This is the basis of the very convenient method of Brown and Hartley (1963, 1966; Hartley, 1970) for determining the positions of S—S bridges in proteins using "diagonal electrophoresis" of peptides on paper. An enzymic hydrolysate of a protein is subjected to electrophoresis at pH 6.5; after drying a strip is cut from the paper and sown on to a new sheet and subjected to electrophoresis at pH 6.5 at right angles to the direction of the first electrophoresis; thus the peptides are distributed according to their mobilities along the diagonal as is

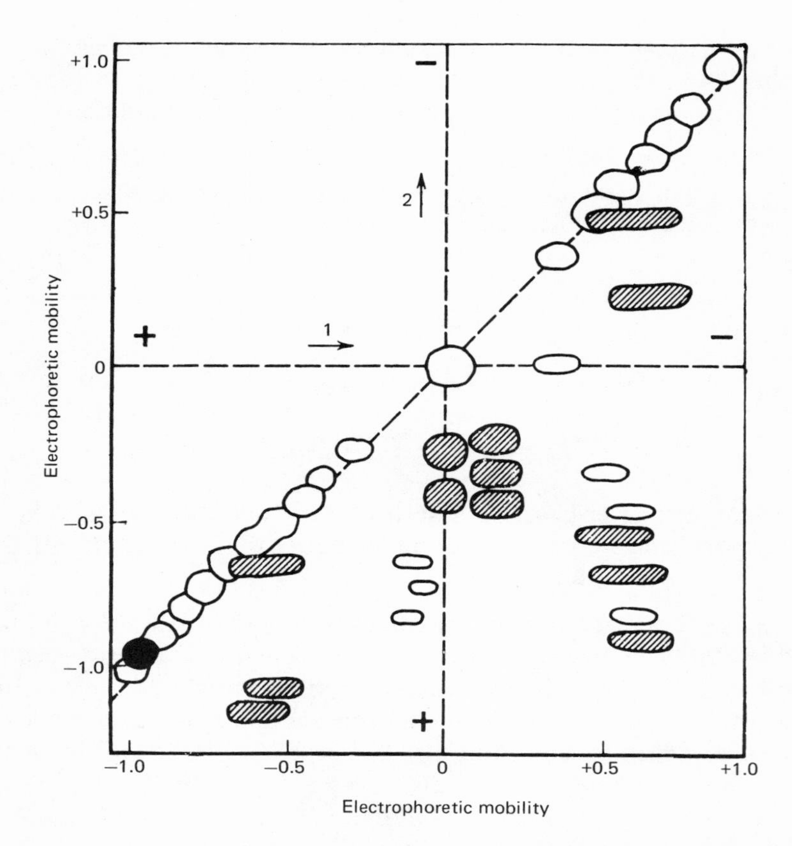

Fig. 8. The diagonal peptide map (at pH 6.5) of a peptic hydrolysate of chymotrypsinogen (Brown and Hartley, 1966). Cysteic acid peptides are shaded. The mobility of peptides is shown relative to the mobility of Dns-OH (−1.0) (the dark spot on the diagram). 1) Direction of the first electrophoresis (before oxidation); 2) direction of the second electrophoresis (after oxidation).

shown in Fig. 8. If after the strip is cut from the first sheet it is exposed to performic acid vapor before the second electrophoresis, then the cysteic acid peptides formed will be located below the diagonal, nearer to the anode. Thus one can easily identify the peptides that are linked in the protein by S–S bridges.

Another method of detecting cystine-containing peptides on peptide maps was proposed by Maeda et al. (1970). Their method is based on reduction of the S–S bonds with borohydride and subsequent spraying of the paper with a solution of 5,5'-dithiobis(2-

nitrobenzoic acid). This reagent allows thiols on the paper to be visualized as yellow spots (see Chapter 3, Section 1c).

15. Concluding Remarks

From the data presented in this chapter it follows that the S—S groups of proteins are stable to varied influences: pH, temperature, metal ions (in low concentration), etc. This property of S—S groups fits their main function, the stabilization of the macromolecular structure of proteins.

The reactivity of the S—S bonds in proteins varies widely; it apparently depends on their localization in the protein macromolecule, accessibility to reagents, the nature of neighboring groups, and, possibly, on the degree of strain (deformation) of the bonds.

An interesting question is whether the S—S groups of proteins can undergo scission under physiological conditions, e.g., by thiol—disulfide exchange with the intracellular thiols. Davidson and Hird (1967) studied the reactivity of the S—S bonds of a number of native proteins toward glutathione (used in physiological concentrations) in a system containing glutathione reductase and NADPH, and so to some extent imitating, as the authors thought, the situation in the living cell. Glutathione reductase and NADPH reduce oxidized glutathione, so that the equilibrium of this system is displaced toward the direction of splitting of the S—S bonds. Nevertheless it appeared that the S—S bonds of the majority of proteins investigated, with the exception of insulin, reacted very little or hardly at all with glutathione at 37°C. The extent of reduction increased markedly on limited proteolysis of the proteins or on raising the temperature to above 50°C. Davidson and Hird concluded that at physiological values of pH and temperature the S—S bonds of native proteins are resistant to the intracellular glutathione.

There are two possible reactions of a nucleophile with a mixed disulfide, according to which sulfur atom is attacked. Both reactions may be expected to occur if the properties of the two halves of the disulfide molecule are similar. Often, however, only one of the two possible reactions is observed. Thus electron-withdrawing substituents improve the leaving ability of a mercaptide ion and direct the nucleophilic attack to the other sulfur atom. This is the basis of the method of Vanaman and Stark (1970) for preparing the S-sulfo and S-cyano derivatives of proteins by expelling a nitrothiophenolate ion with sulfite and cyanide respectively (see Chapter 1, Section 10, and Chapter 2, Section 3). The leaving properties of mercaptide ions can also be improved in other ways, e.g., by protonating them as they leave at the expense of a nearby acidic group (Chapter 2, Section 4). Formation of a favorably sized ring with displacement of a group from the molecule (exocyclic displacement) can provide another basis for the selection of one of the two sulfur atoms of the S—S bond (see Chapter 2, Section 7, for the reaction of dithiothreitol). In proteins it is likely that in addition to such factors, which affect the relative reactivities of the two sulfur atoms, steric and electronic factors may control the relative accessibility of these atoms to approaching nucleophiles. The ability of cyanide to activate papain efficiently (Chapter 6, Section 1a) indicates that it can distinguish between the two sulfur atoms of the bond attacked; Massey's schemes for xanthine oxidase (Chapter 6, Section 2) also postulate such selective attack.

In this chapter much attention has been given to reactions that may be used to split S—S bonds. The scission of these bonds is an important preliminary step in studying protein structure and in determining the amino acid sequences of polypeptide chains. This scission can be achieved by oxidation, reduction, or nucleophilic displacement (with mercaptide ions, sulfite, cyanide, monothiophosphate, etc.). Parker and Kharasch (1959) indicated the possibility of splitting S—S bonds by reaction with arsenic compounds, secondly amines, trialkylphosphines, and certain other nucleophiles; trialkylphosphines have recently been successfully applied for this purpose. Out of all the reactions listed, the reduction of S—S bonds by excess of a thiol is used most frequently in protein chemistry.

Chapter 3

Methods for the Quantitative Determination of SH and S–S Groups in Proteins

1. The Determination of SH Groups

Various methods of determining SH groups in proteins have been proposed, based on the numerous reactions in which these groups take part. A detailed description of these methods is given in a number of reviews (Chinard and Hellerman, 1954; Cecil and McPhee, 1959; Benesch and Benesch, 1962; Leach, 1966). We will limit ourselves to describing the most commonly used and the most exact methods, and also the methods put forward in the last few years and so not mentioned in these reviews. Some of the methods once used to determine SH groups (including those based on the use of nitroprusside, ferricyanide, iodine, and porphyrindin) are hardly used at all now, because they have insufficient specificity, or comparatively low sensitivity, or both. The reagents now most often used for the quantitative determination of protein SH groups are (a) mercaptide-forming compounds, (b) alkylating reagents, and (c) disulfides; recently the use of sulfenyl halides has also been proposed for this purpose.

a. Methods Based on Mercaptide Formation

Spectrophotometric Titration with p-Mercuribenzoate. Of the various methods of titrating SH groups, one of the first places for simplicity, high sensitivity and precision is held by the spectrophotometric method of Boyer (1954); it is based on the increase in optical density in the region 250–255 nm that occurs when p-mercuribenzoate (p-MB) combines with SH groups. The method of Boyer allows measurement not only of the

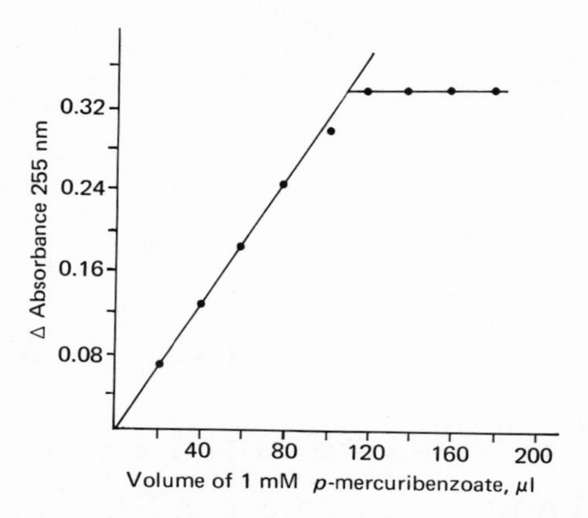

Fig. 9. Spectrophotometric titration of aspartate aminotransferase with p-mercuribenzoate (Torchinskii, 1971). The titration was carried out by adding 20 μl portions of 1 mM p-mercuribenzoate to 2.5 ml of enzyme solution (0.46 mg of protein per ml) in 0.2 M acetate buffer, pH 4.65, containing 0.5% sodium dodecyl sulfate. The reading was taken 15 min after the addition of each portion of p-mercuribenzoate. It was found that there were 4.3 SH groups per enzyme subunit of molecular weight 46,000.

number of SH groups, but also of the rate at which they react with p-MB, as well as correlation of the degree of loss of enzymic activity with the number of SH groups blocked.

The molar extinction of the mercaptide formed by p-MB with cysteine at 250 nm in 0.05 M phosphate buffer (pH 7.0) is 7600; at 255 nm in 0.33 M acetate buffer (pH 4.6) it is 6200. The mercaptides formed by p-MB with various proteins differ somewhat one from another in the magnitude of their molar extinctions. Hence the analysis of the number of SH groups is a protein cannot be based solely on measurement of the increase of optical density in the presence of excess p-MB, but should be carried out in the form of a titration.

The titration is usually carried out by the addition of 1 mM p-MB* (in portions of 5–20 μl) to 2–3 ml of a 0.05–0.3% solution of

* The exact concentration of p-MB is established by iodometric titration, or by determination of the extinction of the solution at 232 nm; according to Boyer ε_M for p-MB at this wavelength in a solution of pH 7.0 is 1.69×10^4.

protein (Fig. 9). Readings are taken at 250 nm (at pH 7.0) or at 255 nm (at pH 4.6) at 10-15 min after the addition of each portion of p-MB. Some proteins have to be incubated for several hours with p-MB to reveal sluggishly reacting SH groups; in this case it is more convenient to set up a series of mixtures that contain the protein and various quantities of p-MB, and to incubate them until a constant value is obtained for the optical density.

One should bear in mind that p-MB itself absorbs strongly at 250-255 nm. Hence equal quantities of p-MB are added to the cuvette with protein solution and to the reference cuvette, which contains buffer solution; this allows measurement of the increase of optical density caused by mercaptide formation with automatic subtraction of the absorption of the p-MB itself. A concentration of SH groups of 0.01 μmole/ml is sufficient for reliable results to be obtained.

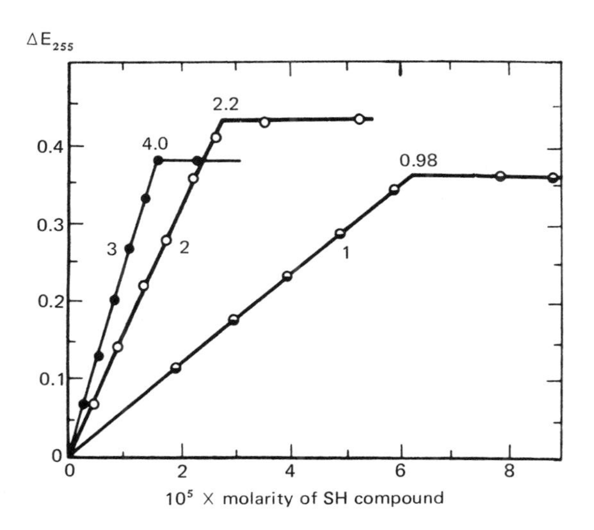

Fig. 10. The increase in optical density at 255 nm on reaction of 60 μM p-mercuribenzoate with (1) cysteine, (2) β-lactoglobulin, and (3) egg albumin in 0.33 M acetate buffer, pH 4.6 (Boyer, 1954). The figures on the curves indicate the number of moles of p-mercuribenzoate that react with 1 mole of SH compound. In the titrations of cysteine and egg albumin the readings were taken 15 min after mixing; in the titration of β-lactoglobulin 20 h later.

The reverse order of titration is also possible, in which a concentrated protein solution is added to 60 μM p-MB in the cuvette (Fig. 10). The presence of EDTA interferes with the titration. When enzymes that contain colored prosthetic groups (FAD, pyridoxal phosphate, etc.) are titrated, it is necessary to take account of the fact that the spectral shifts caused by mercaptide formation can be distorted by the splitting off of these groups from the protein under the influence of p-MB.

In order to determine the so-called "masked" or "buried" SH groups, the titration can be carried out in 8 M urea or in 0.5-1% sodium dodecyl sulfate. Cases are known, however, when even after denaturation of the protein some of its SH groups are unavailable to p-MB. In the case of muscle phosphorylase b, for example, out of the 18 SH groups in the molecule, 6 are titrated in the native protein, and 12 after denaturation; the remaining 6 groups can be blocked only after treatment of the protein with pepsin in 0.05 M HCl (Battell et al., 1968).

The rate and extent of the reaction of the SH groups of many proteins with p-MB is higher at pH 4.6 than at pH 7-8 (Boyer, 1954; Boyer and Segal, 1954; Torchinskii, 1964). This can be explained by the high affinity of p-MB for hydroxide ions, as a consequence of which the reactivity of p-MB increases as the pH is diminished (Benesch and Benesch, 1962). Acidification, however, does not always increase the rate of reaction of SH groups with p-MB. Indeed the SH groups of β-lactoglobulin react with p-MB 150 times more slowly at pH 6.75 than at pH 7.9; the diminution in the rate of reaction is in this case connected with a change in protein conformation (Dunnill and Green, 1966).

Other Spectrophotometric Methods. Bennett and Watts (1958) proposed the use of 1-(4-chloromercuriphenylazo)-2-naphthol (Mercury Orange) as a reagent for determining the total SH groups of soluble and insoluble proteins. Sakai (1968) described the following modification of the method. The Mercury Orange is dissolved in acetone and the solution mixed with an equal volume of 0.1 M phosphate buffer (pH 7) and the mixture added to the solid protein precipitated with acetone. The suspension is incubated at room temperature with mixing for 15-30 min, and then centrifuged. The stained precipitate is washed three times with acetone to remove unspecifically bound Mercury Orange. It is then suspended in acidified acetone (1 drop of 1 M HCl per 5 ml of acetone) to split off the Mercury Orange that had been bound to the SH groups:

$$\text{Protein}-\text{S}-\text{Hg}-\text{C}_6\text{H}_4-\text{N}=\text{N}-\text{C}_{10}\text{H}_6\text{OH} + \text{HCl} \longrightarrow$$

$$\text{Protein}-\text{SH} + \text{Cl}-\text{Hg}-\text{C}_6\text{H}_4-\text{N}=\text{N}-\text{C}_{10}\text{H}_6\text{OH}$$

The quantity of Mercury Orange in the supernatant is determined from its absorption at 470 nm ($E_{470} = 1.833 \times 10^4$ cm^{-1} M^{-1}).

Klotz and Carver (1961) suggested the use of salyrganic acid (mersalyl, the anhydride of o-{[3-(hydroxymercuri)-2-methoxy-propyl]carbamyl}-phenoxyacetic acid) for titrating SH groups. They used as an indicator for the end point of the titration the dye pyridine-2-azo-p-dimethylaniline, which forms a complex that possesses absorption at 550 nm with the excess salyrganic acid. The titration is performed in 0.1 M acetate or phosphate buffer in the pH region of 5.5-6.5 by adding salyrganic acid to a solution of the thiol in the presence of 60 μM dye (see Fig. 11).

Amperometric Titration. The method of amperometric titration of thiols with silver nitrate was first proposed by Kolthoff and Harris (1946) and used for the determination of protein SH groups by Benesch and co-authors (Benesch and Benesch, 1948, 1950; Benesch et al., 1955). This method is also used to determine the S–S bonds of proteins after their scission by sulfitolysis or reduction (see Chapter 3, Section 2c).

The scheme of the apparatus for amperometric titration is shown in Fig. 12. The titration is conducted in a vessel in which a rotating platinum electrode dips, which is connected through a microammeter with a reference electrode. The reference elec-

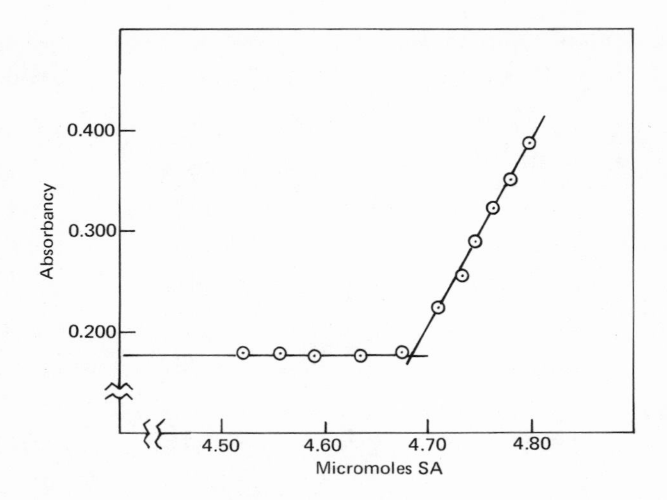

Fig. 11. Typical titration curve of a thiol with salyrganic acid in the vicinity of the equivalence point (Klotz and Carver, 1961). Glutathione (4.686 μmole by weight) was titrated with 1 mM salyrganic acid (SA) in acetate buffer, pH 5.8, in the presence of 60 μM azopyridine indicator. The absorbancy readings (at 550 nm) are corrected for dilution.

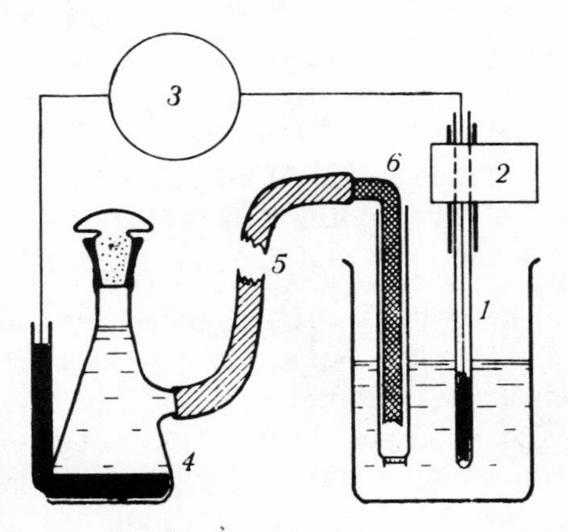

Fig. 12. Diagram of the apparatus for the amperometric titration of SH groups (Kolthoff and Harris, 1946). 1) Rotating platinum electrode; 2) electric motor; 3) microammeter; 4) reference electrode; 5) plastic tube; 6) glass tube, filled with KCl solution in agar gel.

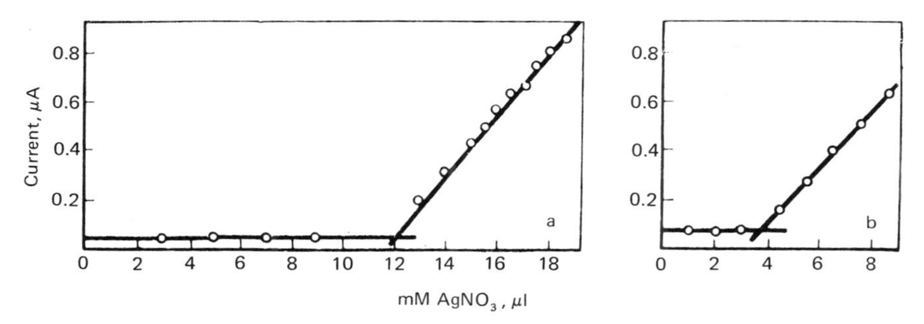

Fig. 13. Amperometric microtitration with 1 mM $AgNO_3$ (Yakovlev and Torchinskii, 1958). a) Titration of 12.05 nmole of glutathione; b) titration of 6.9 μl of blood serum.

trode can be an Hg/HgI_2 electrode (for conducting the titration in an ammonia buffer) or an electrode of Hg/HgO in saturated $Ba(OH)_2$ (for titration in a buffer of tris-HNO_3). The first portions of $AgNO_3$ combine with the SH groups of the protein and do not evoke any current. As soon, however, as the SH groups are blocked, free metal ions appear in the solution (or rather their complexes with ammonia or tris) and these are reduced at the platinum electrode which gives rise to a diffusion current. The strength of this current is proportional to the concentration of metal ions in the solution. The end point of the titration (the point of equivalence) is obtained by plotting the current against the quantity of reagent added. The dropping mercury electrode is sometimes used as an indicator electrode instead of the rotating platinum one; it is less sensitive but more reliable. The titration with it is conducted in a polarograph cell. Stricks and Kolthoff (1956) worked out the construction of a rotating mercury electrode which is 10 times more sensitive than the ordinary mercury electrode.

Gruen and Harrap (1971) have proposed the use of an electrode that responds specifically to Ag^+ ion activity. Such an electrode is based on the selective permeability of a silver sulfide crystal and is commercially available. The titration is therefore carried out in dilute nitric acid (about 10 mM), and in the absence of anything such as tris or ammonia to form complexes with Ag^+.

The sensitivity of amperometric titrations with the rotating platinum electrode is about 0.1-1.0 μmole of SH compound. A micromodification of this method, described by Yakovlev and Torchinskii (1958), allows analysis of quantities of material that contain 1.5-2 nmole of thiol (Fig. 13).

The advantage of amperometric titration over spectrophotometry lies in the possibility of carrying out the analysis in cloudy or colored solutions and even in suspensions. In enzymological investigations, however, this method has not been used as widely as the spectrophotometric method of Boyer. This is apparently due to its lower sensitivity (in its standard form), to the need for special, although not complex, apparatus, and also to the fact that amperometric titration only allows determination of the total number of SH groups in a protein and does not give the opportunity of determining the reactivity and the role of different SH groups, which is achieved by stepwise titration in Boyer's method.

Details of the method and several variants of the apparatus for amperometric titration are described in a number of articles (Kolthoff and Harris, 1946; Kolthoff and Stricks, 1950; Kolthoff et al., 1954, 1965a, b; Benesch et al., 1955; Allison and Cecil, 1958; Ferdinand et al., 1965; Bailey, 1962; Rothfus, 1966; Poglazov et al., 1958; Shol'ts, 1964). A serious drawback of $AgNO_3$ for titrating SH groups is the ability of silver mercaptides to bind excess silver ions with formation of complexes of the type $[(RS-Ag)Ag]^+$. The formation of such complexes explains the raised results obtained by different authors on titrating cysteine, cysteamine, thioglycolate, and hemoglobin (Sluyterman, 1957, 1966; Burton, 1958; Allison and Cecil, 1958; Cole et al., 1958; Börresen, 1963). On the other hand the results of argentometric titration of a number of proteins agree well with the data obtained by other methods. Evidently the degree of formation and the stability of the complexes of mercaptides with silver ions can differ for different thiols and proteins.

Kolthoff et al. (1954) used mercuric chloride for amperometric titration; its affinity for SH groups is higher than that of $AgNO_3$. The mercaptides of mercury can also bind Hg^{2+} ions; the significant difference, however, in stability between the mercaptides of mercury and their complexes with Hg^{2+} ions allows the error connected with this effect to be minimized by performing the titration in the presence of high concentrations of chloride. A more basic complication that arises if $HgCl_2$ is used is due to the ability of mercuric ions to react with either one or two SH groups in a protein (depending on their spatial location in the molecule) with the formation of compounds of the type $R-S-Hg-Cl$ or $R-S-Hg-S-R$, which destroys the stoichiometry of the reaction.

The best reagents for amperometric titration, free from the disadvantages mentioned, are the unifunctional organomercurial compounds: methylmercury iodide, recommended by Leach (1960), and phenylmercury hydroxide, used by Allison and Cecil (1958). For the determination of the SH groups of soluble and insoluble proteins Forbes and Hamlin (1968) recommended a long incubation (about 100 h) with excess methylmercury iodide at pH 7 and subsequent measurement of the excess reagent by potentiometry; their modification allowed determination of about 5 μmole of SH compound.

Radiometric Method. Leach et al. (1966) worked out a radiometric method for the determination of SH groups using [^{14}C]-methylmercury iodide or [^{203}Hg]-phenylmercury acetate. This method is more sensitive than amperometric titration, and permits determination of the SH groups and S–S groups (after scission of the latter with sulfite) in insoluble proteins.

b. Methods Based on Alkylation (see also p. 27)

Determination of the Hydrogen or Iodide Ions Released on Reaction with Iodoacetamide. The course of the reaction of SH groups with iodoacetate or iodoacetamide can be followed by determining the hydrogen or iodide ions released:

$$R-S-H + I-CH_2-CO-NH_2 \rightarrow R-S-CH_2-CO-NH_2 + H^+ + I^-$$

Benesch and Benesch (1962) recommended conducting this reaction at pH 7-7.5 and measuring the quantity of protons liberated with a pH-stat. Watts et al. (1961) worked out a method of determining iodide by measuring the potential at an Ag/AgI electrode. Both methods permit determination not only of the number of SH groups in a protein, but also of their rate of reaction with iodoacetate. Swoboda and Hasselbach (1973) determined the iodide released by its own extinction at 226 nm.

Determination as S-Carboxymethylcysteine. The instability of cysteine residues to acid hydrolysis of the protein has been the reason for working out methods for determining SH groups as S-carboxymethylcysteine or as cysteic acid.

The carboxymethylation of SH groups may be accomplished as follows. The protein is treated with excess iodoacetate at pH 8-9 and 37°C in the presence of 8 M urea and 0.2% EDTA for 2-3 h.

Under such conditions all the SH groups, including "masked" ones, usually react with iodoacetate. To avoid side reactions connected with the formation of iodine by oxidation in the light of the iodide ions released, the carboxymethylation of the protein and the subsequent removal of excess reagent should be conducted in the dark (see, for example, Hirs 1967b). For determination of the total number of half-cystine residues, including those that form S–S bonds or are oxidized to the R–SOH or R–SO_2H forms, the protein is incubated, before addition of iodoacetate, with β-mercaptoethanol or dithiothreitol in an atmosphere of nitrogen. The technique of reduction and subsequent carboxymethylation of proteins is described by Crestfield et al. (1963).

The carboxymethylated protein is separated from iodoacetate, urea, etc., by gel filtration, is freeze-dried, and is hydrolyzed in 5.7 M HCl for 22 h at 110°C. The hydrolysis is conducted in ampoules sealed under vacuum after careful removal of dissolved oxygen. The content of S-carboxymethylcysteine in the hydrolysate is determined with an amino acid analyzer (Spackman et al., 1958). In calculating the number of S-carboxymethylcysteine residues in the protein, correction may be made for its destruction on acid hydrolysis. Cole et al. (1958) found that this was 9% under the conditions described, but Crestfield et al. (1963) found no loss. At pH 3, however, S-carboxymethylcysteine cyclizes (Bradbury and Smyth, 1973) with a half-time of about 30 min at 100°C to reach an equilibrium about half cyclized. Raising the pH slows the reaction; lowering the pH accelerates the reaction, but naturally displaces the equilibrium in favor of opening, since much below pH 3 the opening is accompanied by nett protonation. Cyclization is avoided if 3-bromopropionate is used so that cysteine is analyzed as S-(2-carboxyethyl)-cysteine (see Chapter 1, Section 6c).

Goren et al. (1968) have described the determination of carboxymethylated amino acid residues in proteins, including carboxymethylcysteine, by an isotopic technique.

Determination Using N-Ethylmaleimide. Gregory (1955) proposed a spectrophotometric method of determining SH groups using N-ethylmaleimide (NEM). The method is based on measurement of the absorption maximum of NEM at 302-305 nm, which disappears on combination of the reagent with SH groups.

The reaction is conducted in the spectrophotometer cuvette by adding a concentrated protein solution to 1 mM NEM in 0.1 M phosphate buffer (pH 7.0). The reaction of NEM with glutathione is completed within a few minutes; the reaction with the SH groups of a protein sometimes requires a prolonged incubation. Under these conditions one must consider the possibility of hydrolysis of the NEM, a reaction whose rate increases markedly on raising the pH above 7.

Alexander (1958) used this method to determine the SH groups of protein-free extracts of blood and tissues and of egg albumin. Later the method was used to determine the SH content of a number of other proteins (Leslie et al., 1962a,b), and also for analysis of low-molecular disulfides after their reduction with KCN (De Marco et al., 1966). The use of NEM for the determination of SH groups in proteins has, however, not achieved wide application because the reagent is not specific enough (see Chapter 1), and, more importantly, because of the low sensitivity of the method: the molar extinction coefficient of NEM at 305 nm is 620, i.e., about 12 times less than the molar extinction coefficient of the mercaptides formed on reaction of p-mercuribenzoate with thiols (7600 at 250 nm and pH 7.0).

Determination with 2-Vinylquinoline. Krull et al. (1971) worked out a spectrophotometric method for determining the SH groups in proteins using 2-vinylquinoline. It reacts with these at pH 7.5 to form residues of S-2-(2-quinolylethyl)-cysteine. The number of such residues is calculated from the extinction of the protein solution in 0.1 M acetic acid at 318 nm; the molar extinction coefficient of the residues under these conditions is 10,000 (see Chapter 3, Section 2).

c. Methods Based on Reactions with Disulfides and Sulfenyl Halides

Determination with 5,5'-dithio-bis-(2-nitrobenzoate). As was noted in Chapter 1, disulfides are among the most specific reagents for protein SH groups. Of the various disulfides proposed for the determination of these groups, Ellman's (1959) reagent, 5,5'-dithio-bis-(2-nitrobenzoic acid), has won the greatest popularity. This is because treatment of thiols under mild conditions (pH 8) with only a small excess of reagent leads to stoichiometric release of a colored product (see Chapter 1, Section 10):

$$RS^- + (O_2N\text{-}C_6H_3(COOH)\text{-}S\text{-})_2 \rightarrow {}^-S\text{-}C_6H_3(COOH)\text{-}NO_2 + RS\text{-}S\text{-}C_6H_3(COOH)\text{-}NO_2 \quad \text{(a)}$$

The quantity of nitrothiophenolate anion formed is determined by the increase in extinction at 412 nm (ε_M = 13,600).

When some of the SH groups of a protein have reacted with Ellman's reagent, those remaining can undergo two competing reactions. One is the normal intermolecular reaction with more reagent, i.e., reaction (a) that the first groups to react underwent. Under favorable steric conditions, however, an SH group may react faster intramolecularly with the mixed disulfide already formed, with the resultant liberation of a second equivalent of nitrothiophenolate anion and the formation of a protein disulfide:

$$\text{Protein}(S^-)\text{-}S\text{-}S\text{-}C_6H_3(COOH)\text{-}NO_2 \rightarrow \text{Protein}(S\text{-}S) + {}^-S\text{-}C_6H_3(COOH)\text{-}NO_2 \quad \text{(b)}$$

This reaction has been observed with a number of proteins (Boross, 1969; Connellan and Folk, 1969; Flashner et al., 1972). It should be noted that whether or not reaction (b) competes successfully with the intermolecular reaction, one equivalent of nitrothiophenolate anion is released for each SH group that reacts.

The method is highly sensitive and strictly specific, and can be used for determining SH groups in both native and denatured (e.g., by the presence of sodium dodecyl sulfate) proteins. In native proteins Ellman's reagent often reacts with fewer SH groups than does p-mercuribenzoate.

The pH at which the reaction is carried out is important. In an alkaline medium (pH 8-9) the reaction is much faster, as might be expected, than in a weakly acidic one, and can demonstrate a larger number of SH groups (Jocelyn, 1962; Torchinskii and Sinitsyna, 1970). Brocklehurst and Little (1973) have tabulated the rate constants for the reaction of papain and several other thiols with Ellman's reagent, and also the pH dependence of the reaction.

One of the disadvantages of the method is the fact that prolonged incubation can lead to "autoxidation" of the nitrothiophenolate anion. Difficulties arise in using Ellman's reagent with colored proteins, e.g., heme proteins, which absorb strongly at 412 nm.

One should also bear in mind that before the reaction is carried out it is necessary to remove from the solution any traces of thiols used for the activation or stabilization of some enzymes (this is also important for other methods of determining SH groups).

To avoid the difficulties noted, Butterworth et al. (1967) worked out the following modification of Ellman's method. The protein is incubated with excess of Ellman's reagent at pH 7-8; if necessary the incubation can be carried out in the presence of a thiol. The incubated mixture is gel-filtered to separate the protein, in the form of a mixed disulfide, from the excess reagent and the free nitrothiophenolate anion. Excess dithiothreitol is then added to the thiophenylated protein, and this reduces the mixed disulfide with liberation of a second equivalent of nitrothiophenolate. The quantity of the latter is determined from the increase of absorption at 412 nm. Colored proteins are first precipitated after the dithiothreitol treatment with 60% perchloric acid, and the nitrothiophenolate is measured in the supernatant after its neutralization with 6 M NaOH. The modification described is applicable only to those proteins whose reaction with Ellman's reagent is entirely of type (a). If reaction (b) occurs the modified method will give incorrect results. Brocklehurst et al. (1972a) showed that papain reacts with a single equivalent of Ellman's reagent to form the mixed disulfide by determining not only (1) the nitrothiophenolate produced and (2) the mixed disulfide formed (by the nitrothiophenolate it yielded with dithiothreitol), but also (3) the quantity of reagent used (by determining the excess with dithiothreitol after removal of the protein).

Sedlak and Lindsay (1968) have described a modification of Ellman's procedure that permits determination of the total, protein, and nonprotein SH groups in tissue homogenates.

Among the various disulfides proposed for the spectrophotometric determination of SH groups, β-hydroxyethyl-2,4-dinitrophenyl-disulfide may be noted. Its reaction with protein SH groups proceeds according to the following equation (Bitny-Szlachto, 1965):

$$\text{Protein–S}^- + \text{HOCH}_2\text{CH}_2\text{S–S–C}_6\text{H}_3(\text{NO}_2)_2 \rightleftharpoons$$
$$\rightleftharpoons \text{Protein–S–S–CH}_2\text{CH}_2\text{OH} + {}^-\text{S–C}_6\text{H}_3(\text{NO}_2)_2$$

The increase of absorption at 408 nm, produced by the liberated dinitrothiophenol, gives the number of SH groups that have reacted.

Determination with Pyridine-Disulfides. Grassetti and Murray (1967) proposed the use of 2,2'-dithiodipyridine and 4,4'-dithiodipyridine for the determination of SH groups in proteins and tissue homogenates. These compounds react with SH groups, forming 2- and 4-thiopyridones respectively:

$$\text{(2-Py)–S–S–(2-Py)} + 2\text{RSH} \rightarrow 2\ \text{(2-thiopyridone, N–H, C=S)} + \text{R–S–S–R}$$

The absorption spectra of the thiopyridones is quite different from those of the corresponding pyridine-disulfides, and this allows the reaction to be followed spectrophotometrically. 2-Thiopyridone has a maximum molar extinction coefficient of 7060 at 343 nm; 4-thiopyridone has one of 19,800 at 324 nm. Hence the reaction with 4,4'-dithiodipyridine is more sensitive. The method allows determination of less than 15 nmole of thiol when 4,4'-dithiodipyridine is used.

The favorable equilibrium constant and the high reactivity of the disulfide are doubtless due in this case to the electron-withdrawing properties of the pyridine ring:

$$\text{R–S–S–(2-pyridinium, N}^+\text{–H)} + \text{R'–S}^- \longrightarrow \text{R–S–S–R'} + \text{S=(2-pyridone, N–H)}$$

Whereas with Ellman's reagent the rate falls off greatly as the pH is diminished and RS^- is converted into the inert RSH, with the pyridine-disulfides the drop in $[RS^-]$ is compensated for by the increasing protonation of the reagent which produces its reactive form. Double protonation is likely to make it still more reactive. Grassetti and Murray showed that the reaction was virtually complete in a minute or two with 1 mM reagent down to pH 3.4. The determination can thus be performed over the whole range from this pH up to pH 8. It is extremely useful to have a reagent that can be used under such acidic conditions in which thiols are most resistant to oxidation and other forms of destruction. Brocklehurst and Little (1972, 1973) obtained evidence that the protonated form of 4,4'-dithiodipyridine reacts with cysteine anion about 100 times faster than the neutral form.

Brocklehurst and Little (1972, 1973) found that there is a sharp optimum at pH 3.8 for the reaction of papain with 2,2'-dithiodipyridine. This could be due to the fact that protonation of the reagent (pK 2.5) activates it, whereas papain is activated by deprotonation of His-159 (pK 4, see Chapter 4, Section 2a, and Chapter 6, Section 1a, and Polgar, 1973), but pK values of 2.5 and 4 would be expected to give a broader maximum (see Chapger 1, Section 4) than that observed. Possibly (Brocklehurst, personal communication), efficient binding of the protonated reagent requires deprotonation of Asp-158; the involvement of a second pK of near 4 could explain the sharpness of the optimum. Brocklehurst and Little (1973) have made use of the acidic optimum for specific titration of the SH group of papain. The SH group of denatured papain does not react, and the active enzyme can be titrated even in the presence of a 10-fold molar excess of cysteine or of a 100-fold molar excess of mercaptoethanol. Brocklehurst et al. (1973) have made ingenious use of these findings in order to isolate active papain; they prepared an adsorbent in which the amino group of the glutathione–2-pyridyl disulfide was linked to a solid support. At pH 4 active papain specifically displaced 2-thiopyridone and was itself linked to the adsorbent. When eluted with cysteine it proved to contain 1.0 mole of reactive SH group per mole of protein.

Ampulski et al. (1969) noted an advantage of 4,4'-dithiodipyridine for studying the SH groups of hemoproteins, in that the wavelength of maximum absorption of the reaction product (4-thiopyridine) is far from that of the strong absorption band of heme (412 nm); in the case of hemoglobin the use of 4,4'-dithiodipyridine reduces the background interference almost tenfold in comparison with 5,5'-dithio-bis-(2-nitrobenzoate).

Grassetti and Murray (1969) have described the properties of a number of heterocyclic disulfides (including pyridine, pyrimidine, and thiazole derivatives) whose reactions with thiols can be followed by the appearance of new spectral peaks that are shifted to longer wavelengths and which are due to the absorption of the reaction products (thiones). In most of the thiol–disulfide pairs studied, the thione has appreciable absorption only in the UV region (240-

400 nm). But the product of the reaction of 2,2'-dithio-bis-(5-nitropyridine) with thiols exhibits strong absorption in the visible region of the spectrum; the molar extinction coefficient of this product is 14,000 at its absorption maximum of 386 nm, and is 1000 at 470 nm. This permits application of the reagent for visual detection of small amounts of thiols on chromatograms (but of course special precautions would have to be used to prevent oxidation if small quantities of thiols were to be chromatographed). Spraying of the paper with 0.03% 2,2'-dithio-bis-(5-nitropyridine) readily reveals a spot containing 0.2 μg (about 2 nmole) of cysteine by giving a stable yellow color. Glaser et al. (1970) found that a 0.1% solution of 5,5'-dithio-bis-(2-nitrobenzoic acid) in an alcohol buffer (pH 8.2) mixture can also be used for this purpose, and revealed disulfide-containing peptides on the paper after they had been reduced by spraying with ethanolic sodium borohydride.

Grassetti et al. (1969) have also proposed the use of 6,6'-dithiodinicotinic acid for the quantitative determination of SH groups. It reacts with simple thiols and with the SH groups of proteins as follows:

$$\text{HOOC-C}_5\text{H}_3\text{N-S-S-C}_5\text{H}_3\text{N-COOH} + 2\text{RSH} \rightarrow 2\ \text{HOOC-C}_5\text{H}_3\text{NH(=S)} + \text{RS-SR}$$

The 6-mercaptonicotinic acid formed in this reaction is, to judge from its absorption spectrum at pH 7.2, predominantly in the thiopyridone form. It has absorption maxima at 295 and 344 nm; its molar extinction coefficients at these wavelengths are 19,100 and 10,000 respectively. 6,6'-Dithiodinicotinic acid has the advantage of being more soluble than dithiopyridines in buffer solutions of pH close to neutrality.

Determination Using [^{35}S]-Tetraethylthiuram-Disulfide and Thiamine-Disulfide. Neims et al. (1966) worked out a highly sensitive radiometric method for determining SH groups in proteins, which was based on their reaction with [^{35}S]-tetraethylthiuram-disulfide (disulfuram). This reagent reacts with SH groups at pH 8-9 to form a mixed disulfide and the anion of

[^{35}S]-diethyldithiocarbamate:

$$RS^- + [(C_2H_5)_2NC(=\overset{*}{S})-\overset{*}{S}-]_2 \rightarrow RS-\overset{*}{S}C(=\overset{*}{S})N(C_2H_5)_2 + (C_2H_5)_2NC(=\overset{*}{S})-\overset{*}{S}^-$$

The latter is rapidly destroyed at pH 4.0 to form volatile CS_2, which is absorbed by alkaline piperidine and determined by its radioactivity:

$$(C_2H_5)_2NC(=\overset{*}{S})-\overset{*}{S}^- + H^+ \rightarrow (C_2H_5)_2NH + C\overset{*}{S}_2$$

$$C\overset{*}{S}_2 + C_5H_{10}NH + OH^- \rightarrow C_5H_{10}NC(=\overset{*}{S})-\overset{*}{S}^- + H_2O$$

The sensitivity of the method depends on the specific radioactivity of the [^{35}S]-tetraethylthiuram-disulfide preparation. At a specific activity of 0.25 mCi/mmole the method permits determination of the SH content of about 10 μg of protein. The method can be used to measure the SH groups of colored and insoluble proteins, for which spectrophotometric methods are inapplicable, and also for radioautographic demonstration of the SH groups of tissue proteins.

Kohno (1966) proposed the use of thiamine-disulfide for determining SH groups. When it reacts with thiols free thiamine is formed, and this may be determined as thiochrome by fluorometry. The method allows determination of 0.1-0.2 μmole of SH compound, or of 2 nmole in an ultramicromodification.

Determination with Sulfenyl Halides. The reaction of SH groups with arylsulfenyl chlorides can be used for their determination in proteins (Fontana et al., 1968a). The most convenient reagent for this is p-nitrophenylsulfenyl chloride, which rapidly and stoichiometrically reacts with SH groups in 50% aqueous acetic acid to form mixed disulfides:

$$R'S-Cl + HS-R'' \rightarrow R'-S-S-R'' + HCl$$

These disulfides are stable in acidic media, but are destroyed in alkali, quantitatively liberating nitrothiophenol, which can be determined spectrophotometrically in 0.1 M NaOH (by its absorption at 412 nm), thus permitting measurement of the number of cysteine residues in a protein. By this method 7.8 SH groups were found in reduced ribonuclease (out of 8).

d. Methods Based on Oxidation

Determination Using o-Iodosobenzoate. To determine the SH groups of a protein using o-iodosobenzoate, the protein is incubated at pH 7.0 with a small excess of the reagent. After incubation the quantity of unreacted iodosobenzoate is determined by addition of an acidified solution of KI and titration of the liberated iodine with thiosulfate (for a more detailed description of the method, see Chinard and Hellerman, 1954):

$$HOOC{-}C_6H_4{-}IO + 2I^- + 2H^+ \rightarrow HOOC{-}C_6H_4I + I_2 + H_2O$$

$$I_2 + 2S_2O_3^{2-} \rightarrow S_4O_6^{2-} + 2I^-$$

The course of the oxidation of SH groups by o-iodosobenzoate can also be followed by the drop in absorption of the reagent at 285 nm (Leslie and Varricchio, 1968).

Determination as Cysteic Acid. For determination of the cysteine content of proteins, they are first treated with performic acid. The aim of this treatment is to convert the unstable cysteine residues into cysteic acid, which withstands acid hydrolysis and can then be reliably determined by chromatography of the hydrolysate.

The procedure for treating the protein with performic acid and for the subsequent hydrolysis is described below (Chapter 3, Section 2a). Agreement between the number of cysteic acid residues found in the hydrolysate with the number of SH groups that can be titrated in the protein indicates that the protein possesses no S–S bonds.

2. Determination of S – S Groups

The methods of determining S–S groups in proteins are based on their prior scission by oxidation or reduction. On oxidative scission of S–S groups with performic acid, cysteic acid residues are formed; on reductive scission SH groups are formed, and they may be determined by the methods described above, in particular by amperometric or spectrophotometric titration or as S-carboxymethylcysteine after alkylation with iodoacetate. If thiols are used for the reduction of the S–S groups, then the protein must be carefully separated from the excess of added thiol before the number of SH groups formed can be determined.

a. Determination as Cysteic Acid

The oxidative scission of S–S bonds by performic acid was first applied by Sanger (1947, 1949) in studying insulin and has been successfully used since then in protein chemistry. Two residues of cysteic acid are formed for each S–S bond broken. Determination of the cysteic acid in the protein hydrolysate permits estimation of the total content of cysteine and cystine residues in the protein; they are usually combined under the term "half-cystine residues."

The procedure for oxidizing the protein with performic acid and for determining cysteic acid in the hydrolysate has been developed in a number of studies (Schramm et al., 1954; Moore, 1963; Hirs, 1967a). The usual method is to treat the protein at 0°C for 4 h (or for 20 h if the protein is insoluble) with a solution of performic acid. This is freshly prepared by mixing one volume of 30% hydrogen peroxide with nine volumes of 88% formic acid and allowing to stand for 1-2 h at room temperature (to allow formation of the performic acid) before cooling for the oxidation. The performic acid is destroyed after the oxidation by adding aqueous HBr, and the mixture is evaporated to dryness before the usual procedures of acid hydrolysis and amino acid analysis. Moore recommends carrying out a control analysis of unoxidized protein in order to check that the hydrolysate contains no substances that have the same chromatographic properties on the sulfonic resin used for analysis as cysteic acid. If such substances are present, they may be separated from the cysteic acid by chromatography on a basic resin. Under the standard conditions of hydrolysis there is a loss of 6-8% of the cysteic acid, and correction is made for this.

Treatment of proteins with performic acid under the conditions described leads to the destruction of tryptophan residues, the partial destruction of tyrosine and the quantitative conversion of methionine into methionine sulfone.

Spencer and Wold (1969) found that cystine and cysteine residues are converted into cysteic acid if the acid hydrolysis is conducted in the presence of 0.2–0.3 M dimethylsulfoxide. Starting from this, they worked out a new procedure for determining half-cystine residues in proteins; this procedure does not require prior treatment of the protein with performic acid. Hydrolysis in the presence of dimethylsulfoxide leads to the destruction of some amino acids (tyrosine, histidine, methionine, and, to a lesser extent, serine, threonine, and proline); hence the hydrolysate obtained under these conditions cannot be used for complete amino acid analysis of the protein. The procedure of Spencer and Wold gave 88% of the correct value for the half-cystine of chymotrypsinogen, and about 100% for ribonuclease, insulin, and several other proteins.

b. Determination as S-Sulfocysteine

Inglis and Liu (1970) put forward a new method of determining the content of cystine and cysteine in proteins which did not require their prior oxidation to cysteic acid. The method was based on the conversion of the cystine and cysteine (and their partially oxidized derivatives) present in the acid hydrolysate of a protein into S-sulfocysteine by treatment of the hydrolysate first with dithiothreitol and then with sodium tetrathionate. The quantity of S-sulfocysteine formed is determined on an amino acid analyzer.

The treatment with dithiothreitol and tetrathionate does not affect the determination of other amino acids, with the possible exception of tyrosine. The method gave precise results for half-cystine on ribonuclease, lysozyme, streptococcal protease, and wool. From the data of Inglis and Liu it follows that cystine is more stable to acid hydrolysis than was supposed earlier.

c. Determination as S-β-(4-Pyridylethyl)cysteine

Friedman et al. (1970) described a quantitative determination of cystine plus cysteine residues in proteins as S-β-(4-pyridylethyl)cysteine. In their method the disulfide bonds are first reduced with β-mercaptoethanol, and the SH groups are then alkylated by treatment with 4-vinylpyridine.

They dissolved the protein in a tris-nitrate buffer of pH 7.5 containing 8 M urea. Mercaptoethanol (in about 100-fold excess) was added under nitrogen and the mixture was stirred at room temperature for 16 h. 4-Vinylpyridine in an amount equimolar with the total thiol present was then added and stirring continued for 1.5–2 h. If the protein was insoluble, a 3:1 molar ratio of 4-vinylpyridine was used, and the incubation extended to 4.5 h. After the reaction the solution was adjusted to pH 3 with glacial acetic acid, dialyzed against 0.01 M acetic acid, and freeze-dried. A sample of the dry protein was hydrolyzed and subjected to amino acid analysis.

S-β-(4-Pyridylethyl)cysteine is stable to acid hydrolysis up to 120 h; it is eluted from the sulfonic resin column of an amino acid analyzer before arginine and well resolved from it. The total cystine plus cysteine content of several proteins determined by the method was in good agreement with the values in the literature. The SH group of cysteine was the only group modified when the alkylation was performed as described for soluble proteins. With the extended time of alkylation of 4.5 h, there was some decrease in lysine content.

d. Determination as S-2-(2-Quinolylethyl) cysteine

Krull et al. (1971) have worked out a spectrophotometric method for determining the cysteine plus cystine residues of unhydrolyzed proteins. The method is based on the reduction of S—S bonds with β-mercaptoethanol (under the same conditions as in Section c) and the subsequent reaction of the SH groups with 2-vinylquinoline. This is added in equimolar amount with respect to all SH groups present, and the mixture is incubated for 4 h with mixing. The solution is adjusted to pH 3 with glacial acetic acid and dialyzed until free of all reagents. The content of S-2-(2-quinolylethyl)cysteine residues is calculated from the extinction at 318 nm of a solution of the modified protein in 0.1 M acetic acid. The molar extinction coefficient of these residues is 10,000 in 0.1 M acetic acid, in 0.1 M HCl, and in constant boiling HCl. Proteins that are insoluble after dialysis are hydrolyzed and their extinction read in constant boiling HCl. The total content of cystine plus cysteine residues found by this method for several proteins was in excellent agreement with the values in the literature.

e. Determination as S-(p-Nitrophenethyl) cysteine

Masri et al. (1972) alkylated the SH groups of reduced soluble proteins and keratin with p-nitrostyrene. Acid hydrolysis of the products yielded S-(p-nitrophenethyl)-L-cysteine, which can be determined by amino acid analysis or by spectrophotometry (its absorbance is 9600 M^{-1} cm^{-1} in 6 M HCl at 283 nm). The rate of reaction of p-nitrostyrene with SH groups is 4-5 times less than the rate of reaction of 2-vinylquinoline or of 4-vinylpyridine.

f. Amperometric Titration

Kolthoff and Stricks (1950) described an amperometric titration of cystine by silver nitrate in the presence of sulfite, which breaks S–S bonds according to the following equation:

$$RS{-}SR + SO_3^{2-} \rightleftarrows RS{-}SO_3^- + RS^-$$

Carter (1959) applied the argentometric titration in the presence of sulfite to determination of the S–S content of proteins. According to his method, 0.1 ml of a freshly prepared, saturated solution of Na_2SO_3 is added to 30 ml of a solution of the protein in 8 M urea containing 60 μM EDTA (pH of mixture 8.3) and titration with 1 mM $AgNO_3$ at 37°C is started at once, using a rotating platinum electrode as the indicator electrode. The minimal quantity of protein required for the analysis is 1 mg in 30 ml. Carter determined the content of S–S groups in several proteins (ribonuclease, insulin, trypsin, serum albumin, etc.) and obtained data that agreed well with those found by other methods.

The amperometric titration of the SH groups liberated on scission of S–S bonds by sulfite can also be performed with $HgCl_2$, methylmercury iodide, or ethylmercury iodide (Stricks et al., 1954b; Leach, 1960; Kolthoff et al., 1958, 1965a,b). In this case the dropping mercury electrode is used as indicator, since sulfite forms a complex with mercury ions that is not reduced at a platinum electrode.

Leach (1960) recommended carrying out the mercurimetric titration of proteins in a 0.1 M NH_4Cl-NH_4OH buffer containing 0.5 M KCl, 0.2 M Na_2SO_3, and 8 M urea, at pH 9.2. The titration is carried out in the polarograph cell at 20°C and at a potential of −0.8 V, using 2 mM CH_3HgI in 25% dimethylformamide, or at a potential of −1.0 V with 10 mM $HgCl_2$ in water. According to Leach's results, under these conditions one Hg^{2+} ion combines with two SH groups (see Chapter 2, Section 3); methylmercury iodide reacts with SH groups in the ratio 1:1.

Harrap and Gruen (1971) have worked out a method based on the reduction of the S–S bonds in proteins with $NaBH_4$ in 8 M urea and the subsequent titration of the SH groups formed with 0.1 M $AgNO_3$; they used the silver sulfide specific ion electrode (see Section 1a) as indicator. For titration they took a quantity of protein that contained about 5 μmole of disulfide. Excellent results were obtained with several proteins (ribonuclease, chymotrypsin, serum albumin, etc.); in the case of lysozyme, however, the method gave too high a value, apparently because of unspecific binding of Ag^+ to the reduced protein.

g. Fluorometric Method

Karush et al. (1964) developed a method for the quantitative determination of disulfides using fluorescein mercury acetate (FMA):

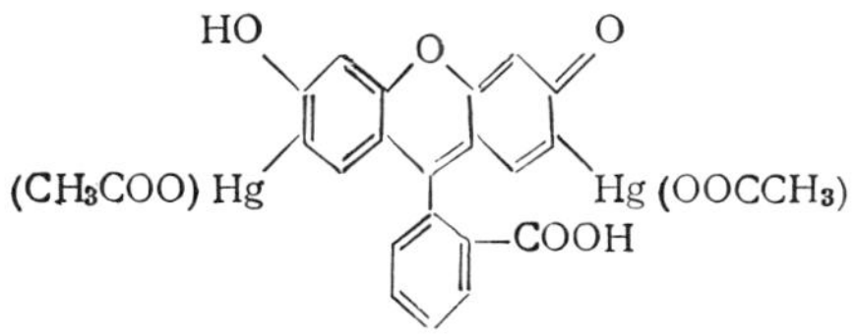

HO O O

(CH_3COO) Hg Hg $(OOCCH_3)$

–COOH

The method is based on the quenching of the fluorescence of FMA when it reacts with disulfides in alkaline media (the excitation maximum of the reagent is at 499 nm and its emission maximum is at 520 nm); the quenching results, apparently, from combination of the reagent with an SH group liberated by alkaline hydrolysis of the disulfide. Indeed, according to Heitz and Anderson (1968), FMA selectively reacts with the SH groups of yeast alcohol dehydrogenase with accompanying quenching of fluorescence.

The method is highly sensitive; on spraying a paper with an alkaline solution of FMA one can visualize spots that contain 0.1 nmole of oxidized glutathione. Methionine and lanthionine give only a weakly positive reaction when they are applied to the paper in an amount of 10 nmole. The strongest unspecific reaction was observed with histidylhistidine, but even this reaction was 100 times less sensitive than the reaction with disulfides.

To determine their S–S content proteins are incubated with 1-10 μM FMA in 1 M NaOH for 30-60 min at room temperature. Using a standard curve obtained from the reaction of FMA with ribonuclease, Karush et al. determined the content of S–S groups in lysozyme, serum albumin, insulin, trypsin, chymotrypsinogen, and purified rabbit antibody. With all the proteins studied, except for the antibodies, they obtained values that agreed with those in the literature. When Habeeb (1966) used this method, he obtained raised values for the S–S contents of γ-globulin and guanidylated albumin. He suggested that the NaOH could liberate from proteins some groups other than SH groups that could also quench the fluorescence of FMA and so interfere with the quantitative determination of disulfides.

h. Methods Based on the Reduction of S–S Groups and the Subsequent Use of Ellman's Reagent

Zahler and Cleland (1968) worked out a sensitive method for determining disulfides, based on their prior reduction with dithiothreitol or dithioerythritol (see the equations for the reaction in Chapter 2, Section 7). After completion of the reduction, the excess dithiol added is bound with arsenite, which forms a stable complex with it (see Table 6, Chapter 1, Section 8). The monothiols formed (which do not give stable complexes with $HAsO_2$) are then titrated spectrophotometrically with 5,5'-dithio-bis-(2-nitrobenzoate).

On the basis of this method Walsh et al. (1970) developed an automatic system for determining cystine-containing peptides in column chromatography of protein hydrolysates. They recommended addition of 0.01-0.02 M EDTA to stabilize the color.

The application of the method of Zahler and Cleland to proteins has encountered difficulties. Some of these are connected with the fact that reduced peptides containing several thiol groups may bind arsenite so strongly that they do not react with 5,5'-dithio-bis-(2-nitrobenzoate). A more reliable method of determining the S–S bonds of proteins may be to reduce them with sodium borohydride before spectrophotometric titration of the SH groups formed; the excess $NaBH_4$ is easily destroyed by acidifying the solution to pH 3 (Chapter 2, Section 9).

Robyt et al. (1971) proposed a method for determining S–S groups on the basis of their conclusion that the nitrothiophenolate anion that arises when Ellman's reagent reacts with an SH group can then react at pH 8 with a disulfide group of a protein to form a mixed disulfide. Since the thiol group simultaneously produced would react with more Ellman's reagent, the disulfide groups of the protein would be progressively converted into mixed disulfides, and since these liberated colored nitrothiophenolate anion on addition of dithiothreitol or on raising the pH to 10.5, they could easily be determined. In the case of an amylase Robyt et al. (1971) appear to have established that the reactions occur, since nitrothiophenolate was released from the protein by dithiothreitol after removal of excess Ellman's reagent. In the case of papain, however, which they also studied, Brocklehurst et al. (1972a) measured the amount of Ellman's reagent that disappeared, as well as the amount of nitrothiophenolate released from the protein previously treated with Ellman's reagent, and concluded that no formation of mixed disulfides had occurred. Brocklehurst et al. (1972a) explained the further release of nitrothiophenolate found by Robyt and coworkers when the pH was raised by a papain-catalyzed alkaline scission of Ellman's reagent. The constant stoichiometry found by Robyt et al. could be explained by the progressive inactivation of the papain in the course of the reaction. It is therefore premature to conclude that the reaction reported provides a general method for estimating S–S groups especially since Telegdi and Straub (1973) observed no reaction between nitrothiophenolate anion and the S–S groups of amylase. Spectrophotometry appears capable of determining whether Robyt's exchange reaction occurs, since mixed disulfides of Ellman's reagent absorb strongly near 325 nm.

i. Determination as Hydrogen Sulfide

Robbins and Fioriti (1963) found that heating protein solutions to 100°C at pH 9.5 in the presence of cadmium hydroxide led to the quantitative liberation of hydrogen sulfide from the cystine and cysteine residues. Neither free methionine nor methionine residues in proteins formed H_2S under these conditions.

Taniguchi (1971) showed that a 0.6% suspension of zinc hydroxide could be used in place of cadmium hydroxide with just as good a result. He worked out a new and very precise micromethod for determining the total content of cystine plus cysteine of proteins. The protein samples, containing 0–1000 nmole of sulfur, are treated with a 0.6% suspension of zinc hydroxide at pH 9.5 and 100°C for 48–96 h under a nitrogen atmosphere. The reaction mixture is acidified with 6 M H_2SO_4; the H_2S liberated is trapped in a 0.6% zinc acetate solution, and then determined by a colorimetric method based on the formation of methylene blue from N,N'-dimethyl-p-phenylenediamine in the presence of H_2S and of ferric ammonium sulfate. Under the conditions described the sulfur of both cysteine and cystine is quantitatively released as H_2S.

j. Direct Spectrophotometric Measurement of Reduction by Dithiothreitol

Iyer and Klee (1973) have published a spectrophotometric method of determining the rate of reduction and the number of

S—S bonds in peptides and proteins. This method is based on the absorbance of the oxidized form of dithiothreitol. The reduced form has no absorbance at wavelengths above 270 nm whereas the oxidized form displays a broad band with a maximum at 283 nm ($\varepsilon_m = 273$). Iyer and Klee recommend measurement of the absorbance increase at 310 nm, because the absorption of the oxidized dithiothreitol at this wavelength is great enough ($\varepsilon_m = 110$) whereas the absorption of the protein chromophores is small enough for them not to interfere appreciably. It is thus relatively easy to measure the concentration of oxidized dithiothreitol at values as low as 0.1 mM (in 1-cm cells) in the presence of protein, using a spectrophotometer equipped with a scale expanded over the absorbance range of 0 to 0.1. The method is rapid and can produce a complete kinetic curve with 0.1 μmole of disulfide. Results with oxidized glutathione, ribonuclease, bovine serum albumin, and α-lactalbumin show that the correct stoichiometry is obtained for the reaction. The reason why the disulfide of dithiothreitol absorbs more at 280–310 nm than does the original protein disulfide is not yet clear; presumably it is connected with the constraint of the disulfide bond into a six-membered ring.

Chapter 4

The Reactivity of SH Groups in Proteins

1. Types of SH Groups

The reactivity of SH groups in proteins varies over a wide range. Hellerman et al. (1943) and later Barron (1951) divided the SH groups of proteins into three types: rapidly reacting, sluggishly reacting, and "masked" or "buried." According to Barron, SH groups of the first type react readily with nitroprusside and with mild oxidizing agents such as ferricyanide, porphyrindin, and o-iodosobenzoate. SH groups of the second type do not give the nitroprusside reaction, and react only with stronger oxidizing agents (e.g., iodine) and with mercaptide-forming reagents (p-mercuribenzoate and organic arsenical compounds). To the third group belong the SH groups that can be detected only after denaturation of the protein, i.e., after destruction of its secondary and tertiary structure.

While the activity of some enzymes, for example succinate dehydrogenase and glyceraldehyde-3-phosphate dehydrogenase, is inhibited by blocking its easily reacting SH groups, inhibition of the activity of others, for example urease, aldolase, malate dehydrogenase, begins only when their slowly reacting or even "masked" SH groups are blocked. By stepwise titration of SH groups by the method of Boyer (Chapter 3) and parallel determination of enzymic activity one can establish which kind of SH group is required for enzymic activity.

Szabolcsi and Biszku (1961) (see also Szájani et al., 1970; Závodszky et al., 1972) found that blocking of the 8-12 most reactive SH groups of aldolase (2-3 per subunit) with p-mercuribenzoate did not affect the activity of the enzyme. The blocking of these groups, however, produced conformational changes in the enzyme that rendered accessible to titration other SH groups that had been unreactive or "buried" in the native protein. The reaction of these newly exposed SH groups with p-mercuribenzoate led to further and greater changes in the conformation of the molecules and to loss of the enzymic activity. It is interesting that the oxidation of 8-10 SH groups in adolase with tetranitromethane led, on the contrary, to complete inactivation of the enzyme (Riordan and Christen, 1968), but there is no evidence whether the thiol groups most reactive to p-mercuribenzoate were those oxidized by tetranitromethane.

Naturally there are no sharp boundaries between the various types of SH group, and it is sometimes difficult to assign a particular group to one type or another. Nevertheless the division of SH groups into types according to their reactivity has proved useful and has been widely used in the literature.

2. Causes of Altered Reactivity of SH Groups

The reactivity of SH groups of native proteins is usually lower than in simple thiols, and rises upon denaturation. There are, however, many exceptions to this rule. Thus the rate of reaction of the catalytically active SH groups of ficin and papain with chloroacetamide is 15-20 times greater than the reaction rates of many simple thiols; the rate constants for ficin, papain, cysteine, and β-mercaptoethylamine are 7.5, 6.5, 0.43, and 0.3 M^{-1} sec^{-1} respectively (Hollaway et al., 1964; Chaiken and Smith, 1969; Lindley, 1960). The SH group of papain reacts with cyanate almost 3000 times faster than the SH group of cysteine (Sluyterman, 1967). The rate constants of the reactions of 2-chloromercuri-4,6-dinitrophenol with the SH groups of papain and of thioglycolic acid are 1.4×10^7 and 3×10^5 $M^{-1} \cdot sec^{-1}$ respectively (Gutfreund and McMurray, 1970). The SH group of the active site of streptococcal protease is also highly reactive; it competes succesfully for iodoacetate with 100- to 1000-fold excess of β-mercaptoethanol or cysteine; the rate of its reaction with chloroacetamide is 50-100 times more than that of the SH groups of glutathione (see Chapter 1, Section 6c) (Liu et al., 1965; Gerwin, 1967). The catalytically active SH group of glyceraldehyde-3-phosphate dehydro-

TABLE 11. Rate Constants for the Reactions of the SH Groups of Some Enzymes and of Cysteine with DL-α-Iodopropionic Acid at 25°C (Wallenfels and Eisele, 1970)

	$10^3 \times k$ ($M^{-1} \cdot sec^{-1}$)	
	pH 7.0	pH 8.5
Cysteine	3.2	45
Papain	2500	700
Yeast alcohol dehydrogenase	32	10
Yeast glyceraldehyde-3-phosphate dehydrogenase	40	270
Lactate dehydrogenase	–	0.2

genase reacts faster than β-mercaptoethanol or cysteine at neutral pH with iodoacetate and with 5,5'-dithio-bis-(2-nitrobenzoate) (Boross et al., 1969; Boross, 1969).

The reactivity of the SH groups of a number of enzymes and of cysteine with respect to α-iodopropionate are given in Table 11.

It is not only catalytically active SH groups that can possess high reactivity. Thus four SH groups of muscle phosphorylase b react with chlorodinitrobenzene at pH 8 five times faster than cysteine and 12 times faster than β-mercaptoethanol, although these groups do not take part in the catalytic action of the enzyme or in the binding of its substrates (Gold, 1968). One of the 17 SH groups of phosphofructokinase reacts with 5,5'-dithio-bis-(2-nitrobenzoate) 2×10^4 times faster in the native conformation than after denaturation; this SH group also fulfills no catalytic function, since after its blocking only 25% of the activity of the enzyme was lost (Kemp and Forest, 1968).

Thus, in discussing the causes of altered reactivity of the SH groups of proteins, it is useful to consider separately possible causes of increased reactivity, i.e., mechanisms for activating SH groups, and possible causes of diminished reactivity, i.e., mechanisms of "masking" of SH groups.

a. Mechanisms of Activation of SH Groups

Two different mechanisms have been proposed to account for the raised reactivity of some SH groups in proteins.

Fig. 14. Scheme of activation of the SH group of an enzyme by formation of a hydrogen bond with the imidazole ring of histidine (Rabin and Whitehead, 1962).

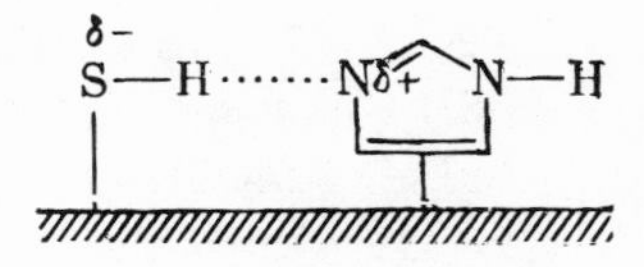

Rabin and colleagues (Rabin and Watts, 1960; Rabin and Whitehead, 1962; Watts and Rabin, 1962; Whitehead and Rabin, 1964; see also Olson and Park, 1964; Husain and Lowe, 1968) postulated that SH groups situated in the active sites of enzymes may be activated as a result of the formation of hydrogen bonds with neighboring functional groups of the protein, such as imidazole or carboxylate anion (Fig. 14); Hammond and Gutfreund (1959) first postulated hydrogen bonding between SH and carboxyl groups in the active site of an enzyme in their study of ficin. The participation of the proton of the SH group in hydrogen-bond formation leads to an increase of electron density at the sulfur atom and consequently to an increase in its nucleophilic properties. This should be accompanied by a rise in the pK of the SH group and a fall in the pK of the proton-accepting group (e.g., imidazole) since both ionization of the SH group and protonation of the imidazole involve breaking of the hydrogen bond.

In agreement with this hypothesis Rabin and colleagues found that the rates of alkylation by iodoacetamide of the SH groups situated in the active sites of alcohol dehydrogenase and of ATP-creatine-phosphotransferase (creatine kinase) were unaltered over a wide range of pH (from 4 to 9–10 for the former enzyme, and from 6 to 10 for the latter). This fact can be explained by an increase in the pK of the SH group to above 10, since ionization of the SH group at lower pH values should be accompanied by a marked increase in the rate of alkylation (see Chapter 1, Section 6c). The rate of alkylation of these SH groups was also independent of the ionic strength, whereas a secondary salt effect would be expected if mercaptide ion participated in the reaction (Watts and Rabin, 1962).

One would expect that the reactivity of an SH group activated by formation of a hydrogen bond with some basic group in the protein would be intermediate between the reactivity of an un-ionized SH group and that of a mercaptide ion. And in fact Whitehead and Rabin (1964) found that the rate constant for alkylation by iodoacetamide of the reactive SH group of yeast alcohol dehydrogenase was 0.4–0.65 $M^{-1} \cdot sec^{-1}$, which is much lower than the rate constants for RS^- ions formed from simple thiols (800 $M^{-1} \cdot sec^{-1}$). Further, Little and Brocklehurst (1972; Brocklehurst and Little, 1973); found that the rate of reaction of papain with Ellman's reagent fell from 284 to 40 $M^{-1} \cdot sec^{-1}$ as a group with a pK of 8 was protonated, so the acid form could be a hydrogen-bonded SH group. But activation of SH groups by hydrogen bonding cannot be the sole factor in cases, such as that of ficin, where the reactivity of the SH group of the enzyme is higher than that of a simple RS^- ion. And the fact that the pK values of the SH groups are not raised (8.55 in ficin and 8.0 in papain) is hard to reconcile with their involvement in hydrogen bonding. Hollaway et al. (1964) explained the high rate of reaction of the SH group of ficin with chloroacetamide on the hypothesis that a molecule of the reagent interacts simultaneously with both the SH group and with a cationic group of the protein located nearby. As shown in Fig. 15a, the cationic group that participates in forming the activated complex assists in breaking the bond between the carbon and chlorine atoms, i.e., it lowers the energy of the transition state. The mechanism of activation of SH groups proposed by Hollaway and co-authors postulates a highly specific interaction between the thiol reagent and the immediate environment of the SH group in the protein. It is natural to expect that such an interaction could occur with only a limited group of reagents. In fact the SH group of ficin, which reacts so unusually rapidly with chloroacetamide, reacts with another reagent, N-ethylmaleimide, considerably more slowly than cysteine. Evidently the difference in mechanisms be-

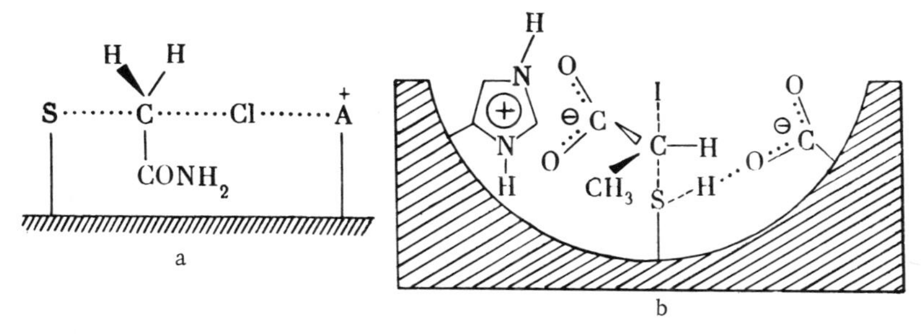

Fig. 15. The proposed transition states in alkylation reactions of the SH groups of ficin and papain. Two possible mechanisms for acceleration of the reaction. a) Interactions of chloroacetamide in the active site of ficin (Hollaway et al. 1964); A is a cationic group of the protein. b) Interactions of L(–)-α-iodopropionic acid in the active site of papain (Wallenfels and Eisele, 1968).

tween the reactions of the SH group with chloroacetamide (nucleophilic displacement of a halogen atom at tetrahedral carbon) and with N-ethylmaleimide (addition to an activated double bond) determines the different dependence of the rates of these reactions upon the nature of the environment of the SH group in the protein, although steric factors could also be involved.

The influence of the immediate environment of the SH group on its reactivity is shown clearly in the stereospecific alkylation by compounds that contain an asymmetric center. Wallenfels and Eisele (1968) found that L(−)-α-iodopropionic acid reacts with the SH group of papain 12 times faster than its D(+)-antipode. The inverse ratio of rates was found for the reactions of the L(−)- and D(+)-antipodes of α-iodopropionamide. Wallenfels and Eisele explained the high rate of reaction between the SH group of papain and L(−)-α-iodopropionate (the rate constant is 1000 times higher than for the reaction with cysteine) by supposing that this reagent is specifically orientated in the active site of the enzyme thanks to the concerted action of a cationic (possibly imidazole) and an anionic group of the protein; besides this they postulated hydrogen bonding between the SH group and a carboxyl group of the protein (Fig. 15b). The combination of these factors leads, according to their calculations, to a diminution in the free energy of activation ($\Delta G^{\neq}$) by 4.5 kcal/mole (in comparison with the alkylation of cysteine).

Polgar (1973) recently extended such investigations of alkylation, and showed that not only must a group of pK 8.4 be protonated but one of pK 4 must be deprotonated. He identifies the latter with histidine-159, and discusses possible reasons for its low pK (cf. Little and Brocklehurst, 1972). Since he finds no detectable isotope effect in D_2O, the proton must have passed from the SH group to the histidine before the rate-determining step.

Brocklehurst, Suschitzky and colleagues (unpublished) have found that 2-mercaptomethyl-4,5-benzimidazole shares with papain (see Chapter 3, Section 1c and Chapter 6, Section 1a) the possession of an optimum rate of reaction with 2,2'-dithiodipyridine near pH 4, and that it loses this property when one or two more methylene groups are inserted between the SH group and the imidazole ring. This model compound may assist the study of how imidazole can activate a neighboring SH group.

The mechanism of activation of the SH groups of an enzyme postulated by Hollaway and co-authors agrees with the results of model experiments by Heitmann (1968b), who studied the reactivities of SH groups introduced into anionic and cationic micelles. The anionic micelles were formed from N-dodecanoyl-DL-cysteinate or from a mixture of it with N-dodecanoylglycine; the cationic micelles from hexadecyltrimethylammonium bromide. The incorporation of SH groups into anionic micelles was accompanied by their "masking," i.e., by a sharp drop in their reactivity. On incorporating N-dodecanoylcysteinate into cationic micelles the reactivity of the SH groups, on the other hand, was greatly increased. The rate constants for the reactions of RS^- ions with iodoacetamide and with p-nitrophenylacetate were 60-100 and 100-200 times greater, respectively, in cationic micelles than in solution. The effect is still greater if the rate constants are calculated with respect to the total concentration of SH groups, since their degree of ionization is markedly increased in the cationic micelles. Heitmann suggested that the ionized SH group is situated on the surface of the micelle in an environment of the positively charged groups of the detergent. The reagent attacking the SH group, iodoacetamide for example, is orientated on the micelle surface according to its dipolar structure in the way shown in Fig. 16. Thus the bond between the carbon and iodine atoms becomes the object of synchronous attack by the RS^- ion and by the cationic group. Participation of the latter in stabilizing the transition state greatly accelerates the reaction. The cationic group can facilitate breakage of the carbon—oxygen bond in the acetylation of the RS^- ion by p-nitrophenylacetate in an analogous manner (Fig. 17).

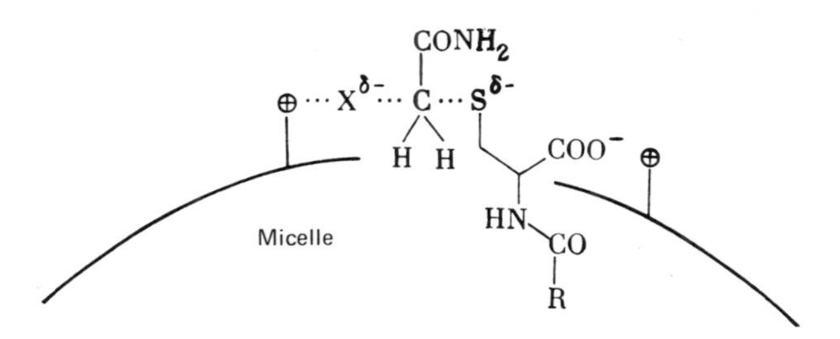

Fig. 16. The proposed transition state for alkylation of an SH group on the surface of a cationic micelle (X = Cl or I) (Heitmann, 1968b).

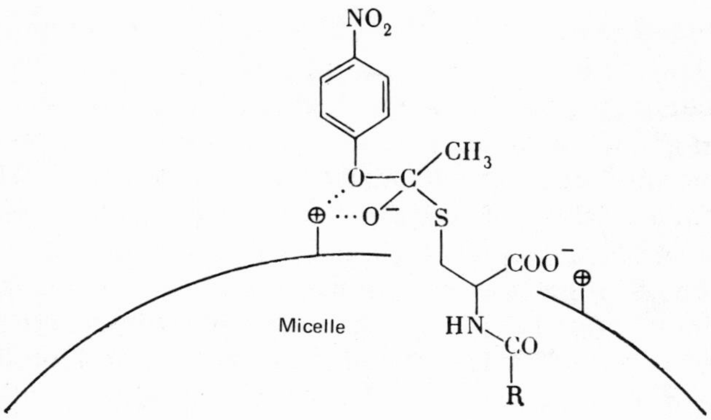

Fig. 17. Proposed tetrahedral intermediate in the acylation of a SH group on the surface of a cationic micelle (Heitmann, 1968b).

b. Mechanisms of "Masking" SH Groups.

The Formation of Intramolecular Bonds

By "masking" of SH groups in proteins is meant the slowing or abolition of their reaction with some kind of specific thiol reagent. This phenomenon was observed much earlier than the activation of SH groups, and it is encountered incomparably more often. Two hypotheses have been advanced to explain the mechanism of such "masking":

a) the steric inaccessibility of the SH groups, i.e., their screening by the neighboring amino acid residues;

b) the formation of intramolecular chemical bonds by the SH groups ("chemical masking").

The first of these hypotheses was put forward as early as the thirties by Mirsky and Pauling (1936; Mirsky, 1941). According to it, the sulfhydryl and other functional groups that are characterized by a diminished reactivity are located inside the protein globule and are thus sterically inaccessible to chemical agents. The long known fact of the increase of reactivity of the SH groups on denaturation, i.e., on unfolding of the globule, argues in favor of this viewpoint. The comparison of data on the reactivities of SH groups with the results of x-ray crystallographic studies on protein crystals that locate these groups precisely in the molecule can give a direct check of the hypothesis. Such a comparison is so far possible only for oxyhemoglobin. According to Cullis et al. (1962) and Smith and Perutz (1960) horse hemoglobin contains two pairs of SH groups, of which only one (belonging to cysteine-93 of the β-chains) reacts readily with various thiol reagents. The other pair (belonging to residue α-104) belongs to the maked category. In complete accord with these data Perutz (1965; Perutz et al., 1968) has found that the reactive SH groups of hemoglobin are located on the surface of the molecule, while the others are buried inside the tetrameric macromolecule of the protein.

The example of hemoglobin supports the hypothesis of steric inaccessibility of masked SH groups. But there are also data in the literature that are hard to explain on this hypothesis. Already Anson (1945) had pointed out that postulation of steric inaccessibility does not explain why the SH groups of native egg albumin do not react with nitroprusside, ferricyanide, oxygen, and dilute H_2O_2, since they do react with iodoacetamide, iodine, and concentrated H_2O_2. On the other hand even the SH groups of denatured egg albumin do not react with mild oxidizing agents under some conditions.

The absence of correlation between the molecular dimensions of thiol reagents and the accessibility of protein SH groups to them was noted in a number of other investigations. Thus Smith (1958) found that the SH group of papain, although unavailable for reaction with nitroprusside or porphyrindin, reacts unusually quickly with the rather large molecule of p-mercuribenzoate. Clearly such differences cannot be explained by simple screening of the SH group, and important supplementary factors must be involved in determining the peculiar reactivities of the SH groups of proteins. Among such factors is the influence of the functional groups that are located close to the SH group in the tertiary, secondary, or primary structure of the protein. Such groups may:

a) facilitate or hinder, according to their charge, the approach of ionized reagents to the SH groups;
b) alter the degree of ionization of the SH groups;
c) participate in stabilizing the transition state of the reaction of the SH groups with thiol reagents;
d) form intramolecular bonds of various kinds with the SH groups.

The influence of the charge of the side chains of neighboring amino acid residues on the interaction of SH groups with ionized reagents can be illustrated by the example of alkylation reactions. The alkylation of SH groups by iodoacetate should proceed faster if there is a positively charged group in the vicinity, since it both facilitates the approach of the iodoacetate anion and lowers the pK of the SH group. Conversely, a negatively charged group near to the SH group should slow reaction with the iodoacetate anion both by electrostatic repulsion of the anion and by raising the pK of the SH group. These considerations explain the difference in the kinetics of the reactions of the SH group in the active site of alcohol dehydrogenase with the anion of iodoacetate and with the neutral molecule of iodoacetamide (see Chapter 1, Section 6c).

As already noted, Heitmann (1968b) found that placing SH groups in anionic micelles led to a large fall in their reactivity with iodoacetate and with p-nitrophenylacetate. The SH groups are located on the surface of the micelles in an environment of negatively charged groups, which, in Heitmann's opinion, do not merely lower the degree of ionization of the SH groups, but also directly hinder their reaction with the two reagents. The results of experiments with micelles (which may be regarded as models of a protein) show clearly that the cause of masking is not solely the burying of SH groups inside the protein globule: an SH group on the surface may appear masked if one or more negatively charged groups are close to it.

In addition to purely ionic interactions, the cause of masking of SH groups in proteins may be their participation in various kinds of intramolecular chemical bonding. What kinds of such bonding can SH groups form?

Linderström-Lang and Jacobsen (1940, 1941) expressed the thought that the lowered reactivity of some SH groups in proteins could be explained by their forming thiazolidine or thiazoline rings:

```
—CH—NH—C—       —CH—NH—C—OH       —CH—N=C—
 |       ||  ⇄    |      / \    ⇄    |    /
CH2—SH   O       CH2—S             CH2—S
```

Several authors (Calvin, 1954; Basford and Huennekens, 1955; Garfinkel, 1958; Goodman and Salce, 1965; Jones and Nelson, 1969) obtained evidence in favor of the formation of thiazoline structures in strongly acid solutions by glutathione, pantetheine and N-acetyl-β-mercaptoethylamine. But all attempts to observe such structures in proteins have been unsuccessful. Salce and Goodman (1961), on the basis of study of the ultraviolet spectra, the kinetics of oxidation, and the dissociation of model thiazolines and thiazolidines,reckoned that it was most unlikely that thiazolines could play an important role in protein structure.

Several investigators have suggested that thioester bonds may be present in proteins. This hypothesis arose at the very dawn of modern ideas of protein structure (Mastin and Schryver, 1926; Chibnall, 1942). Later Smith (1958) came out in support of it. He proposed that a thioester bond occurs in the active site of papain and of some other thiol enzymes, and that it plays an important part both in the catalytic action of these enzymes and in stabilizing their macromolecular structure. In Smith's opinion, the intramolecular thioester bond, formed by SH and COOH groups of the protein, arose and was maintained at the expense of the "energy of folding" (conformational energy) of the native enzyme molecule. Smith tried to explain the peculiar reactivity of the SH group of papain, which reacts rather rapidly with p-mercuribenzoate but not at all with nitroprusside and porphyrindin, by the presence of a thioester at the active center. Smith's explanation proved, however, to be incorrect. Sanner and Pihl (1962, 1963) showed that p-mercuribenzoate and cystamine, which rapidly block the catalytically active SH group of papain, are incapable of reacting with low-molecular thioesters, which clearly contradicts the hypothesis of Smith. The suggestion of a preformed thioester in the active site of papin was also not confirmed by X-ray analysis of the structure of this enzyme (Drenth et al., 1968).

A reason for the masking of SH groups could be their participation in the formation of hydrogen bonds of the type $RS^{-}\cdots H-D$ (where H—D is a proton-donating group). Such bonding should diminish the reactivity of mercaptide ions. The presence of such bonds in glutathione and some other thiols was postulated by Cecil (1950; 1951) and also by Benesch and Benesch (1953) on the basis of experiments in which urea increased the reactivity of SH groups. The investigations of Wright (1958), Lindley (1962), and Edsall (1965), however, did not support the indications that hydrogen bonding of the mercaptide ion occurred in cysteine or glutathione (see Chapter 1, Section 2), and consequently there is little basis for expecting such bonds to be formed in proteins. In the physiological range of pH most SH groups are in the un-ionized state; under these conditions one could picture the formation of hydrogen bonds of the type RS−H ··· A (where A is an acceptor) in which the SH group participates in the role of proton donor. But such bonds, as was shown in the preceding section, should lead not to the masking of SH groups, but conversely to their activation.

Finally, a reason for the masking of SH groups in proteins could be their involvement in hydrophobic interactions. The possibility of the participation of un-ionized SH groups in hydrophobic interactions has been discussed in the literature only relatively recently. Klotz (1960), and also Némethy and Scheraga (1962) considered SH groups as nonpolar on the grounds that methylmercaptan forms a crystal hydrate. Némethy and Scheraga (1962) calculated that the standard free energy for formation of a hydrophobic bond between the side chain of cysteine and the side chains of nonpolar amino acids at 25°C is 0.6 kcal/mole. It should be noted that the dipole moment of the bond S−H is 0.7 D, which is significantly less than the dipole moment of the bonds O−H (1.51) and N−H (1.31) (Smyth, 1955; Minkin et al., 1968). Cecil (1963; Cecil and Thomas, 1965) considered that the behavior of the masked SH groups of hemoglobin and other proteins could best be explained by participation of these groups in hydrophobic interactions and presented experimental data that indirectly supported this point of view. He found that a number of alcohols, which weakened hydrophobic bonds in proteins, increased the rate of reaction of the inactive SH groups of carboxyhemoglobin with $HgCl_2$ in the presence of sulfite. The effect increases in the series from methanol to butanol, i.e., with lengthening of the hydrocarbon chain of the alcohol. One could imagine, however, that the increase in reactivity of the SH groups is a secondary phenomenon evoked by conformational changes that occur under the influence of the alcohols. Direct support for the involvement of the SH groups of hemoglobin in hydrophobic bonds was obtained by Perutz et al. (1968). From the atomic model of the macromolecule of horse oxyhemoglobin that they constructed it follows that the unreactive SH groups of the cysteine-104 (G11) residues of the α chains participate in forming the hydrophobic contacts $\alpha_1-\beta_1$ between subunits. Human hemoglobin has another unreactive cysteine at β-112, and the leucine at this position of the horse protein is also involved in $\alpha_1-\beta_1$ contacts.

Data were obtained in studies of the SH groups of fumarase that provided evidence that these groups were located in a hydrophobic environment within the molecule of the enzyme. Robinson et al. (1967) found that the rate of reaction of these groups with p-mercuribenzoate increased in the presence of low concentrations of aliphatic alcohols in direct proportion to the number of methylene groups in their hydrocarbon chains. The results of experiments with methyl-, ethyl-, n-propyl-, and n-butylmercury nitrates (250 μM) were highly suggestive; these compounds react with the SH groups of the enzyme the faster, the longer the alkyl radical; the rate constants are 6.4, 8.6, 17.8, and 34×10^{-4} sec^{-1} respectively. The rate of inactivation of fumarase by iodoacetate at pH 6.5 is minimal at 23.5°C and increases on lowering the temperature; this indicates the importance of hydrophobic bonds.

Several authors (Heitz et al., 1968; Fonda and Anderson, 1969) have found that the rate of reaction of the SH groups of yeast alcohol dehydrogenase and of kidney D-amino acid oxidase with a series of N-alkylmaleimides increases steeply with lengthening of the N-alkyl group of the reagent. These data, like the results of the experiments with fumarase, indicate only the nonpolar character of the environment of the SH groups, which evidently assists the binding of reagents that contain the longer alkyl chains; it still does not follow from this that the SH groups themselves are involved in hydrophobic bonds.

Evidence for the participation of SH groups in hydrophobic bonds was obtained by Heitmann (1968a) in model experiments with micelles. He made a comparison of the stabilities of micelles formed with different N-acylamino acids, in which the N-acyl groups contained long hydrocarbon chains. The stability of these micelles depended markedly on hydrophobic bonding. It appeared that micelles formed by N-acylcysteine were significantly more stable than those formed by the corresponding derivatives of glycine and serine. On this basis Heitmann concluded that the SH group of N-acylcysteine participates in the formation of hydrophobic bonds in the micelle. According to his data the replacement of a hydrogen atom on the α-carbon of N-acylglycine by the $-CH_2SH$ group (i.e., by the side chain of cysteine) has the same influence on micelle stability as lengthening the hydrocarbon chain of the N-acyl residue by one methylene group; the $-CH_2OH$ group of serine does not have a similar effect.

Tanford (1973, personal communication) believes, however, that Heitmann's results can have another interpretation. He thinks that the amide group of N-acyl cysteines would be very hydrophilic and would lie outside the hydrophobic core of the micelles. If so, the cysteinyl side chain could not be incorporated into this hydrophobic core. Thus a possible alternative explanation for Heitmann's results would be that the SH group lies between the external carboxyl groups and thereby diminishes the repulsion between them. Because of the large size of the sulfur atom an SH group might be more effective in such surface interaction than an OH group.

3. Conclusions

Summarizing the data presented in this chapter, it should be noted that of the various types of intramolecular interactions in proteins into which SH groups could enter, the most probable are hydrophobic ones. SH groups can, in addition, form charge-transfer complexes with tyrosine and tryptophan, but the occurrence of such complexes in proteins, like their involvement in hydrogen bonding, has not yet been demonstrated experimentally.

Attempts to observe in proteins labile covalent bonds involving SH groups have been unsuccessful. The only proven type of intramolecular covalent bond in whose formation SH groups participate are those mediated by a metal ion, i.e., mercaptide bonds that participate in forming chelate complexes. Such bonds are found in a number of metalloenzymes (see Chapter 6, Section 3); they can be considered as a particular case of the masking of SH groups.

The building of three-dimensional atomic models of thiol enzymes on the basis of the results of X-ray analysis, which is being done in several laboratories, will probably throw light on a number of the questions discussed in this chapter. The comparison of the reactivity of SH groups with their location and microenvironment in the protein molecule will permit a better understanding of the mechanisms of both their activation and their masking. The data of Perutz, cited above, on the localization of the SH groups in the hemoglobin molecule, represent the first step in this direction. But the data already at hand allow us to conclude that the reasons for the changed reactivities of SH groups in proteins may be, in addition to purely steric factors, the peculiarities of their ionic environment and hydrophobic interactions.

PART II

THE ROLE OF SH AND S–S GROUPS IN ENZYMES AND OTHER PROTEINS

Chapter 5

The Essential SH Groups of Enzymes and Methods for Their Identification

1. The Concept of Essential Groups

It is usual to divide the SH groups of enzymes, like other functional groups, into essential and nonessential. The former are those groups whose blocking or destruction by various reagents is accompanied by a change in catalytic activity, usually by its loss or diminution, but more rarely by its increase.

The essential SH groups of enzymes can be divided into three categories according to their role:

I. Groups of the active site: a) catalytic, i.e., catalyzing the conversions and breakdown of the intermediate enzyme–substrate complexes; b) binding, i.e., participating in the formation of bonds between enzymes and substrates or cofactors (metal ions and coenzymes); these groups are part of the contact or substrate sites.
II. Groups located outside the active site, but participating in in its maintenance of an active conformation.
III. Groups that participate in binding allosteric effectors, i.e., that are part of the allosteric or regulatory site.

The division of the SH groups of enzymes into essential and nonessential is somewhat arbitrary. Firstly, the so-called nonessential groups (whose blocking in experiments *in vitro* does not affect the catalytic activity) may be important for the functioning of the enzyme in the cell, ensuring, for example, the attachment of the enzymes to intracellular organelles and the formation of elaborate multimolecular complexes. Secondly, the inhibition of enzymic activity on chemical modification of so-called essential SH groups may be sometimes due, not to damage of the functioning of the blocked group, but to the influence of the attached inhibitor (its hydrophobic or hydrophilic radical or charged group) on the position and reactivity of neighboring amino acid residues. In this connection the history of the study of the role of the reactive SH groups of hemoglobin is instructive. The study of the influence of p-mercuribenzoate on the properties of hemoglobin led Riggs (1953, 1959; Riggs and Wolbach, 1956) to consider that the SH groups played an important role in the interaction between the heme groups and in the Bohr effect. The subsequent investigations of Taylor et al. (1963, 1966), however, showed that blocking of the same SH groups (which belong to residues 93 of the β-chains) with disulfides had no effect on the interactions between the heme groups; hemoglobin so modified did not differ qualitatively from normal hemoglobin in its kinetic properties, although it was different from hemoglobin treated with p-mercuribenzoate. It thus became clear, from consideration of the results obtained on blocking the SH groups with varied organomercurial, disulfide, and alkylating, reagents, that the SH groups were not directly involved in the heme–heme interactions or in the Bohr effect, and that the changes observed in the properties of the modified hemoglobins were due to the structural characteristics of the molecules of the bound reagent (size, charge, presence of heavy atom, etc.) rather than to the destruction of the SH groups *per se* (Amiconi et al., 1971; Giardina et al., 1971; Simon et al., 1971).

In some cases an inhibitor molecule, when bound to an SH group, may sterically hinder the access of substrate to the active site, and so inhibit activity, even though the SH group itself plays no part in substrate binding or in catalysis. This conclusion follows, for example, from the findings of Kress et al. (1966), who found that the degree of inhibition of muscle ATP-AMP-phosphotransferase (myokinase) by organomercurial compounds increased with enlargement of the reagent molecules; ethylmercury chloride, phenylmercury chloride, and p-acetamidophenylmercury acetate inhibited the enzyme by 22, 55, and 75% respectively.

The significance of the size and charge of the reagent molecule is also clear in the results of Chung et al. (1971) with isocitrate dehydrogenase from *Azotobacter vinelandii.* Blocking of the one most reactive SH group of this enzyme with iodoacetate, p-mercuribenzoate, N-ethylmaleimide, or 5,5'-dithio-bis-(2-nitrobenzoate) leads to an almost complete loss of catalytic activity. Replacement, however, of the nitrothiobenzoate group by cyanide, with formation of a thiocyanoalanyl residue, leads to regeneration of 30–50% of the activity. The thiocyanate group is uncharged and less bulky than the reagents listed above, and apparently allows substrate to bind to the active site of the enzyme. Similarly oxidation of the SH group of aspartate transcarbamylase is accompanied by loss of catalytic activity and lowered affinity for substrate analogs (Benisek, 1971), but its conversion into –S–Me or –SCN is not (Jacobson and Stark, 1973).

The division of the SH groups of the active site of an enzyme (like that of other functional groups) into catalytic and binding groups is also somewhat arbitrary; frequently a group that assists in binding a substrate also fulfills a catalytic function, and vice versa. Analysis of the kinetics of the modified enzyme is the criterion for differentiating these two types of group. If blocking of the group is accompanied by a marked rise in the Michaelis or substrate constant (K_m or K_s) and by little change in the maximal rate of reaction (V) one can postulate that the group blocked takes part only in substrate binding; a significant drop in V or in the rate constants of the intermediate steps of the reaction indicates modification of catalytic function. One reason why this criterion may be unreliable has been explored by Fastrez and Fersht (1973); if correct and incorrect binding of a substrate compete with each other, and if incorrect binding greatly predominates, then K_m is determined by the incorrect binding, and any factor that affects correct binding will appear in V and not in K_m.

The assignment of SH groups to one of the categories indicated above, i.e., establishment of the actual role of a particular essential SH group of an enzyme, is thus not a simple task; a wide range of direct and indirect methods may contribute to its solution.

2. Direct Methods of Identifying SH Groups in the Active Sites of Enzymes

One of the more direct and reliable methods of identifying the SH groups located in the active sites of enzymes is their labeling with substrates, pseudosubstrates, and specific inhibitors. This approach is used when one can expect formation of a covalent bond between the SH group and the substrate. For this it is necessary to find conditions in which the enzyme–substrate complex is stable. The isolation of such a complex and subsequent hydrolysis of the protein permits identification of the amino acid residue that had formed a covalent bond with a substrate molecule or some part thereof. The introduction of radioisotopes or of fluorescent or colored groups into the substrate or pseudosubstrate molecule greatly facilitates the separation and identification of the reacting amino acid residue.

Various means are used to stabilize the intermediate enzyme–substrate complexes. One of these is to treat the enzyme with a mild denaturing agent in the presence of saturating substrate concentrations; such treatment does not in favorable cases destroy the bonds between enzyme and substrate, but prevents the further conversion of the enzyme–substrate complex into enzyme and products. Sometimes the complex may be selectively chemically modified.

An enzyme–substrate intermediate that involved a protein SH group was first isolated in the study of glyceraldehyde 3-phosphate dehydrogenase by Krimsky and Racker (1955). They incubated NAD^+-free dehydrogenase with acetyl phosphate or with 1,3-diphosphoglycerate, after which they precipitated the enzyme with $(NH_4)_2SO_4$ in the cold. They found that the isolated enzyme–substrate complex possessed the properties of a thioester, but they mistakenly supposed that the acyl residue was attached to the SH group of enzyme-bound glutathione. Later Park et al. (1961; Harris et al., 1963) isolated $[^{14}C]$-acetyldehydrogenase after reaction of the enzyme with another pseudosubstrate, p-nitrophenyl-$[^{14}C]$acetate; after denaturation and peptic hydrolysis they isolated from the acetyl-dehydrogenase a peptide fragment that contained a residue of S-acetylcysteine, and thus identified the catalytically active group of glyceraldehyde-3-phosphate dehydrogenase (see Chapter 6, Section 1). The use of pepsin may be noted, since it acts under acidic conditions in which thioester bonds are relatively stable and in which any nearby nucleophilic groups may be protonated until the proteolysis removes them from the vicinity of the labile bond.

Identification of the SH group in the active site of acetoacetyl-CoA thiolase (EC 2.3.1.9) was achieved by incubation of the enzyme with excess [^{14}C]-acetyl-CoA in the cold, and subsequent precipitation of the protein. This proved to contain covalently bound [^{14}C]-acetyl groups, and the following peptide was isolated from its tryptic hydrolysate:

$$\overset{[^{14}\text{C}]\text{Ac}}{\text{Val}-\text{Cys}}-\text{Ala}-\text{Ser}-\text{Gly}-\text{Met}-\text{Lys}$$

This peptide was identical in amino acid composition and sequence with the labeled peptide isolated from the thiolase after blocking its reactive SH groups with iodo[^{14}C]acetate (Gehring and Harris, 1968).

In identifying the SH group in the active site of liver transglutaminase, p-nitrophenyl-[^{14}C]-trimethylacetate was used as a pseudosubstrate. It reacts with the enzyme to form a stable [^{14}C]-trimethylacetyl-enzyme. This derivative was hydrolyzed with proteases, and from the hydrolysate it was possible to isolate a radioactive peptide, which contained the trimethylacetyl group bound by thioester linkage to the SH group of a cysteine residue. Folk and Cole (1966; Folk et al., 1967) showed that the acyl group was attached to the same SH group that reacts with iodoacetamide when this reagent inactivates the enzyme.

The complexes of some transferases with the fragments of the substrate molecules that they transfer can sometimes be isolated by conducting experiments in the absence of acceptors for these fragments. Grazi et al. (1965; Grazi and Rossi, 1968) successfully used this approach to isolate a stable amidine complex of kidney transamidinase (L-arginine: glycine-amidinotransferase, EC 2.1.4.1) and showed that the SH group of the protein participated in formation of this complex. It should be noted that Walker (1957, 1958) had earlier, on the basis of indirect evidence, put forward the idea that the intermediate compound of transamidinase with its substrate was an S-substituted derivative of thioisourea. Grazi et al. isolated the covalent enzyme—amidine complex (compound I in the scheme below) by incubation of the transamidinase with [^{14}C]-arginine (containing the label in the amidine group) in the absence of acceptors of the amidine residue followed by gel filtration; the complex so obtained preserved its ability to donate the labeled amidine group to glycine or ornithine, but at neutral pH it underwent a slow, spontaneous hydrolysis to form urea and free enzyme. A much sturdier complex could be obtained by stabilizing it with

0.1 M HCl. After acid hydrolysis of this complex, 2-aminothiazoline-4-carboxylic acid (II) was isolated and identified, and it contained in position 2 the carbon atom of the amidine group. Oxidation of this compound with bromine gave carbamoylcysteic acid (III), from which cysteic acid, NH_3, and radioactive CO_2 were formed by alkaline hydrolysis. The scheme of identification of the cysteine residue in the active site of the transamidinase is given below:

$$\text{Enzyme}-\text{SH} + [^{14}\text{C}]\text{-Arginine} \longrightarrow \text{Ornithine}$$

$$\downarrow$$

$$\text{Enzyme}-\text{S}-\overset{*}{\text{C}}{\scriptstyle\swarrow}^{\text{NH}}_{\searrow \text{NH}_2} \qquad (I)$$

$$\downarrow \text{Acid hydrolysis}$$

$$\begin{array}{l} \text{H}_2\text{C}-\text{S}\searrow^{*} \\ \;\;|\qquad\quad\;\, \text{C}-\text{NH}_2 \\ \text{HC}-\text{N}\nearrow \\ \;\;| \\ \text{COOH} \end{array} \qquad (II)$$

$$\downarrow \text{Oxidation}$$

$$\underset{(IV)}{\text{NH}_3 + \overset{*}{\text{C}}\text{O}_2 + \begin{array}{l} \text{H}_2\text{C}-\text{SO}_3\text{H} \\ \;| \\ \text{HC}-\text{NH}_2 \\ \;| \\ \text{COOH} \end{array}} \xleftarrow[\text{hydrolysis}]{\text{Alkaline}} \begin{array}{l} \text{H}_2\text{C}-\text{SO}_3\text{H} \\ \;| \\ \text{HC}-\text{NH}-\overset{*}{\text{C}}\text{O}-\text{NH}_2 \\ \;| \\ \text{COOH} \end{array} \qquad (III)$$

The isolation of enzyme–substrate (or pseudosubstrate) compounds permits identification of an essential SH group of the active site only when this group forms a sufficiently stable covalent bond with the substrate, and this is by no means always the case.

In the last few years a new, more general, and most promising approach has been developed for identifying functional groups in the active sites of enzymes. This is the use of irreversible substrate-like inhibitors. These are reagents that are structurally similar to the substrate, and consequently have selective affinity for the substrate-binding site of the enzyme, and also possess additional reactive groupings, that can block (usually alkylate or acylate) functional groups of the enzyme that are situated in the active site of the enzyme or close to it. The first step in the action of such reagents is binding to the active site of the enzyme. The modification reactions that proceed after such binding therefore possess an essentially intramolecular character and may be rapid. The method has received the name of "affinity labeling" (see the reviews by Singer, 1967; Baker, 1967; Torchinskii, 1968). Examples of inhibitors of this type are the 6-chloro derivative of inosine-5-phosphate, which reacts, apparently, with an SH group of the active site of inosinephosphate dehydrogenase (Brox and Hampton, 1968), and an analog of ATP, 6-mercapto-9-β-D-ribofuranosylpurine-5'-triphosphate, which suppresses the ATPase activity of myosin, forming a disulfide with a cysteine residue in the active site (Murphy and Morales, 1970). Pinkus and Meister (1972) have used a chloroketone analog of glutamine, L-2-amino-4-oxo-5-chloro[5-^{14}C]pentanoic acid, for identifying a nucleophilic group at the active site of carbamoyl phosphate synthetase from *Escherichia coli*; they found that this analog occupies the site of the glutamine that serves as donor of its amide nitrogen in the reaction catalyzed by the enzyme, and that it forms a covalent bond with an SH group.

It is of interest that some antibiotics have the properties of irreversible substrate-like inhibitors: cycloserine – with respect to alanine aminotransferase; azaserine – with respect to the enzymes that catalyze transfer of the amide group of glutamine in the processes of biosynthesis of nucleotides, aminosugars, etc. (p. 162). The formulas of azaserine and the glutamine it imitates are here compared:

$$\underset{\text{Azaserine}}{N_2{=}CH{-}COOCH_2{-}\underset{\substack{|\\ NH_2}}{CH}{-}COOH} \qquad \underset{\text{Glutamine}}{NH_2{-}OC{-}CH_2{-}CH_2{-}\underset{\substack{|\\ NH_2}}{CH}COOH}$$

French et al. (1963) found that azaserine (O-diazoacetylserine), which is a structural analog of glutamine, binds at the substrate site of 5'-phosphoribosyl-formylglycineamide: L-glutamine amido-ligase (EC 6.3.5.3) and alkylates, with release of N_2, the SH group of a cysteine residue which is located in the active site of the enzyme. After hydrolysis of enzyme previously inhibited with [^{14}C]-azaserine, a peptide fragment was isolated which contained a covalently bound residue of azaserine:

$$-\text{Ala}-\text{Leu}-\text{Gly}-\text{Val}-\underset{\substack{|\\ CH_2-CO-O-CH_2-\underset{\substack{|\\ NH_2}}{CH}-COOH}}{\text{Cys}}-$$

Peptides containing reactive SH groups have been isolated from the active sites of various thiol enzymes (Tables 12 and 13). The bifunctional reagent 1,3-dibromo[^{14}C]acetone has been successfully

TABLE 12. **Sequences Adjacent to the Cysteine Residue in the Active Sites of Some Dehydrogenases**

Enzyme	Source	Amino acid sequence	References
Alcohol dehydrogenase	Yeast	−Lys−Tyr−Ser−Gly−Val−Cys−His−Thr−Asp−Leu−His−Ala−	Harris, 1964
	Horse liver	−Val−Ala−Thr−Gly−Ile−Cys−Arg−Ser−Asp−Asp−His−Val−	Harris, 1964; Li and Vallee, 1964
Lactate dehydrogenase	Frog muscle	−Val−Ile−Ser−Gly(Gly, Cys)Asn−Leu−Asn(Ala, Thr)Arg−	Fondy and Everse, 1964
	Chicken muscle	−Val−Ile−Ser−Gly−Gly−Cys−Asn−Leu−Asp−Thr−Ala−Arg−	Fondy et al., 1965
	Ox heart	−Val−Ile−Asn(Asp, Gly_3, Ser_2, Cys)Asn−Leu−Ala−Arg−	Gold and Segal, 1965
	Pig heart	−Val−Ile−Gly−Ser−Gly−Cys−Asn−Leu−Asp−Ser−Ala−Arg−	Holbrook et al., 1967
Glyceraldehyde-3-phosphate dehydrogenase	Yeast, and muscle from rabbit, pig, monkey, badger, and lobster	−Val−Ser−Asn−Ala−Ser−Cys−Thr−Thr−Asn−Cys−Leu−Ala−Pro−	Harris et al., 1963; Allison and Harris, 1965; Harris and Perham, 1968; Perham, 1969; Davidson et al., 1967
α-Glycerophosphate dehydrogenase	Rabbit muscle	−Phe−Ala−Val(Val_{0-1})−Asp−Thr−Cys−Ser−Gly(Glu, Gly)Ile−	Sajgo and Telegdi, 1968

The reactive cysteine residue is underlined.

TABLE 13. Amino Acid Sequences Close to the Cysteine Residue in the Active Sites of Plant Proteases

Enzyme	Sequence	Reference
Papain	−Pro−Val−Lys−Asn−Gln−Gly−Ser−Cys−Gly−Ser−Cys−Trp−	Light et al., 1964; Husain and Lowe, 1965, 1968b
Ficin	−Pro−Ile−Arg−Gln−Gln−Gly−Gln−Cys−Gly−Ser−Cys−Trp−	Wong and Liener, 1964; Husain and Lowe, 1970b
Bromelain	−Asn−Gln−Asp−Pro−Cys−Gly−Ala−Cys−Trp−	Husain and Lowe, 1970b

The catalytically active cysteine residue is underlined.

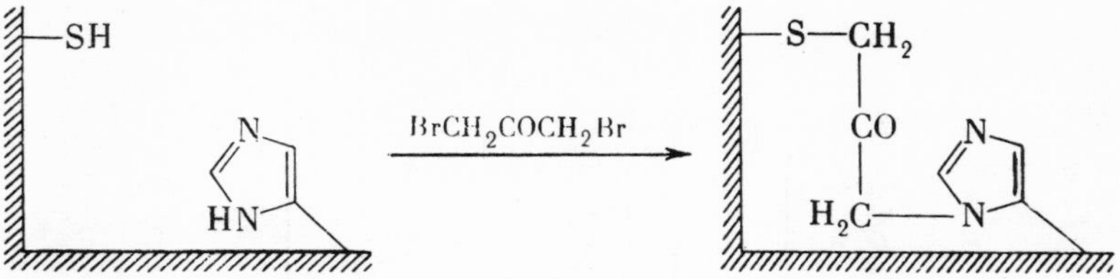
SH
N
HN
BrCH₂COCH₂Br
S—CH₂
CO
N
H₂C—N

Fig. 18. The interaction of 1,3-dibromoacetone with SH and imidazole groups in the active site of papain.

used in studying the active site of papain and other plant proteases; it alkylates the SH group and the imidazole group of a histidine residue that is nearby in the tertiary structure (Fig. 18) (Husain and Lowe, 1968a,b, 1970b).

In some two-component enzymes the functional groups of the active site are already "labeled" by nature itself by covalently bound prosthetic groups. Such a situation exists with cytochrome c. After its enzymic hydrolysis a hemopeptide (i.e., a fragment of the polypeptide chain to which heme is attached) was isolated, and it contained two cysteine residues, linked by thioether bonds to the vinyl groups of heme (Tuppy and Paleus, 1955; Tuppy, 1959):

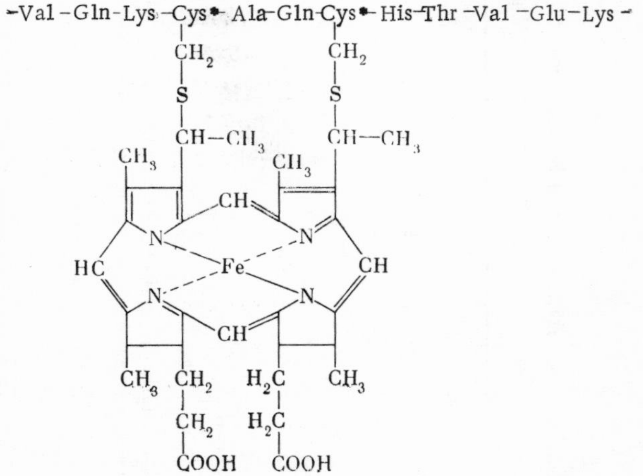
-Val-Gln-Lys-Cys*-Ala-Gln-Cys*-His-Thr-Val-Glu-Lys-
CH₂
S
CH—CH₃
CH₃
CH
N
Fe
HC
COOH

The symbol Cys represents only the backbone of Cys, as the modified side chain is shown in the formula.

Similarly it is by means of a thioether bond to cysteine that FAD is bound to the apoenzyme in liver monoamine oxidase, and possibly in some other flavoproteins. This was shown by Walker et al. (1971) who isolated the following flavin peptide from a digest of the enzyme (R represents the rest of the FAD molecule):

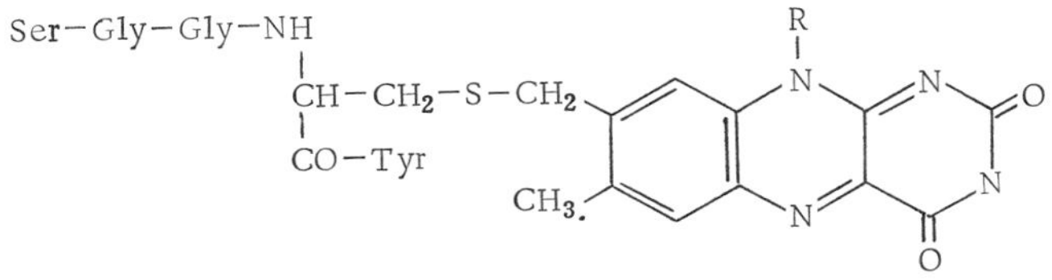

Ser–Gly–Gly–NH
|
CH–CH_2–S–CH_2 R N N O
|
CO–Tyr CH_3 N N O

The X-ray analysis of the structure of protein crystals with isomorphous replacements by atoms of heavy elements may be counted among the direct methods for identifying the functional groups that are located at the active site. Each year this method gains importance in studying enzymes and their active sites. Several thiol enzymes are now being studied in this way: glyceraldehyde-3-phosphate dehydrogenase, alcohol dehydrogenase, lactate dehydrogenase, and papain. It should be noted, however, that in spite of the successes in the development of direct methods of identifying functional groups, for the moment the less reliable indirect methods retain their value (and will retain it in the future), since they can give valuable preliminary information about the essential groups of enzymes and about the reactions in which they take part.

3. Indirect Methods of Identifying SH Groups in the Active Sites of Enzymes

Among indirect methods of identifying SH groups are the following:

a. Kinetic analysis of the dependence of enzymic activity on the pH of the medium, particularly the separate determination of the dependence on pH of the Michaelis constant (K_m) or substrate constant (K_s) and of the maximal velocity (V); the curves expressing this dependence often reflect the ionization of a catalytically active group (or groups) and allow determination of the pK of this group. Identification of SH groups from pK values, however, is seriously hindered by the fact that the pK of the SH groups in proteins varies over fairly wide limits (usually from 7.5 to 10) and partly overlaps with the pK values of other functional groups (imidazole, phenolic, and amino groups).

b. Study of the effect on enzymic activity of reagents that selectively block SH groups. Mercaptide-forming, alkylating, and disulfide reagents find widespread application for this purpose (see Chapter 1, Section 12). The advantages and disadvantages of various reagents for SH groups are discussed in Chapter 1. Here we will only note that it is useful to determine not merely the degree of inactivation of the enzyme under the influence of a particular thiol reagent, but also the number of SH groups modified. When using alkylating reagents it is essential to keep in mind the possibility of modification of other functional groups, which can be detected by amino acid analysis.

An advantage of indirect methods is that they can be applied to relatively small quantities of enzyme. Such methods, however, are not always sufficiently specific, and, what is particularly important, they cannot differentiate between SH groups that are located in the active site of the enzyme and those that take part solely in maintaining the catalytically active three-dimensional structure of the protein macromolecule. Experiments on the "protection" of SH groups by substrates, coenzymes, and their analogs can contribute toward the solution of this problem.

In analyzing the protective action of substrates and coenzymes the following circumstances should be taken into account. Substrates and coenzymes can prevent loss of enzymic activity not only by direct protection (screening) of functional groups, but also as a consequence of their general stabilization of the structure of the enzyme. Since the functional groups that are part of the active site often belong to different polypeptide chains (or different parts of one chain), it is not difficult to see how binding of substrate or coenzyme to these groups can form supplementary bonds between the chains and thus prevent secondary conformational changes that follow modification of amino acid residues. Hence to establish that a substrate or coenzyme protects essential SH groups it is not enough to carry out measurements of enzymic activity alone; a direct measurement of the number of these groups before and after action of the specific reagent is needed. Even a drop in the number of SH groups that can be titrated in the presence of substrate or coenzyme is not unambiguous evidence that these groups are actually in the active site, because formation of enzyme–substrate or apoenzyme–coenzyme complexes can be accompanied by conformational changes in the protein molecule, which can lead to a change in the environment, and hence in the reactivity, of functional groups that are outside the active site. Cases have been described in the literature in which the reactivity of SH groups in the presence of substrates is not lowered, but raised (Yankeelov and Koshland, 1965; Birchmeier and Christen, 1971).

Comparative studies of enzymes derived from different species can help in substantiating conclusions about the role of SH groups, since any functionally important residue should be conserved and have similar properties in the enzymes from all species. The enzymes compared must, of course, be performing an identical function and should have arisen by divergent evolution, i.e., they should have homologous amino acid sequences. The work of Anderson (1972) provides an example. A sulfhydryl group in rabbit muscle aldolase is alkylated by iodoacetamide with loss of enzymic activity. The presence of the substrate, fructose-1,6-diphosphate, protects this group from alkylation and the enzyme from inactivation. Nevertheless the enzyme from sturgeon muscle has no cysteine residue homologous with the substrate-protected residue of the rabbit muscle enzyme. The conclusion must be that the loss of activity is caused by the alkyl group, and that the sulfhydryl group can have no function that is essential for catalysis.

Salter et al. (1973) have confirmed the presence of an SH group in the region of the active site of rabbit muscle aldolase. They found that the substrate analog chloroacetol phosphate, $ClCH_2-CO-CH_2-O-PO_3H_2$, occupies the binding site for dihydroxyacetone phosphate and selectively alkylates an SH group that is situated nearby.

These considerations mean that finding that substrates or coenzymes protect some functional groups is only a preliminary indication that these groups are at or near to the active site. A supplementary, but still indirect, criterion for establishing the role of an SH group in an enzyme is the extent of loss of enzymic activity when it is blocked. The modification of a group that has a key catalytic function is accompanied, as a rule, by an almost complete disappearance of catalytic activity toward all substrates of the enzyme. Modification of a group that is part of the contact site, or that participates in maintaining the native structure of the enzyme, often leads to only partial loss of enzymic activity, and this loss may be expressed to different degrees with different substrates (Weiner et al., 1966).

In cases where a substrate or coenzyme screens functional groups, but without forming a covalent bond with them, labeling of these groups may sometimes be achieved as follows. The enzyme is first treated, in the presence of the substrate, with a nonradioactive reagent, which blocks all the functional groups of a given type, with the exception of those that are protected by substrate and are consequently presumed to be located in the active site. The excess reagent and the substrate are then removed and the enzyme is again treated with the reagent, but this time with labeled reagent. After hydrolysis of the protein the labeled modified amino acid residues are identified, i.e., the residues protected by substrate (the method of differential labeling). It is of course important to remember that since substrate molecules usually dissociate freely from enzymes, complete protection of screened functional groups is unlikely to be achieved. As the enzyme is saturated with substrate so the fraction of the time for which these groups are accessible to the modifying reagent is diminished, and this can be enough to achieve the selective effect.

We have described above the difficulties that arise in interpreting the data obtainable by indirect methods. One must also recognize that these methods have yielded extensive and valuable information about a large number of thiol enzymes. It was by these methods that the presence in the active site of a reactive SH group first indicated for glyceraldehyde-3-phosphate dehydrogenase, transamidinase, papain, ficin, and some other enzymes; in many cases these findings have since been confirmed by direct methods.

Chapter 6

The Role of SH Groups in Enzymes

1. The Role of SH Groups in Catalysis

a. SH Groups of Proteins as Nucleophilic Catalysts

Only a few enzymes are known in which the direct participation of SH groups in catalysis has already been unambiguously demonstrated. Glyceraldehyde-3-phosphate dehydrogenase (Gra-3-P-dehydrogenase, EC 1.2.1.12) is one of the most studied of these. It is one of the enzymes of glycolysis and plays an important part in the carbohydrate metabolism of most living organisms. In the presence of NAD^+ and phosphate it catalyzes the reversible oxidative phosphorylation of glyceraldehyde-3-phosphate to form 1,3-diphosphoglyceric acid.

Gra-3-P dehydrogenase was the first enzyme for which direct evidence was obtained for an intermediate covalent compound of thioester type between the protein and the substrate. Later the formation of acyl enzyme has been demonstrated in studies of the mechanisms of a number of proteases and esterases (trypsin, chymotrypsin, cholinesterase, etc.) where a serine residue acts as acyl acceptor from the substrate. In Gra-3-P dehydrogenase a cysteine residue fulfills this role. Before presenting data concerning the function of SH groups in this enzyme, it is necessary to consider its mechanism of action.

Warburg and Christian (1939), who had first isolated Gra-3-P dehydrogenase from yeast, proposed that in the oxidation of glyceraldehyde-3-phosphate there is first a nonenzymic addition of phosphate to the aldehyde group, which is then oxidized by NAD^+ with formation of 1,3-diphosphoglyceric acid:

$$\begin{array}{l} CH_2OPO_3H_2 \\ | \\ CHOH \\ | \\ C{-}H \\ \| \\ O \end{array} + H_3PO_4 \rightleftarrows \begin{array}{l} CH_2OPO_3H_2 \\ | \\ CHOH \\ | \\ CHOH \\ | \\ OPO_3H_2 \end{array} + NAD^+ \xrightleftharpoons{\text{Enzyme}} \begin{array}{l} CH_2OPO_3H_2 \\ | \\ CHOH \\ | \\ C{=}O \\ | \\ OPO_3H_2 \end{array} + NADH + H^+$$

This scheme was subjected to experimental testing in the fifties, and, in spite of the efforts of Warburg to defend it, was rejected by most authors. The basic argument advanced by Warburg to substantiate the scheme was that the addition of phosphate raised the initial rate of enzymic oxidation of the poor substrate glyceraldehyde (Warburg et al., 1954, 1957). But other authors who carried the reaction out at higher concentrations of NAD^+ and enzyme found no difference in the initial rate of this reaction in the presence and absence of phosphate (Velick and Hayes, 1953; Koeppe et al., 1956). The interpretation of experiments with phosphate is complicated by the fact that it apparently affects the binding of NAD^+ and the conformation of the protein.

The experiments of Velick and Hayes (1953) and also of Segal and Boyer (1953) provided a convincing refutation of Warburg's scheme. These authors, using "substrate quantities" of the enzyme, were able to observe separately two stages of the reaction; in the first stage, which they carried out in the absence of phosphate, at most two to three moles of NADH were formed per mole of enzyme; the reaction stopped at this and restarted only on addition of phosphate. It followed from these experiments that the oxidation of Gra-3-P and the reduction of NAD^+ occur in the first stage of the enzymic reaction and that the phosphorylation occurs in the second. Segal and Boyer (1953; Boyer and Segal, 1954) postulated the following scheme for the action of Gra-3-P dehydrogenase, which is now accepted (with slight modifications) by most authors:

$$R{-}\overset{\overset{\displaystyle O}{\|}}{C}{-}H + HS{-}\text{Protein}{-}NAD^{+} \rightleftarrows R{-}\underset{\underset{\displaystyle H}{|}}{\overset{\overset{\displaystyle OH}{|}}{C}}{-}S{-}\text{Protein}{-}NAD^{+} \quad (1)$$

$$R{-}\underset{\underset{\displaystyle H}{|}}{\overset{\overset{\displaystyle OH}{|}}{C}}{-}S{-}\text{Protein}{-}NAD^{+} \rightleftarrows R{-}\overset{\overset{\displaystyle O}{\|}}{C}{\sim}S{-}\text{Protein}{-}NADH + H^{+} \quad (2)$$

$$R{-}\overset{\overset{\displaystyle O}{\|}}{C}{\sim}S{-}\text{Protein}{-}NADH + NAD^{+} \rightleftarrows NADH +$$

$$+ R{-}\overset{\overset{\displaystyle O}{\|}}{C}{\sim}S{-}\text{Protein}{-}\underline{NAD^{+}} \quad \text{(see p. 191)} \quad (3)$$

$$R{-}\overset{\overset{\displaystyle O}{\|}}{C}{\sim}S{-}\text{Protein}{-}NAD^{+} + HOPO_3^{2-} \rightleftarrows R{-}\overset{\overset{\displaystyle O}{\|}}{C}{-}O{\sim}PO_3^{2-} + HS{-}\text{Protein}{-}NAD^{+} \quad (4)$$

According to this scheme Gra-3-P dehydrogenase with the participation of its bound NAD^+ forms in the first stage [reactions (1) and (2)] an intermediate compound with the acyl residue of the substrate (a phosphoglyceryl enzyme). This compound, after oxidation of its bound NADH, then undergoes phosphorolysis to form 1,3-diphosphoglyceric acid and to regenerate the original form of the enzyme [reactions (3) and (4)]. It should be noted that the immediate hydrogen donor in the reaction catalyzed by Gra-3-P dehydrogenase is not the aldehyde itself or its hydrate, but a hemimercaptal (thiohemiacetal), which is oxidized to a thioester (see also Trentham, 1971).

Convincing evidence in favor of the intermediate formation of an acyl enzyme was obtained by Krimsky and Racker (1955) who isolated the acyl enzyme, after they had worked out conditions for stabilizing it, after reaction of Gra-3-P dehydrogenase with 1,3-diphosphoglycerate or with acetyl phosphate. They showed that the isolated acyl enzyme (like thioesters) could react with hydroxylamine, forming a hydroxamic acid, and was also rapidly reduced by NADH with aldehyde formation.

Even in the thirties it was known that blocking the SH groups of Gra-3-P dehydrogenase with various thiol reagents destroyed its enzymic activity (Green et al., 1937; Rapkine, 1938; Cori et al., 1948; Barron and Dickman, 1949). The discovery of the participation of thiol compounds, particularly coenzyme A, in acyl group transfer (Lynen and Reichert, 1951) led to the thought that the function of the SH groups in the dehydrogenase could be the formation of a thioester bond with the acyl group of the substrate molecule. This hypothesis was developed and given a broad base in the work of Holzer (Holzer and Holzer, 1952), Boyer (Segal and Boyer, 1953; Boyer and Segal, 1954; Koeppe et al., 1956), and Racker and Krimsky (1952; 1958; Krimsky and Racker, 1952, 1954; Racker, 1954, 1955). Holzer, working in Lynen's laboratory in 1952, and Segal and Boyer somewhat later, postulated the participation of the SH groups of Gra-3-P dehydrogenase in forming the acyl enzyme compounds on the basis of experiments on the protection of these groups by glyceraldehyde-3-phosphate.

Segal and Boyer (1953) found that two of the SH groups of the dehydrogenase could be protected from alkylation with iodoacetate by the substrate glyceraldehyde-3-phosphate. The protection of these two groups ensured almost complete retention of enzymic activity. Segal and Boyer also supported the hypothesis of S-acyl enzyme formation on the basis of the known ability of aldehydes to react readily with various thiols to form hemimercaptals (Schubert, 1936a). Krimsky and Racker (1954) showed that acetyl phosphate protects Gra-3-P dehydrogenase from inactivation by N-ethylmaleimide. Koeppe et al. (1956) presented convincing support for the role of SH groups as acceptors of acyl groups when they showed that the formation of the acyl enzyme by reaction of the dehydrogenase with acetyl phosphate was accompanied by a fall in the number of free SH groups that reacted with p-mercuribenzoate.

Krimsky and Racker (1954; Racker, 1954, 1955) postulated that the SH group of Gra-3-P dehydrogenase that forms a thioester belongs to enzyme-bound glutathione, which they had observed in highly purified preparations of the dehydrogenase. Koeppe et al. (1956) showed, however, that the glutathione content of the enzyme was insignificant (0.12-0.23 mole per mole of enzyme) and that it was apparently combined with the protein as a mixed disulfide. Experiments with p-nitrophenylacetate helped in clarifying the nature of the catalytically active SH group of the dehydrogenase.

Park et al. (1961; Olson and Park, 1964) found that Gra-3-P-dehydrogenase is five times better than chymotrypsin and 700 times better than cysteine or glutathione in catalyzing the hydrolysis of p-nitrophenylacetate. This hydrolysis appears to take place at the active site of the enzyme, since it is inhibited in the presence of glyceraldehyde 3-phosphate, acetyl phosphate, NAD^+, and thiol reagents.* The hydrolysis occurs in two steps, of which the second is rate-determining:

$$CH_3\overset{O}{\overset{\|}{C}}-O-C_6H_4-NO_2 + HS-E \rightleftarrows CH_3\overset{O}{\overset{\|}{C}}\sim SE + NO_2-C_6H_4-OH$$

$$CH_3\overset{O}{\overset{\|}{C}}\sim SE + H_2O \rightleftarrows E-SH + CH_3\overset{O}{\overset{\|}{C}}-OH$$

After reaction with p-nitrophenyl[^{14}C]acetate the [^{14}C]-acetyl enzyme was isolated, and it contained 3-4 acetyl groups per molecule of protein. The [^{14}C]-acetyl-dehydrogenase was subjected to hydrolysis with pepsin and oxidation with performic acid. From the products Harris et al. (1963; Perham and Harris, 1963) isolated the following peptide fragment, which contained a residue of S-[^{14}C]-acetylcysteine:

```
                                  Ac─┐                     O3H
                                     │                      │
Lys–Ile–Val–Ser–Asn–Ala–Ser–Cys–(Thr,Thr,Asn)–Cys
```

Since the second cysteine residue in this sequence becomes unmasked when the protein is denatured it must be converted by oxidation or alkylation into a derivative that is stable to atmospheric oxygen to allow isolation of the peptide. The structure of this fragment agreed well with the results of analysis of alkylated peptides that had been isolated from Gra-3-P dehydrogenase after blocking its reactive SH groups with iodo[^{14}C]acetate or colored

*Olson and Park (1964) suggested that the high rate of esterolysis of p-nitrophenylacetate in the active site of Gra-3-P dehydrogenase is explained by the cooperative action of the SH group and an imidazole group, which firstly activates the SH group by hydrogen bonding with it and secondly catalyzes hydrolysis of the thioester. The findings of Behme and Cordes (1967), however, are against this, since the acylation of the enzyme depends on a group of pK 8.1, and the deacylation on a group of pK 8.7. A pK value of 8.1 is characteristic of a free SH group not involved in hydrogen bonding, and a value of 8.7 is not characteristic of an imidazole. As discussed, hydrogen bonding between SH and imidazole groups should lead to a rise in the pK of the SH group and a fall in the pK of the imidazole.

derivatives of N-ethylmaleimide (Harris et al., 1963; Perham and Harris, 1963; Segal and Gold, 1963; Gold and Segal, 1964). Davidson et al. (1967) and Harris and Perham (1968) established the complete primary structure of Gra-3-P dehydrogenases from lobster and pig muscle; they found that both enzymes consist of four identical polypeptide chains (of molecular weight 36,000), each of which contains four (in the case of pig) or five (in the case of lobster) residues of cysteine. Only one of these residues (Cys-149) reacts with iodoacetate or p-nitrophenylacetate in the native enzyme (see also Wassarman and Major, 1969). The remaining cysteine residues, including the nearby Cys-153, are relatively inert and are evidently in a masked state; possibly they participate in maintaining the native conformation of the enzyme (see Chapter 6, Section 3).

There has been discussion of the possible cooperative participation of an SH group and an imidazole group in the catalytic action of Gra-3-P dehydrogenase. On the basis of much indirect evidence Olson and Park (1964; see also Colowick et al., 1966) proposed that the stage of deacylation of the S-acyl enzyme was catalyzed by an imidazole group located close to the SH group in the macromolecular structure of the protein. The scheme they proposed was

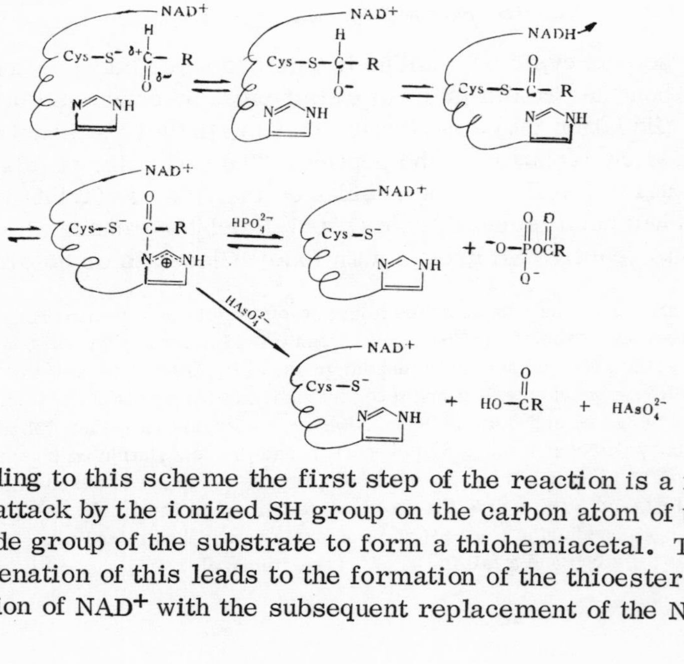

According to this scheme the first step of the reaction is a nucleophilic attack by the ionized SH group on the carbon atom of the aldehyde group of the substrate to form a thiohemiacetal. The dehydrogenation of this leads to the formation of the thioester and the reduction of NAD^+ with the subsequent replacement of the NADH formed by the more tightly bound NAD^+. At this stage imidazole enters the reaction by catalyzing transfer of the acyl group to various acceptors, such as phosphate and arsenate, with the intermediate formation of an acylimidazole. Reverse of this stage explains the transacylase activity of the dehydrogenase.

In support of the scheme of Olson and Park it was found that the photooxidation of a single residue of histidine (per subunit) selectively diminishes the rate of deacylation of the S-acetyldehydrogenase by 50–60%, but has no effect on the rate and degree of acetylation of the SH group by p-nitrophenylacetate (Bond et al., 1970; Francis et al., 1973; see also Friedrich et al., 1964). It is possible, however, that the inhibition of deacylation is a result of a limited conformational change produced by the destruction of the histidine residue. Behme and Cordes (1967) found that the deacylation step depends on a group of pK 8.7, which is not characteristic of imidazole. Thus if an imidazole group is involved in the reaction, it must be one with somewhat unusual properties. Behme and Cordes, and also Lindquist and Cordes (1968) advanced the alternative hypothesis that the deacylation of the S-acetyldehydrogenase proceeds with general base catalysis. The question of the mechanism of deacylation still remains open. According to x-ray data (at 3 Å resolution), His-176 in the lobster enzyme is positioned so that it could participate in the enzymic reaction as an acid–base catalyst (Buehner et al., 1973).

The successful investigation of the role of SH groups in Gra-3-P dehydrogenase has stimulated study of the role of these groups in a number of other aldehyde dehydrogenases (Rothschild and Barron, 1954; Albers and Koval, 1961). Jacoby (1958; Nirenberg and Jacoby, 1960) proposed that the aldehyde dehydrogenases of yeast, liver, and *Pseudomonas fluorescens* contain in their active sites two SH groups that are close together in the tertiary structure of the proteins. This proposal was based on the following findings: 1) the aldehyde dehydrogenases are inactivated by low concentrations of arsenite, which has, as is known, a particularly high affinity for dithiols; 2) dithiols, for example 2,3-dimercaptopropanol, unlike monothiols, readily reactivate enzyme poisoned with arsenite; 3) there are competitive relations between substrates and arsenite.

The interaction between SH groups of enzymes and carbonyl groups of substrates was postulated also in the study of glutamic semialdehyde reductase (Smith and Greenberg, 1957), yeast pyruvate decarboxylase (Stoppani et al., 1953; Schellenberger, 1967),

3-α-hydroxysteroid dehydrogenase (Tomkins, 1956), ribulose-1,5-diphosphate carboxylase (Racker and Krimsky, 1958; Rabin and Trown, 1964), β-hydroxy-β-methylgutaryl-coenzyme A reductase (Kirtley and Rudney, 1967), and acetoacetyl-CoA thiolase (Gehring et al., 1968). This interaction leads, apparently, as in the case of Gra-3-P dehydrogenase, to the formation of enzyme–substrate compounds of the hemimercaptal or thioester type. Boyko and Fraser (1964) postulated formation of a thioester in the course of action of glycyl-tRNA synthetase, but this suggestion has not been convincingly established.

Lipmann et al. (1970; Gevers et al., 1969) obtained evidence that thioester intermediates were involved in the synthesis of peptide antibiotics by *Bacillus brevis*. It appears that gramicidins and tyrocidins are built up on the SH groups of cysteine residues of the enzyme system, one group for each activated amino acid. A pantetheine residue apparently transfers the growing chain from one (SH) site to the amino group of the residue on the next site (see Laland and Zimmer, 1973, for review, and p. 184).

Protein sulfhydryl groups can act as reactive acceptors in enzymic transfer reactions not only of acyl but also of amidine, phosphate, and other residues. Chapter 5, Section 2 gives the scheme of action of transamidinase, whose SH group takes part in the transfer of the amidine residue from arginine onto ornithine or glycine. French et al. (1963) found that the active site of bacterial amidoligase (EC 6.3.5.3), which forms formylglycineamidine-ribonucleotide from glutamine and formylglycineamide-ribonucleotide, contains an SH group, which apparently acts as a nucleophilic group in displacing the amide nitrogen from the carbonyl group of glutamine in the course of the catalyzed reaction. An SH group in the active center of carbamoyl phosphate synthetase of *Escherichia coli* plays a similar role (Pinkus and Meister, 1972); possibly such participation of an SH group is a general feature of glutamine amidotransferases (reviewed by Akhtar and Wilton, 1972, and Buchanan, 1973). They include CTP-synthetase, one of the most studied, and also the plasma enzyme responsible for the cross-linking reaction between glutamine and lysine residues of fibrin in which ammonia is evolved and a blood-clot stabilized (see Holbrook et al., 1973).

A special group of thiol enzymes is formed by the plant proteases: papain, chymopapain, ficin, bromelain, and asclepain.*

*Pinguinain apparently also belongs to this group. It was recently purified from the tropical plant *Bromelia pinguin L.*; in several of its properties it differs from bromelain (Toro-Goyco et al., 1968).

Papain (EC 3.4.4.10) is the most studied representative of this group. Even in the first papers devoted to the study of this enzyme it was noticed that papain was activated by H_2S and by cyanide, and that it was inhibited by various thiol reagents (iodine, hydrogen peroxide, quinone, iodoacetate, Cu_2O, and the organic mercurials C_6H_5HgCl and C_6H_5HgOH) (Bersin and Logemann, 1933; Bersin, 1933, 1935; Hellerman and Perkins, 1934). Smith and colleagues (Kimmel and Smith, 1954; Stockell and Smith, 1957; Smith and Parker, 1958; Smith and Kimmel, 1960; Smith et al., 1962; Light et al., 1964; Chaiken and Smith, 1969; Mitchel et al., 1970) have studied the chemical properties of papain and its mechanism of action over the course of a number of years. Drenth et al. (1968, 1970) have published findings obtained by x-ray structural analysis of papain crystals at a resolution of 2.8 Å. As a result of these investigations the primary and macromolecular structure of papain (molecular weight 23,000) has been elucidated and the catalytically active cysteine residue identified. The papain molecule contains three disulfide bridges and one SH group, which is partly masked in the native enzyme and is observed after this is treated with thiols, borohydride, or cyanide. The activity of papain preparations before activation is directly proportional to its content of free SH groups. Klein and Kirsch (1969a,b) showed that the inactive form of papain is a mixed disulfide formed between the SH group of the enzyme and cysteine. Reduction or nucleophilic scission of the disulfide bond liberates the SH group of papain, together with a stoichiometric quantity of cysteine, or of cysteine whose sulfur atom is covalently bound to the attacking nucleophilic reagent. Activation of papain with $K^{14}CN$ goes according to the following scheme and leads to the formation of $[^{14}C]$-2-iminothiazolidine-4-carboxylic acid.

$$\text{Papain}-S-S-CH_2-CH(\overset{+}{N}H_3)(COO^-) \xrightarrow{\;^{*}CN^-\;}$$

$$\text{Papain}-SH \; + \; S(-\overset{*}{C}\equiv N)-CH_2-CH(\overset{+}{N}H_3)COO^- \longrightarrow \underset{\text{(ring: }S-\overset{*}{C}(H_2N)\overset{+}{=}NH-CH(COO^-)-CH_2-S)}{}$$

Brocklehurst and Kierstan (1973) have expressed the opinion that the mixed disulfide of papain and cysteine is formed in the course of the purification of the enzyme by the method of Kimmel and Smith (1954) and that it is not an inactive precursor of the enzyme *in vivo* There is a natural propapain, which like the activated enzyme has one SH and three S—S groups, but differs in which of the seven half-cystine residues is present as the free SH group (see Chapter 7, Section 1).

Papain is inhibited by some carbonyl reagents, e.g., by phenylhydrazine; the mechanism of this inactivation has long remained mysterious. The most satisfactory explanation is that of Klein and Kirsch (1969b) who suggest that inactivation by phenylhydrazine is due to the ability of this substance to oxidize the SH group of papain. If this reaction is carried out in the presence of cyanide, then oxidative coupling of cyanide to the SH group occurs, so that the cysteine residue in the active site is transformed into a residue of β-thiocyanoalanine (which then apparently forms an adduct with excess phenylhydrazine):

$$RSH \xrightarrow[C_6H_5-NH-NH_2 \text{ or } H_2O_2]{CN^-} R-S-CN \xrightarrow{C_6H_5-NH-NH_2} R-S-C\begin{matrix} \diagup NH_2 \\ \diagdown\!\!\diagdown N-NH-C_6H_5 \end{matrix}$$

Allison and Swain (1973) believe that aerobic oxidation of phenylhydrazine produces H_2O_2 and phenyldiimide. The peroxide oxidizes the SH group to —SOH, which may react with cyanide to form —SCN, whereas the di-imide can add to a thiol to form —S—NH—NH—Ph.

As early as 1937, Weiss (1937) put forward the hypothesis that papain and other thiol enzymes catalyze hydrolytic and synthetic reactions by forming intermediate acylthioenzyme complexes. Investigations of the kinetics of hydrolysis by papain of various low-molecular substrates (esters and amides of N-acylated amino acids) confirmed the hypothesis that an acyl enzyme was formed in the course of the hydrolysis (Smith et al., 1955a; Whitaker and Bender, 1965; Lake and Lowe, 1966; Kirsch and Katchalski, 1965; Kirsch and Igelström, 1966). This hypothesis is consistent with the observation that papain catalyses ^{18}O exchange between the carboxyl group of N-acylated amino acids and water (Grisaro and Sharon, 1964). Like trypsin and chymotrypsin, papain catalyses the transfer of the acyl residude of the substrate to various acceptors in addition to water, including hydroxylamine, aniline, phenylhydrazine, peptides, and amino acids (Fruton, 1950; Durell and Fruton, 1954; Kimmel and Smith, 1957). The action of papain can therefore be presented in the following scheme:

$$\underset{\substack{| \\ SH}}{E} + R-CO-X \rightleftharpoons \underset{\substack{| \\ SH \\ \text{Enzyme-substrate} \\ \text{complex}}}{E} \cdot R-CO-X \rightarrow \underset{\substack{\text{Acyl enzyme} \\ +XH}}{E-S-CO-R} \rightarrow \underset{\substack{| \\ SH}}{E} + R-CO_2H$$

The formation of an intermediate thioester between the SH group of papain and the acyl residue of a pseudosubstrate (methylthionohippurate or trans-cinnamoylimidazole) was confirmed spectrophotometrically (Lowe and Williams, 1964, 1965; Bender and Brubacher, 1964; Brubacher and Bender, 1966). A number of authors have found that the curve of rate of acylation of papain by various substrates against pH is bell-shaped; its sides correspond with the ionization of groups of pK_1 of 4-4.5 and of pK_2 of 8-8.5 (Whitaker and Bender, 1965; Lucas and Williams, 1969). The pK_2 was ascribed to an SH group; it is close to the pK values found in studying the kinetics of alkylating the SH group with irreversible inhibitors (Bender and Brubacher, 1966; Wallenfels and Eisele, 1968).

The fact that the rate of acylation of papain falls as its SH group dissociates (forming a more powerful nucleophile) seems at first paradoxical. But as described in Chapter 1, Section 4, this may only mean that the RS^- form is required together with the protonated form of some group largely unprotonated. The kinetics alone cannot show where the extra proton is required, whether on the thiol or elsewhere (in the transition state or earlier). And since general acids could protonate a group that was being expelled and thus improve its leaving quality, or could activate the peptide carbonyl group of the substrate to facilitate mercaptide attack, the need for an acidic group to be protonated would not be surprising.

The identity of the group that catalyzes the deacylation of acyl papain has evoked lively discussion. Brubacher and Bender (1966), and also Kirsch and Igelström (1966), supposed that an ionized carboxyl group of the enzyme played this part. Lowe and Williams (1965), however, pointed out that imidazole is a much better catalyst of thioester hydrolysis than carboxylate ions. Later Husain and Lowe (1968a,b), using the bifunctional reagent 1,3-dibromoacetone, showed that there are SH and imidazole groups close together in the active site of papain. These authors proposed that the findings on the mechanism of action of papain and the dependence of its inhibition on pH can be explained by postulating in the active site of the enzyme a thiol—imidazole pair, joined by a hydrogen bond. The closeness of cysteine-25 and histidine-159 in the active site has been confirmed by x-ray analysis. This analysis showed that the distance between the S atom and N^{π} of histidine-159 is 3.4 Å; a strong interaction between the SH group and the imidazole ring at this distance is unlikely, and the hydrogen bond, if formed, should be very weak (Drenth et al., 1968, 1970).

The mechanism of action of papain, as proposed by Lucas and Williams (1969) is shown below:

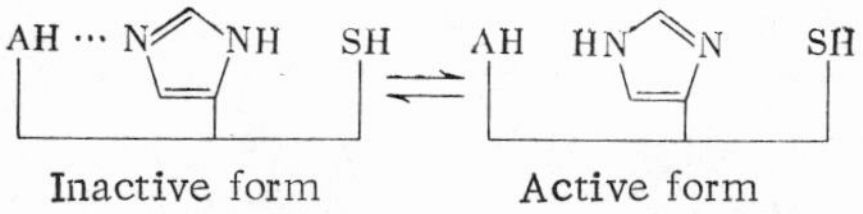

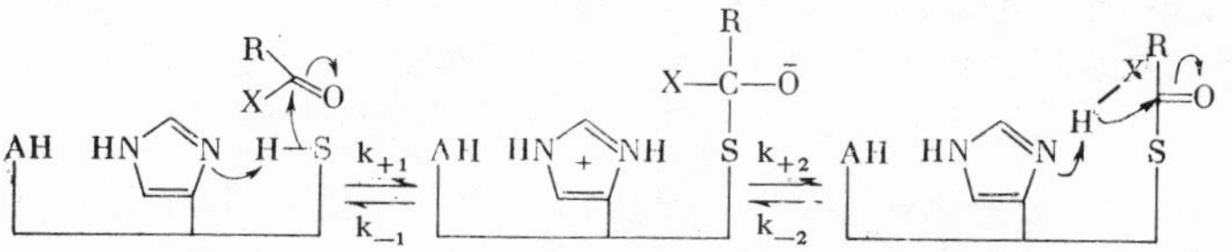

AH ··· N NH SH ⇌ AH HN N SH

Inactive form Active form

AH HN N H–S (R, X, O) $\underset{k_{-1}}{\overset{k_{+1}}{\rightleftharpoons}}$ AH HN + NH S–C(R)(X)–Ō $\underset{k_{-2}}{\overset{k_{+2}}{\rightleftharpoons}}$ AH HN N H (X, R, O) S

AH = Aspargine residue

According to their scheme free imidazole acts as a general base, removing the proton of the SH group at the acylation stage (k_{+1} in their scheme) and removing the proton of water or of some other nucleophilic group in the deacylation step (k_{-2} in the scheme) which is the microscopic reverse of acylation. Lucas and Williams consider that the apparent pK of the imidazole is diminished to about 4.5 by hydrogen bonding of the tertiary nitrogen atom of the imidazole ring with the amide group of an asparagine residue which is also in the active site. Sluyterman and Wolthers (1969) have proposed a different mechanism, also based on diminution of the pK of the histidine residue, shown in the following scheme:

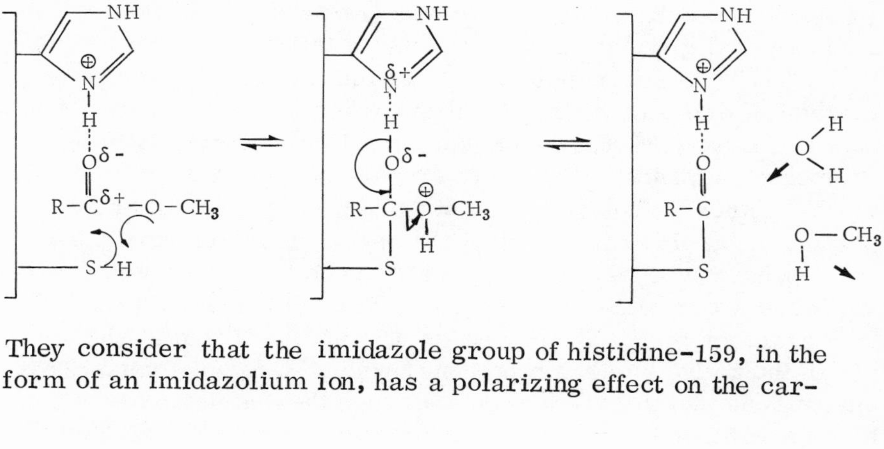

NH, ⊕N–H···$O^{\delta-}$, R–$C^{\delta+}$–O–CH_3, S–H ⇌ NH, $N^{\delta+}$···H, $O^{\delta-}$, R–C–$\overset{\oplus}{O}$(H)–CH_3, S ⇌ NH, ⊕N–H···O=C(R)–S, H_2O, CH_3–O–H

They consider that the imidazole group of histidine-159, in the form of an imidazolium ion, has a polarizing effect on the carbonyl group of the substrate, while the SH group releases its proton to the leaving group of the substrate, simultaneously attacking the carbonyl carbon and forming a thioester.

In a detailed kinetic study Lowe and Yuthavong (1971) showed that an intermediate exists between the enzyme–substrate complex and the acyl-enzyme. Although they were careful to point out that there was no compelling evidence that this was the tetrahedral intermediate envisaged in schemes like that of Lucas and Williams, they considered that this was the most probable form. They demonstrated general-base catalysis of formation of the intermediate and general-acid catalysis of its breakdown; these are just the features of the scheme of Lucas and Williams. The kinetics of the reaction of papain with disulfides (Little and Brocklehurst, 1972; see also Brocklehurst and Little, 1973) also supports activation of the SH group by a nearby histidine residue (see also Chapter 4, Section 2, and Polgar, 1973).

Some of the most interesting findings of Lowe and Yuthavong (1971) were the indications that binding of the substrate by hydrogen bonding of the peptide bond adjacent to the one to be broken forces the substrate into a strained conformation. This puts the peptide bond to be split under strain, possibly forcing it out of a planar state and so destroying its resonance. This would give its carbon atom enhanced electrophilicity and thus reactivity with the sulfhydryl group.

In concluding this section we shall note that papain and Gra-3-P-dehydrogenase are two of the few thiol enzyme for which participation of the SH group in the catalytic act is rigorously proven; in both enzymes the SH groups play a similar role of accepting the acyl part of the substrate molecule. The SH group of acetoacetyl-CoA thiolase also apparently acts as an acceptor and carrier of acyl groups (Lynen, 1970). The roles of the SH groups that are supposed to be parts of the active sites of a number of other thiol enzymes is much less clear. Hypotheses on the mechanisms of action of glutamate dehydrogenase (Hellerman et al., 1958), β-methylaspartase (Bright, 1974; Williams and Libano, 1966), sphingosine 1-phosphate aldolase (Stoffel, 1973), and ATP:creatine phosphotransferase (Watts and Rabin, 1962), which consider the SH groups to participate in these reactions as nucleophilic catalysts, have not yet received convincing experimental support.

b. Dithiol–Disulfide Conversions in the Active Sites of Oxidizing Enzymes

In 1951 Barron (1951) put forward the view that the SH groups of oxidative enzymes might act as intermediate electron carriers from substrates to such acceptors as NAD^+. Some years later this surmise received experimental backing in studies on dihydrolipoate dehydrogenase (NADH:lipoamide oxidoreductase, EC 1.6.4.3). This enzyme forms part of the α-ketoglutarate and pyruvate dehydrogenase complexes (see Section c) and catalyzes the following reaction:

$$\mathrm{Lip(SH)_2} + \mathrm{NAD^+} \rightleftarrows \mathrm{Lip}\begin{smallmatrix}\diagup \mathrm{S} \\ \diagdown \mathrm{S}\end{smallmatrix} + \mathrm{NADH} + \mathrm{H^+}$$

The lipoamide dehydrogenase is a flavoprotein and does not itself contain protein-bound lipoic acid. Nevertheless after preincubation with NADH its activity is strongly inhibited by arsenite, and the inhibition is reversed by 2,3-dimercaptopropanol (Massey and Veeger, 1961; Searls et al., 1961). The treatment of the enzyme with NADH leads to the appearance of two extra titratable SH groups per molecule of FAD (Searls et al., 1961; Palmer and Massey, 1962). In view of the known specificity of arsenite for dithiols these facts suggest that the active site of the dehydrogenase contains a dithiol grouping, which arises by the specific interaction of reduced substrates with an unusually reactive disulfide group of the oxidized enzyme. This suggestion is consistent with many other spectrophotometric and kinetic findings. One of these findings is that although the enzyme can be reduced to a form in which four electrons are accepted (reduction of both the flavin and the disulfide), normal reduction is associated with the acceptance of two electrons. In this state the enzyme is red and has a spectrum like that of flavin semiquinones (flavins reduced by one electron) with raised extinction at 530 nm. Massey (Massey and Veeger, 1961; Massey, 1963) considers that one of the two electrons goes to the flavin and the other to the disulfide. Thus one SH group would be formed and the other sulfur atom would go into the free radical form which would interact with the flavin semiquinone and stabilize it. In agreement with this hypothesis, p-mercuriphenylsulfonate and phenylmercury acetate, which block SH groups, destabilize the semiquinone (Massey et al., 1960; Massey, 1963; Casola and Massey, 1966).

It should be noted that attempts to observe the signal of a free radical in the EPR spectrum of the reduced enzyme have been unsuccessful (Searls et al., 1961); the absence of a signal could be due to strong interaction between the sulfur atom and the FAD semiquinone (possibly a charge-transfer complex is formed between them).

The following is the scheme of the action of lipoamide dehydrogenase postulated by Massey (Massey and Veeger, 1961; Massey and Gibson, 1962; Massey, 1963):

(I) —S—S— ··· FAD ···
Lip $(SH)_2$ Lip S_2
—SH HS— ··· FAD ··· (II)
NADH NADH
(III) —SH $\dot{S}$— ··· FAD$\dot{H}$ ···
R—N (pyridine ring) (IV)
NAD$^+$ H$^+$ NAD$^+$ H$^+$
(V) —S HS— ··· FAD ···
—S $\dot{S}$— ··· FAD$\dot{H}$ ···

According to this scheme, the disulfide group of the protein is the first acceptor of electrons and protons from dihydrolipoate; in this reaction it is reduced to form a dithiol (structures I → II). The dithiol transfers one hydrogen atom (with its electron) to the FAD molecule, which goes into the semiquinone form (structure III). Then a molecule of NAD$^+$ combines with one of the thiol groups, and is finally reduced to NADH with regeneration of the original form of the enzyme (structures IV and V). A somewhat different scheme was put forward by Searls et al. (1961), but their scheme also envisaged participation of the disulfide group of the protein in the catalytic cycle.

Lipoamide dehydrogenase from *Escherichia coli* has been shown to contain a single active disulfide and three SH groups in each of the subunits of the native enzyme. In the amino acid sequence the cysteine residues that form the disulfide group of the active site are separated by four residue (Burleigh and Williams, 1972; Brown and Perham, 1972): Val–Cys–Leu–Asn–Val–Gly–Cys–Ile–Pro–Ser–Lys. This sequence appears to be the same in the enzyme from pig heart, except for replacement of the initial Val by Thr, and the homology extends for another four residues toward the N-terminus (Brown and Perham, 1974).

Many data suggest that a reactive S–S group participates in the electron transfer catalyzed by glutathione reductase (EC 1.6.4.2) (Massey and Williams, 1965) and by thioredoxin reductase (Thelander, 1967, 1968; Zanetti and Williams, 1967). Both enzymes resemble lipoamide dehydrogenase in some respects. Like it they contain two molecules of FAD per protein molecule and have similar and characteristic absorption spectra in the visible region; they also catalyze the reduction of S–S bonds in their substrates using a nicotinamide coenzyme. In distinction from lipoamide dehydrogenase and glutathione reductase, whose substrates are low-molecular compounds, thioredoxin reductase catalyzes the reduction of the S–S bond of the relatively large molecule of the protein thioredoxin, whose molecular weight is 12,000. The dithiol form of thioredoxin is the hydrogen donor in the reduction of ribonucleotides to 2-deoxyribonucleotides and in some other reductions. Titration of thioredoxin reductase under anaerobic conditions with NADH or dithionite shows that the enzyme can accept four electrons per molecule of FAD. Since a flavin can accept at most two electrons, some other group in the enzyme must accept the other two. When thioredoxin reductase is titrated with substrates, no red-colored intermediate like that characteristic of lipoamide dehydrogenase (a stabilized flavin semiquinone) can be observed, so the mechanisms of action of these enzymes are probably different. Thelander (1970), and also Ronchi and Williams (1972), found that the cysteine residues that can form an S–S group in the active site of thioredoxin reductase from *Escherichia coli* are separated in the primary structure by two other residues: –Ala–Cys–Ala–Thr–Cys–Asp–Gly–Phe–. They are thus the same distance apart as those of thioredoxin, which are at positions 32 and 35 in a sequence of 108 residues.

Massey and co-authors (Miller and Massey, 1965; Brumby et al., 1965; Massey et al., 1969) have proposed that the active sites of the metalloflavoproteins xanthine oxidase (EC 1.2.3.2) and dihydroorotate dehydrogenase (EC 1.3.3.1) contain a special form of labile, reactive disulfide, one of whose sulfur atoms is combined with ferric iron, while the other belongs to a cysteine residue. They consider that this structure is responsible for the sulfide and iron that are released from these enzymes on acid or heat denaturation, and that the increase of titratable SH groups on treatment with NADH or substrates is explained by its reduction:

$$\begin{array}{l} >Fe^{3+}-S \\ \quad\quad\quad\;\; | \\ -CH_2-S \end{array} \underset{}{\overset{\pm H}{\rightleftharpoons}} \begin{array}{l} >Fe^{3+}-S\cdot \\ \\ -CH_2-SH \end{array}$$

Phenylmercury acetate inactivates xanthine oxidase only in the presence of substrate. This points to reduction of the S–S bond by the substrate, but one cannot exclude the possibility that the phenylmercury acetate inactivates the enzyme by reacting with masked SH groups which become accessible as a result of a conformational change evoked by the substrate.

Massey and Edmondson (1970; Edmondson et al., 1972) have also postulated another kind of persulfide in xanthine oxidase. By using an adsorbent that contained a purine analog they separated the enzyme by affinity chromatography into two forms. The one not bound to the adsorbent was inactive with respect to xanthine, although fully active in catalyzing the NADH–ferricyanide reaction. Both forms possessed four atoms of labile iron, four atoms of labile sulfide, and one atom of molybdenum per molecule of flavin. The form active to xanthine, however, alone released thiocyanate when treated with cyanide, and simultaneously lost its xanthine oxidase activity. The authors explained this by postulating that it contains an $R-S-S^-$ group which reacted with CN^- to yield $R-S^-$ and SCN^-.

Massey and colleagues consider that the persulfide group may attack xanthine to expel hydride which is transferred to the Fe/S part of the molecule. Hydrolysis of the bond formed would release uric acid and regenerate persulfide. The large number of Fe–S groups can explain the fact that several molecules of xanthine can be oxidized per molecule of enzyme in the absence of other oxidizing agent.

Transfer of reducing equivalents from persulfide region of the molecule to the Fe–S chromophore, possibly via the molybdenum, may also explain the reactivation of cyanide-treated material with sulfide. Edmondson and Massey propose the following cycle, where C is the sum of the other redox groups in the molecule (Fe–S, flavin, and molybdenum):

$$\left[\begin{matrix} -C_{ox} \\ -S{-}S^- \\ -S^- \end{matrix}\right] \xrightarrow{CN^- \;\to\; SCN^-} \left[\begin{matrix} -C_{ox} \\ -S^- \\ -S^- \end{matrix}\right] \rightleftharpoons \left[\begin{matrix} -C_{red} \\ -S \\ | \\ -S \end{matrix}\right] \xrightarrow{O_2 \;\to\; H_2O_2} \left[\begin{matrix} -C_{ox} \\ -S \\ | \\ -S \end{matrix}\right]$$

$$\xleftarrow{\quad HS^- \quad} \text{(return to first form)}$$

Thus xanthine oxidase appears to have two forms of persulfide group, one associated with its iron and providing labile sulfide, and the other associated with its xanthine site and labile to cyanide. Possibly they are given these very different properties by their different environments. Massey and colleagues stress that parts of these ideas on the structures of the groups are still speculative.

c. Thiol Cofactors

Among thiol cofactors are glutathione, lipoic acid, coenzyme A, and acyl-carrier protein. The role of SH groups in these compounds is analogous in many ways to the role of the SH groups of the cysteine residues of proteins that carry out catalytic functions in the active sites of some enzymes (glyceraldehyde-3-phosphate dehydrogenase, papain, etc.). Because of this we consider it useful to review the functions of thiol cofactors in a section on the role of SH groups in catalysis.

Glutathione. De Rey-Pailade (1888) first isolated from yeast a sulfur-containing substance which he called "philothion." Thirty-three years later this substance was rediscovered, purified, and obtained in crystalline form by Hopkins (1921), and called "glutathione" by him. As noted in the Introduction, the discovery of glutathione and the study of its role served as a stimulus to the broader investigation of the SH groups of proteins and enzymes. The structure of glutathione was finally established in 1929 by three groups of authors (Pirie and Pinhey, 1929; Hopkins, 1929; Kendall et al., 1929) and confirmed by its synthesis in 1935 (Harrington and Mead, 1935). Glutathione consists of the tripeptide γ-glutamyl-cysteinyl-glycine:

$$HOOC{-}\underset{}{\overset{NH_2}{\overset{|}{C}H}}{-}[CH_2]_2{-}CO{-}NH{-}\overset{CH_2-SH}{\overset{|}{C}H}{-}CO{-}NH{-}CH_2{-}COOH$$

and it exists in two forms: reduced (G–SH) and oxidized (G–S–S–G).

A very large amount of work has been devoted to searches for the physiological functions of glutathione (see the reviews of Barron, 1951; Knox, 1960; Waley, 1966, and the monographs edited by Colowick et al., 1954; Crook, 1959); nevertheless its role in the cell is not yet entirely clear. In view of its almost universal distribution in the tissues of animals and plants, and in microorganisms, and of its relatively high concentration in many cells (from 0.3 to 7 mM) one may agree with Barron (1951) that the role of glutathione is to protect the SH groups of intracellular enzymes from oxidation and from blocking by heavy metal ions and other poisons. There are data in the literature that glutathione takes part in the detoxication of foreign organic compounds by combining with those of them that contain groups sufficiently electrophilic to react with SH groups. These reactions, catalyzed by glutathione-S-transferases, are apparently the first step in the biosynthesis of mercapturic acids (the S-substituted derivatives of N-acetylcysteine) (Boyland and Chasseaud, 1969; Wit and Leeuwangh, 1969; Royland and Speyer, 1970). The scheme of formation of a mercapturic acid from 1,2-dichloro-4-nitrobenzene and glutathione is given here as an example (according to Boyland and Chasseaud, 1969):

$$NO_2-C_6H_3(Cl)-Cl + HS-G \xrightarrow[\text{S-Aryl transferase}]{\text{Glutathione}} NO_2-C_6H_3(Cl)-S-G$$

$$\downarrow \gamma\text{-Glutamyl transferase}$$

$$NO_2-C_6H_3(Cl)-S-CH_2-CH(NH-COCH_3)-COOH \leftarrow NO_2-C_6H_3(Cl)-S-CH_2CH(NH_2)-CO-NH-CH_2COOH$$

Specific coenzymic functions for glutathione have been found in only a small number of enzymic reactions. Lohmann (1932) found that glutathione is the coenzyme of glyoxylase, which catalyzes the conversion of methylglyoxal into lactic acid. It was later shown that glyoxylase consists of two components: glyoxylase I

(EC 4.4.1.5) and glyoxylase II (EC 3.1.2.6) (Hopkins and Morgan, 1948). Racker (1951) postulated the following mechanism for the glyoxylase reaction:

$$
\begin{array}{ccccccccc}
CH_3 & & CH_3 & & CH_3 & & CH_3 & & CH_3 \\
| & & | & & | & & | & & | \\
C{=}O & & C{-}OH & & C{-}OH & & HC{-}OH & \xrightarrow{H_2O} & HC{-}OH \\
| & \rightleftarrows & \| & \rightarrow & \| & \rightleftarrows & | & & | \\
C{=}O & & C{=}O & & C{-}OH & & C{=}O & & COOH \\
| & & & & | & & | & & \\
H & & + & & S{-}G & & S{-}G & & + \\
 & & G{-}SH & & a & & b & & G{-}SH
\end{array}
$$

On the basis of a kinetic study Cliffe and Waley (1961) proposed that the hemimercaptal (a), formed by nonenzymic combination of glutathione with methylglyoxal, is the true substrate of glyoxylase I, which converts it into the thioester (b), S-lactoylglutathione. Glyoxylase II (thiolesterase) catalyses the hydrolysis of the thioester to form lactic acid and free glutathione.

Glutathione plays a similar part as coenzyme of formaldehyde dehydrogenase (EC 1.2.1.1). The substrate of this enzyme is also, apparently, a hemimercaptal, which undergoes dehydrogenation by reaction of NAD^+; the thioester so formed (S-formyl glutathione) gives formic acid and free glutathione on hydrolysis (Strittmatter and Ball, 1965; Rose and Racker, 1962):

$$
\begin{array}{ccccccc}
H & & H & & \left[\begin{array}{c} H \\ | \\ C{=}O \\ | \\ S{-}G \end{array}\right] & & HCOOH \\
| & & | & \underset{\longleftarrow}{\xrightarrow{NAD^+,\ enzyme}} & & \xrightarrow{H_2O} & + \\
H{-}C{=}O & \rightleftarrows & H{-}C{-}OH & & & & G{-}SH \\
+ & & | & & & & \\
G{-}SH & & S{-}G & & & &
\end{array}
$$

Another type of enzymic reaction, in which glutathione acts as a coenzyme, is that of cis–trans isomerizations (Edward and Knox, 1956; Knox, 1960; Lack, 1961). Maleylacetoacetate isomerase (EC 5.2.1.2) and maleylpyruvate isomerase (EC 5.2.1.4), which catalyze these reactions, are active only in the presence of glutathione, which apparently adds reversibly to the ethylenic grouping of the substrate. Glutathione is also the coenzyme of indolylpyruvate keto-enol tautomerase (Spencer and Knox, 1962). It has also been suggested to be the coenzyme for the oxygenase that catalyses the oxidation of sulfur to sulfite in *Thiobacillus thioparus* (Suzuki and Silver, 1966).

Lipoic Acid. Lipoic acid was discovered in the late forties by several groups of workers (O'Kane and Gunsalus, 1948; Stokstad et al., 1949; Kidder and Dewey, 1949; Kline and Barker, 1950). Reed, Gunsalus, and co-authors (Reed et al., 1951, 1953) first obtained it in a crystalline form, and they obtained about 30 mg of the pure acid from ten tons of the water-insoluble residue of ox liver. Isolation of lipoic acid in a pure form made possible determination of its structure, which was soon confirmed by synthesis (Reed et al., 1953; Brockman et al., 1954; Bullock et al., 1952, 1954; Walton et al., 1955). Lipoic acid is 1,2-dithiolan-3-valeric acid:

```
       H2
       C
      / \
   H2C   CH—[CH2]4—COOH
    |     |
    S  —  S
```

Lipoic acid is widely distributed in the tissues of animals and plants and in microorganisms. Its only firmly established physiological function is the important one of participating in the oxidative decarboxylation of α-ketoacids (Reed, 1960; Schmidt et al., 1965), although Bradley and Calvin (1955) proposed that it takes part in the transfer of electrons from chlorophyll to pyridine nucleotides in plants. Even before the discovery of lipoic acid, Peter and co-workers (Peters et al., 1946; Stocken and Thompson, 1946) had postulated that a dithiol grouping participated in the oxidation of pyruvate in brain. They did this because they observed that compounds of tervalent arsenic severely inhibit the pyruvate dehydrogenase system and that dithiols, but not monothiols, reverse this inhibition. After the discovery of lipoic acid, direct evidence was obtained that it formed part of the pyruvate and α-ketoglutarate dehydrogenase complexes isolated from *Escherichia coli* and also from pig heart, ox liver and pigeon breast muscle. Reed et al. (1958; Koike and Reed, 1960; Suzuki and Reed, 1963) showed that the splitting off of lipoate under the influence of a specific hydrolytic enzyme (lipoamidase) led to the inactivation of these complexes, and that the reincorporation of lipoate into them on incubation with ATP and a lipoate-activating enzyme led to a restoration of activity.

The oxidative decarboxylation of pyruvate and of α-ketoglutarate that is catalyzed by the dehydrogenase complexes takes place according tothe following series of reactions (Massey, 1963; Reed, 1960):

$$R{-}COCO_2H + TPP{-}E_1 \rightarrow [RCHO{-}TPP]{-}E_1 + CO_2 \qquad (1)$$

$$[RCHO{-}TPP]{-}E_1 + \underset{S{-}S}{\frown}{-}[CH_2]_4{-}\underset{\underset{O}{\|}}{C}{-}E_2 \rightarrow$$

$$\underset{HS\quad S{-}\underset{\underset{O}{\|}}{C}R}{\frown}{-}[CH_2]_4{-}\underset{\underset{O}{\|}}{C}{-}E_2 + TPP{-}E_1 \qquad (2)$$

$$\underset{HS\quad S{-}\underset{\underset{O}{\|}}{C}R}{\frown}{-}[CH_2]_4{-}\underset{\underset{O}{\|}}{C}{-}E_2 + HS{-}CoA \rightarrow \underset{HS\quad SH}{\frown}{-}[CH_2]_4{-}\underset{\underset{O}{\|}}{C}{-}E_2 + R\underset{\underset{O}{\|}}{C}{-}S{-}CoA \qquad (3)$$

$$\underset{HS\quad SH}{\frown}{-}[CH_2]_4{-}\underset{\underset{O}{\|}}{C}{-}E_2 + FAD{-}E_3{<}\begin{matrix}S\\|\\S\end{matrix} \rightarrow$$

$$\underset{S{-}S}{\frown}{-}[CH_2]_4{-}\underset{\underset{O}{\|}}{C}{-}E_2 + \cdot FADH{-}E_3{<}\begin{matrix}S\cdot\\SH\end{matrix} \qquad (4)$$

$$\cdot FADH{-}E_3{<}\begin{matrix}S\cdot\\SH\end{matrix} + NAD^+ \rightarrow FAD{-}E_3{<}\begin{matrix}S\\|\\S\end{matrix} + NADH + H^+ \qquad (5)$$

$$\overline{R{-}COOOH + CoA{-}SH + NAD^+ \rightarrow R{-}CO{-}S{-}CoA + CO_2 + NADH + H^+}$$

The pyruvate and α-ketoglutarate dehydrogenase complexes consist of three enzymes: a decarboxylase (E_1), which catalyzes reaction (1) and contains thiamine pyrophosphate (TPP) as coenzyme; lipoate reductase–transacylase (lipoate acetyl transferase) (E_2), which catalyzes reactions (2) and (3); and lipoamide dehydrogenase (E_3), which catalyzes reactions (4) and (5).

From this scheme it follows that the enzyme-bound lipoic acid is involved in the formation and transfer of the acyl group, and also in the transfer of electrons and protons. The interaction of the so-called "active aldehyde," 2-(α-hydroxyethyl)-thiaminepyrophosphate, formed in reaction (1), with lipoate leads to reductive acylation of the latter:

$$\underset{\text{"Active acetaldehyde"}}{HO(H_3C)\overset{\ominus}{C}{-}C{=}\overset{\oplus}{N}{<}(\text{thiazole ring, }S)} + \underset{R{-}\frown}{S{-}S} \longrightarrow \overset{\ominus}{C}{=}\overset{\oplus}{N}{<}(\text{thiazole ring, }S) + CH_3\underset{}{\overset{\overset{O}{\|}}{C}}{-}S\underset{R}{\frown}SH$$

From the S-acyldihydrolipoate the acyl group is then transferred to coenzyme A, and the dihydrolipoate formed then undergoes oxidation under the action of the flavoprotein dihydrolipoate dehydrogenase, whose mechanism of action was discussed in Chapter 6, Section 1b.

In contrast with other thiol coenzymes (glutathione, coenzyme A), lipoic acid is firmly bound to the enzyme and does not dissociate even under harsh treatment. Reed and colleagues (Reed, 1960; Nawa et al., 1960; Daigo and Reed, 1962) found that lipoic acid was bound by an amide bond to the ε-amino group of a lysine residue in the protein molecule; they isolated the corresponding peptide fragments from the pyruvate dehydrogenase complex of *E. coli:*

```
            Lipoyl
              |ε
−Gly−Asp−Lys−Ala−
```

and from the α-ketoglutarate dehydrogenase complex

```
            Lipoyl
              |ε
−Thr−Asp−Lys−Val−
```

In the opinion of Reed (1966) the combination of lipoic acid with the side chain of lysine (the total length of this chain with its bound valeric acid residue is 14 Å) gives the dithiolan ring a certain mobility which is necessary for its sequential interaction with the 2-(α-hydroxyethyl)-thiaminepyrophosphate of the decarboxylase, the coenzyme A, and the prosthetic group of the lipoamide dehydrogenase (Fig. 19).

Coenzyme A and Acyl-Carrier Protein. Coenzyme A, discovered by Lipmann (1945, 1946; Lipmann et al., 1947) consists of residues of 3',5'-diphosphoadenosine and 4'-phosphopantetheine:

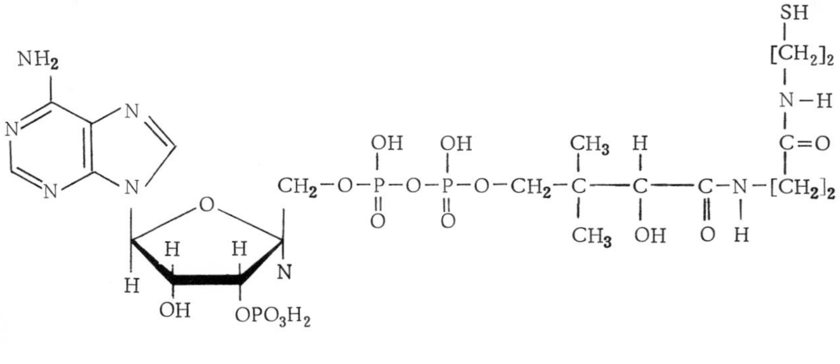

```
                                                                          SH
                                                                          |
  NH2                                                                   [CH2]2
                                                                          |
                                                                         N−H
                                                                          |
                          OH    OH          CH3   H                      C=O
                          |     |           |     |                      |
   (adenine)   CH2−O−P−O−P−O−CH2−C−−−−C−−−−C−N−[CH2]2
                          ||    ||          |     |     ||  |
                          O     O          CH3   OH     O  H
      (ribose ring: H, H, H, N; OH, OPO3H2)
```

4'-Phosphopantetheine, in its turn, consists of residues of 4'-phosphopantothenic acid and cysteamine.

The physiological role of coenzyme A is the activation and transfer of acyl residues. Lynen et al. (1951) first showed that the so-called "active acetate," formed in the oxidation of carbohydrates and fats, was the thioester of acetate with coenzyme A. In the view of Lipmann (1948-1949) there are two different types of activation of acetate: activation of the "head" and activation of the "tail." By activation of the "head" is meant raised ability of acetate to enter reactions of combination of its carboxyl group. This type of activation is caused by the particular properties of thioesters, reviewed in Chapter 1, Section 3. By activation of the "tail" is meant raised reactivity of the methyl carbon of acetate, which is a consequence of resonance stabilization of the carbanion formed by dissociation of a proton from the α-carbon atom:

$$\left[R-\overset{H}{\underset{\cdot\cdot}{C}}-\overset{:O:}{\overset{\|}{C}}-\underset{\cdot\cdot}{\overset{\cdot\cdot}{S}}\,CoA \longleftrightarrow R-\overset{H}{C}=\overset{:\ddot{O}:}{C}-\underset{\cdot\cdot}{\overset{\cdot\cdot}{S}}\,CoA \right]^{\ominus}$$

An example of a reaction in which both types of activation are important is the formation of acetoacetyl-CoA from two molecules of acetyl-CoA under the influence of the enzyme acetoacetyl-CoA

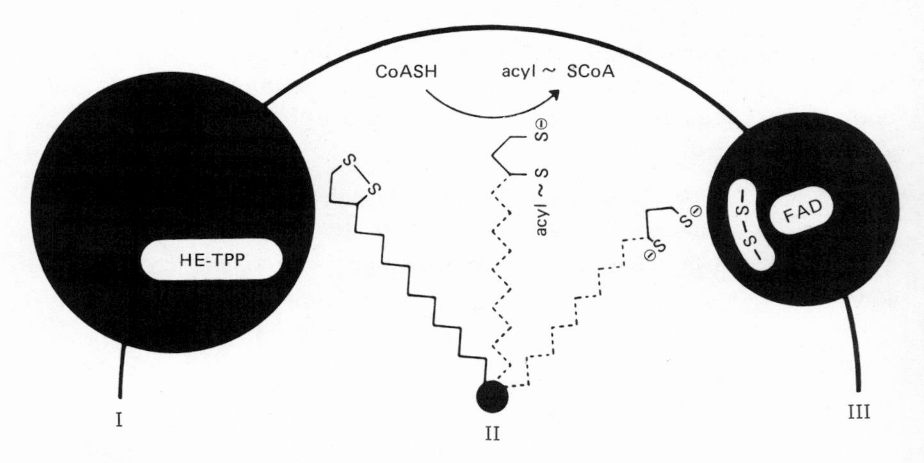

Fig. 19. Scheme of the functioning of bound lipoic acid in the process of oxidative decarboxylation of α-ketoacids (Reed, 1968). I) Decarboxylase, containing 2-(α-hydroxyethyl)-thiaminepyrophosphate (HE-TPP); II) lipoate reductase–transacylase, containing bound lipoic acid; III) lipoamide dehydrogenase, containing FAD and an active S–S group.

thiolase:

$$\mathrm{CoA{-}S{-}\underset{\underset{O}{\|}}{C}{-}\overset{\ominus}{C}H_2\ \ \underset{\underset{O^{\delta-}}{\|}}{\overset{\overset{CH_3}{|}}{C^{\delta+}}}{-}S{-}CoA\ (H^+)} \rightleftharpoons \mathrm{CoA{-}S{-}\underset{\underset{O}{\|}}{C}{-}CH_2{-}\underset{\underset{O}{\|}}{C}{-}CH_3 + CoA{-}SH}$$

As can be seen from this scheme, one of the molecules of acetyl-CoA acts in this reaction as a nucleophilic reagent, attacking with its activated methylene group (the "tail") the electrophilic carbonyl carbon of the other molecule.

Coenzyme A plays an important part in the reactions of oxidation of aldehydes and α-ketoacids, the β-oxidation of fatty acids, and also in the synthesis of mevalonic acid and of some phospholipids (Jaenicke and Lynen, 1960; Goldman and Vagelos, 1964).

The participation of coenzyme A in the oxidative decarboxylation of pyruvate and α-ketoglutarate can be seen from the equations on p. 176 As a result of these reactions acetyl-CoA is formed by oxidation of pyruvate, and succinyl-CoA by oxidation of α-ketoglutarate, and these are further metabolized in the tricarboxylic acid cycle.

Fatty acids undergo β-oxidation in the form of activated acyl derivatives of coenzyme A. The formation of these derivatives takes place with involvement of ATP:

$$\mathrm{R{-}CH_2{-}CH_2{-}CH_2{-}COOH + ATP + CoA{-}SH} \rightleftharpoons$$
$$\mathrm{R{-}CH_2{-}CH_2{-}CH_2{-}CO{-}S{-}CoA + AMP + Pyrophosphate}$$

As a result of the sequential reactions of dehydrogenation, hydration, and dehydrogenation, acetyl-CoA is formed together with the fatty acid residue shortened by two carbon atoms (also in the form of its coenzyme A derivative):

$\mathrm{R{-}CH_2{-}CH_2{-}CH_2{-}CO{-}S{-}CoA}$
↓ FAD → $FADH_2$ — Acyl dehydrogenase
$\mathrm{R{-}CH_2{-}CH{=}CH{-}CO{-}S{-}CoA}$
↓ H_2O — Enoyl hydratase
$\mathrm{R{-}CH_2{-}CHOH{-}CH_2{-}CO{-}S{-}CoA}$
↓ NAD^+ → NADH — β- + H^+ Hydroxyacyl dehydrogenase
$\mathrm{R{-}CH_2{-}CO{-}CH_2{-}CO{-}S{-}CoA}$
↓ CoA−SH — Thiolase
$\mathrm{R{-}CH_2{-}CO{-}S{-}CoA}$ + $\mathrm{CH_3{-}CO{-}S{-}CoA}$

The enzymes for the first three of these reactions are specific for the CoA-derivatives, even though the thioester bond plays no important part in the reaction catalyzed. Consequently the β-ketoacid is produced in the form of its CoA derivative, and this makes the thiolase reaction possible, in which, as we have seen, the thioester nature of the substrates is essential.

The acetyl groups thus emerge from the breakdown of fatty acids in the form of acetyl-CoA, i.e., already activated for the next reaction. Acetyl-CoA, both from this source and derived from pyruvate by oxidative decarboxylation, condenses with oxaloacetate, the reaction being catalyzed by citrate synthase (an example of "tail" activation):

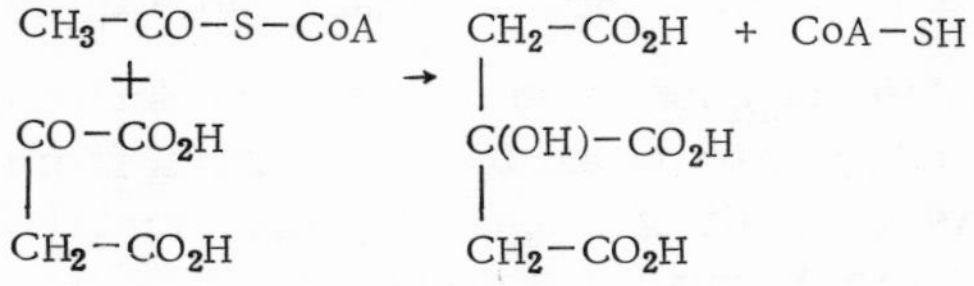

$$\begin{array}{lcl} CH_3-CO-S-CoA & & CH_2-CO_2H \;+\; CoA-SH \\ + & \rightarrow & | \\ CO-CO_2H & & C(OH)-CO_2H \\ | & & | \\ CH_2-CO_2H & & CH_2-CO_2H \end{array}$$

Citrate

Thus coenzyme A is liberated and the citrate formed is metabolized in the Krebs cycle. In this reaction the proton removed from the methyl group of acetyl-CoA is on the side away from the oxaloacetate, as shown by the inversion of configuration of a chiral methyl group (-CHDT) during the reaction, which has the following stereochemistry (Eggerer et al., 1970; Retey et al., 1970):

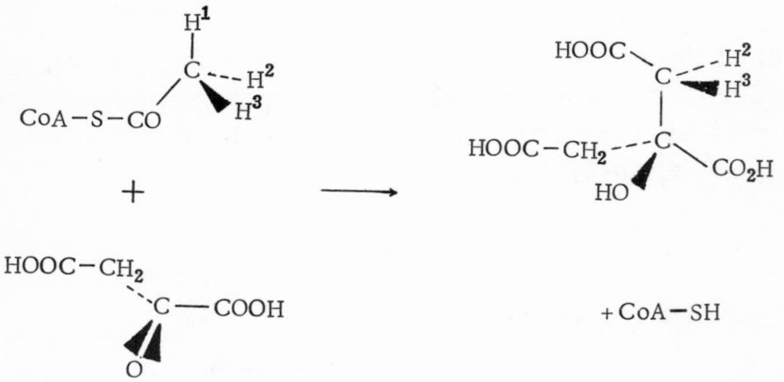

H^1, C, H^2, H^3, CoA−S−CO

\+

HOOC−CH_2, C—COOH, O

⟶

HOOC, C, H^2, H^3, HOOC−CH_2, C, CO_2H, HO

\+ CoA−SH

This shows that the stabilized carbanion has no free existence apart from the enzyme. Probably a basic group on the enzyme is correctly placed to assist in removing the proton from the methyl group of the acetyl-CoA.

Significant progress has been made in the last few years in working out the mechanism of biosynthesis of fatty acids. It has proved to be not simply, as was first thought, the reverse of the process of β-oxidation. Lynen (1961) observed that the purified fatty acid synthetase multienzyme complex catalyzes the condensation of acetyl-CoA and malonyl-CoA with the formation of enzyme-bound acetoacetate. On the basis of this fact, and of the impossibility of finding the intermediates of fatty acid biosynthesis in free form Lynen proposed that these intermediates are bound to a protein SH group and undergo a series of enzymic conversions while they remain attached to this group.

An "acyl-carrier protein" (ACP, molecular weight 9000) was isolated in the laboratories of Vagelos and Wakil in fractionation of the fatty acid synthetase of *Escherichia coli* and it was shown that the intermediate acyl compounds that are involved in fatty acid biosynthesis are actually bound by a thioester bond to the SH group of ACP (Alberts et al., 1964; Majerus et al., 1964; Wakil et al., 1964; Vagelos et al., 1966). The prosthetic group of ACP proved to be 4'-phosphopantetheine, which is bound by a phosphoester link to the hydroxyl group of a serine residue of the protein (Majerus and Vagelos, 1965; Majerus et al., 1965a,b; Pugh and Wakil, 1965; Vanaman et al., 1968):

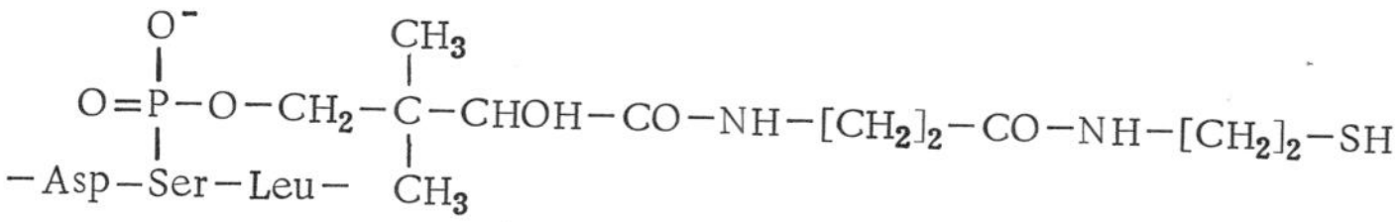

$$\begin{array}{ll} \quad O^- & \quad CH_3 \\ \quad | & \quad | \\ O{=}P{-}O{-}CH_2{-} & C{-}CHOH{-}CO{-}NH{-}[CH_2]_2{-}CO{-}NH{-}[CH_2]_2{-}SH \\ \quad | & \quad | \\ {-}Asp{-}Ser{-}Leu{-} & \quad CH_3 \end{array}$$

Thus ACP may be regarded as a protein analog of coenzyme A; in both cofactors the function of acceptor of acyl residues is fulfilled by the SH group of 4'-phosphopantetheine. Indeed an enzyme has been found in E. coli which catalyzes the transfer of 4'-phosphopantetheine from coenzyme A onto apo-ACP to form holo-ACP and liberate adenosine-3',5'-diphosphate (Elovson and Vagelos, 1968; Prescott et al., 1969).

ACP plays an important part in the biosynthesis of fatty acids in bacteria, yeasts, plants, and animals (Vagelos et al., 1966; Willecke et al., 1969; Larrabee et al., 1965; Simoni et al., 1967; Wells et al., 1967; Matsumura and Stumpf, 1968); in yeasts and animal tissues the synthesis of fatty acids is carried out in the cytoplasm by a stable multienzyme complex, the 4'-phosphopantetheine is combined with a protein component of this complex which is similar in function to the ACP of *E. coli*, but apparently different in molecular weight and amino acid composition (Willecke et al., 1969; Lynen, 1970). Acyl derivatives of ACP may also be involved in the acylations of phospholipid synthesis (Goldfine et al., 1967; Goldfine and Ailhaud, 1971; Van den Bosch and Vagelos, 1970).

The first reactions of bacterial fatty acid synthesis are the formation of acetyl-ACP and malonyl-ACP from the corresponding CoA derivatives by specific transacylases:

$$\text{Acetyl-S-CoA} + \text{ACP-SH} \rightleftharpoons \text{Acetyl-S-ACP} + \text{CoA-SH}$$

$$\text{Malonyl-S-CoA} + \text{ACP-SH} \rightleftharpoons \text{Malonyl-S-ACP} + \text{CoA-SH}$$

Malonyl-CoA is formed by carboxylation of acetyl-CoA under the influence of the biotin enzyme acetyl-CoA carboxylase:

$$CO_2 + \text{ATP} + CH_3\text{-CO-S-CoA} \overset{Mn^{2+}}{\rightleftharpoons} \text{HOOC-}CH_2\text{-CO-S-CoA} + \text{ADP} + P_i$$

The first reaction of the multi-enzyme complex is the condensation catalyzed by β-ketoacyl-ACP synthetase:

$$\text{Acetyl-S-ACP} + \text{Malonyl-S-ACP} \longrightarrow \text{Acetoacetyl-S-ACP} + \text{ACP-SH} + CO_2$$

Alberts et al. (1965) have put forward the hypothesis that in this reaction the acetyl group of acetyl-ACP is first transferred to the SH group of the synthetase, and then combines with the methylene group of malonyl-ACP. According to Lynen (1970), yeast fatty acid synthesis starts with the transfer of an acetyl group from acetyl-CoA onto the SH group of the 4'-phosphopantetheine combined with a protein component of the complex (yeast ACP). The acetyl residue is transferred thence onto the SH group of the active center of the condensing enzyme (β-ketoacyl synthetase). There follow transfer of the malonyl residue from malonyl-CoA onto the liberated SH group of the 4'-phosphopantetheine, and condensation of the bound acetyl and malonyl residues, forming an acetoacetyl residue combined with the SH group of the 4'-phosphopantetheine.

Since the condensation is accompanied by decarboxylation, its equilibrium is far further in the direction of synthesis than that of the reaction of acetoacetyl thiolase in fatty acid oxidation. This shift is achieved at the expense of the ATP used in the synthesis of malonyl-CoA. But the properties of thioesters contribute in the same way (Chapter 1, Section 3 and above) to this reaction.

The further reactions of fatty acid synthesis are similar to those of fatty acid oxidation in reverse. The β-ketoacyl group is reduced to a β-hydroxyacyl group, this is dehydrated to form an $\alpha\beta$-unsaturated acyl group, which is finally reduced to a saturated fatty acyl group. In this way a butyryl-ACP molecule is formed. In both reductions the hydrogen donor is NADPH; this is true also in animal tissues where NADPH is the reducing agent available in the cytoplasm. In the yeast system described above the butyryl group when formed is transferred to the second SH group to allow acceptance of a further malonyl residue by the 4'-phosphopantetheine, so that the whole process can be repeated. Finally a fatty acid with a long chain is formed, and liberated from the ACP by transfer to coenzyme A.

These reactions are summarized by Ayling et al. (1972) in the following scheme, where the upper SH group represents that of ACP and the lower one a cysteine residue:

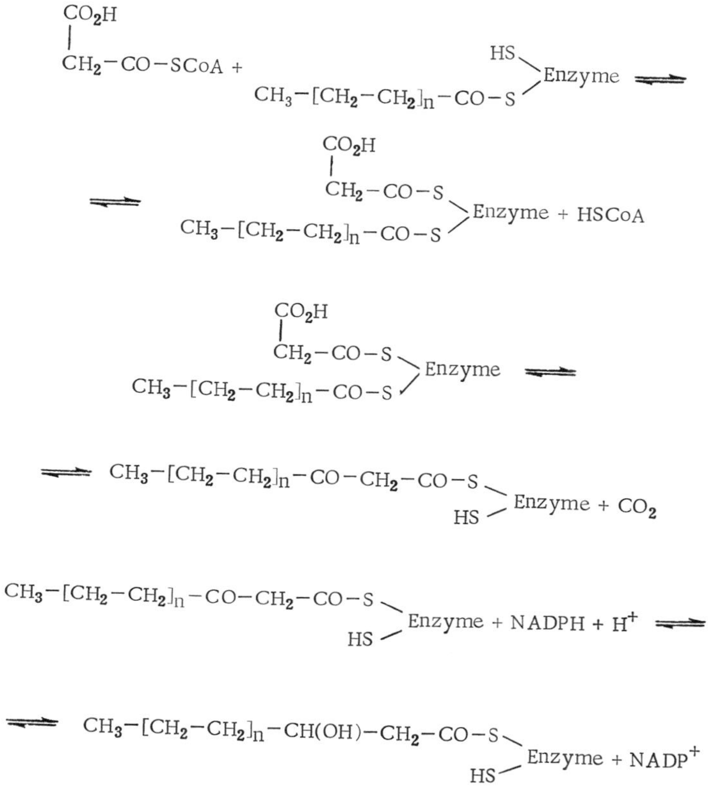

$$\begin{array}{l} CO_2H \\ | \\ CH_2{-}CO{-}SCoA \end{array} + \begin{array}{l} \qquad\qquad\qquad\qquad HS \\ CH_3{-}[CH_2{-}CH_2]_n{-}CO{-}S \end{array}\!\!>\!Enzyme \rightleftharpoons$$

$$\rightleftharpoons \begin{array}{r} CO_2H \qquad\quad\; \\ | \qquad\qquad\quad \\ CH_2{-}CO{-}S \\ CH_3{-}[CH_2{-}CH_2]_n{-}CO{-}S \end{array}\!\!>\!Enzyme + HSCoA$$

$$\begin{array}{r} CO_2H \qquad\quad\; \\ | \qquad\qquad\quad \\ CH_2{-}CO{-}S \\ CH_3{-}[CH_2{-}CH_2]_n{-}CO{-}S \end{array}\!\!>\!Enzyme \rightleftharpoons$$

$$\rightleftharpoons \begin{array}{r} CH_3{-}[CH_2{-}CH_2]_n{-}CO{-}CH_2{-}CO{-}S \\ HS \end{array}\!\!>\!Enzyme + CO_2$$

$$\begin{array}{r} CH_3{-}[CH_2{-}CH_2]_n{-}CO{-}CH_2{-}CO{-}S \\ HS \end{array}\!\!>\!Enzyme + NADPH + H^+ \rightleftharpoons$$

$$\rightleftharpoons \begin{array}{r} CH_3{-}[CH_2{-}CH_2]_n{-}CH(OH){-}CH_2{-}CO{-}S \\ HS \end{array}\!\!>\!Enzyme + NADP^+$$

$$CH_3-[CH_2-CH_2]_n-CH(OH)-CH_2-CO-S\text{ (HS)}>\text{Enzyme} \rightleftharpoons$$

$$\rightleftharpoons CH_3-[CH_2-CH_2]_n-CH=CH-CO-S\text{ (HS)}>\text{Enzyme} + H_2O$$

$$CH_3-[CH_2-CH_2]_n-CH=CH-CO-S\text{ (HS)}>\text{Enzyme} + NADPH + H^+ \longrightarrow$$

$$\longrightarrow CH_3-[CH_2-CH_2]_{n+1}-CO-S\text{ (HS)}>\text{Enzyme} + NADP^+$$

$$CH_3-[CH_2-CH_2]_{n+1}-CO-S\text{ (HS)}>\text{Enzyme} \rightleftharpoons \text{(HS)}\ CH_3-[CH_2-CH_2]_{n+1}-CO-S>\text{Enzyme}$$

It has been established that β-ketoacyl-ACP synthetase and β-hydroxyacyl-ACP dehydratase from *E. coli* are absolutely specific for the thioesters of ACP and do not catalyze reactions of the CoA derivatives. β-Ketoacyl-ACP reductase catalyzes the reduction of acetoacetyl-CoA at a rate 60 times slower than that of acetoacetyl-ACP. Figure 20 shows some of the details of fatty acid synthesis catalyzed by the multienzyme complex from pigeon liver. The attachment of the malonyl group to 4'-phosphopantetheine not only activates its "tail" for condensation, but evidently allows access of the growing chain to each of the enzymes of the complex.

In conclusion it may be noted that the use of two different thiols (coenzyme A and acyl carrier protein) as activator-acceptors of fatty acyl residues may facilitate the separate control of the two processes of the breakdown and synthesis of fatty acids in the cell, processes that go through a number of similar intermediate reactions.

A protein subunit similar to ACP and also containing phosphopantetheine has been found in a bacterial citrate lyase by Dimroth et al. (1973). The enzyme apparently functions as follows:

Acetyl–enzyme + Citrate → Citryl–enzyme + Acetate

Citryl–enzyme → Acetyl–enzyme + Oxaloacetate

Thus here too the function of the phosphopantetheine residue is to carry acyl groups. It also seems to be involved, presumably in a

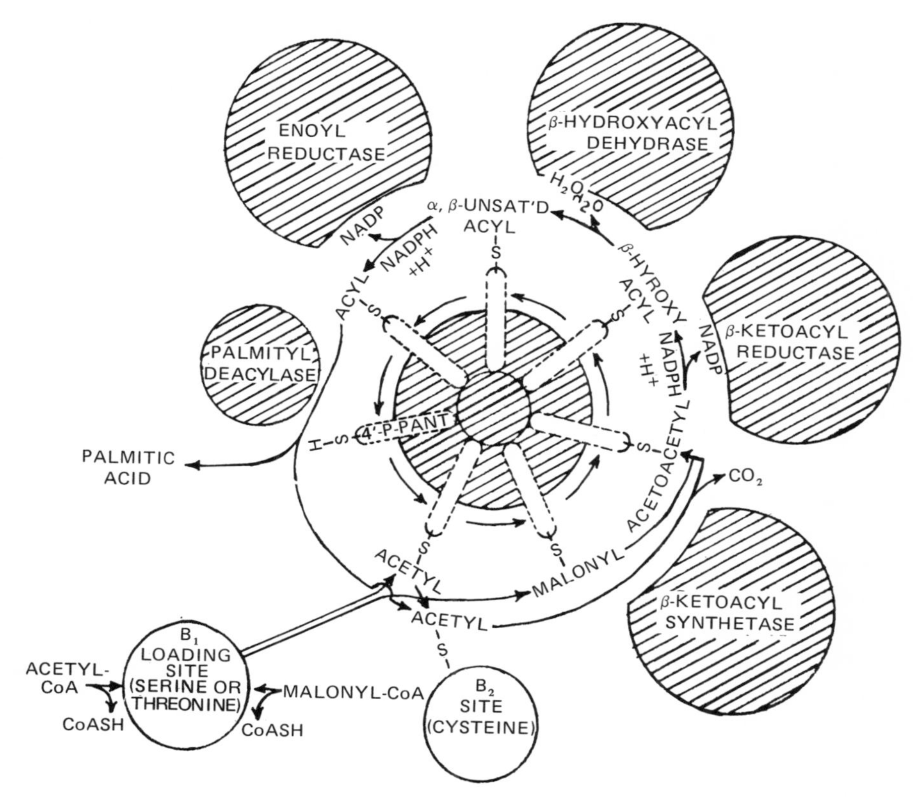

Fig. 20. A mechanism proposed for fatty acid synthesis by pigeon liver multienzyme complex (Phillips et al., 1970). One complete cycle of reactions (from acetyl-CoA and malonyl-CoA to butyryl enzyme) is shown. Six additional cycles would occur in the synthesis of palmitic acid. The two reductions and the dehydration reaction are depicted as occurring on a rotating (or freely moving) 4'-phosphopantetheine prosthetic group.

similar role, in the synthesis of antibiotic peptides by an enzyme system from *Bacillus brevis* (Gilhuus-Moe et al., 1970; Lipmann et al., 1971; Laland and Zimmer, 1973; see p. 162).

2. The Role of SH Groups in the Binding of Substrates and Cofactors (Metal Ions and Coenzymes)

The SH groups of the active sites of enzymes, besides participating directly in the catalytic act, may also play a part in establishing bonds between the apoenzyme and the molecules of substrate or coenzyme. In this case the SH groups are part of the so-called contact region of the active center ("binding site"). The division of SH groups into catalytically active groups and groups that ensure substrate or cofactor binding is arbitrary to the extent that the division of the active site into catalytic and contact regions is itself arbitrary. The formation of bonds from the SH or other functional groups of the enzyme to some group of the substrate or coenzyme often leads to changes in the electron densities and conformation of the substrate or coenzyme that significantly facilitate the chemical reaction carried out by the enzyme.

Various methods, described in Chapter 5, are used to discover participation of SH groups in substrate or cofactor binding. One of such methods is investigation of ability of substrates, coenzymes, and their analogs, to protect SH groups from blocking by thiol reagents. The first observations of this ability were made in the thirties for the succinate oxidase system (Hopkins and Morgan, 1938) and for glyceraldehyde-3-phosphate dehydrogenase (Rapkine, 1938; Rapkine et al., 1939). Hopkins and his co-workers showed that succinate and fumarate, and also malonate which is closely related in structure, protect partially purified preparations of succinate oxidase from inactivation by oxidized glutathione. These findings were later fully confirmed in studies on purified succinate dehydrogenase, isolated from the succinate oxidase complex. It proved that succinate and its analogs protect the SH groups of succinate dehydrogenase not only from the action of oxidizing agents, but also from alkylating agents and compounds of tervalent arsenic; the protective action is weaker in relation to mercuric ions and p-mercuribenzoate (Stoppani and Brignone, 1957).

It has now been established for a fairly large number of enzymes that substrates, coenzymes, or their analogs have a protective action against SH reagents. To these enzymes belong succinate, malate, lactate, isocitrate, 3-phosphoglycerate, glyceraldehyde-3-phosphate, betaine aldehyde, β-hydroxyisobutyrate, glutamate, homoserine, 3α-hydroxysteroid, aldehyde, and alcohol dehydrogenases, dihydrofolate reductase, cytochrome reductase, glyoxylate reductase, glutamic semialdehyde reductase, D-amino acid oxidase, imidazolylacetate monooxygenase, fumarase, ribulose-1,5-diphosphate carboxykinase, pyruvate decarboxylase, citrate lyase, aldolase, glutamine and tryptophan synthetases, adenosine-triphosphatase (myosin), β-amylase, pyridoxamine-pyruvate aminotransferase, transamidinase, ATP-creatine and ATP-arginine phosphotransferases, phosphofructokinase, phosphoenolpyruvate carboxykinase, valyl-, glycyl-, isoleucyl- and tryptophanyl-tRNA synthetases, luciferase, collagen-proline hydroxylase, thiolase, etc.

In Chapter 5 we described the difficulties in interpreting experiments on the protection of SH groups. A fall in the number of titratable SH groups of an enzyme in the presence of substrate or coenzyme can be taken as a preliminary indication of the presence of SH groups in the active site.* Less convincing are experiments that measure only the degree of loss of activity under the influence of the thiol reagent, without direct determination of the number of SH groups in the presence and absence of substrate or coenzyme. This is because substrates and coenzymes can diminish the degree of inactivation by a thiol reagent not only by screening the SH group, but also by stabilizing the structure (conformation) of the protein molecule. Even the splitting of a coenzyme from a specific protein or dissociation of an enzyme–pseudosubstrate complex under the influence of thiol reagents must not be taken unreservedly as the result of direct displacement of the coenzyme or pseudosubstrate from the SH group, since the displacement could also have been due to changes in the structure of the protein evoked by blocking of its SH groups.

Difficulties have arisen, for example, in discussing coenzyme protection of the SH groups of alcohol dehydrogenase (EC 1.1.1.1). Theorell and Bonnichsen (1951) first proposed participation of the SH groups of the liver enzyme in the binding of NADH on the basis of spectrophotometric studies. They showed that addition of p-mercuribenzoate to the complex of the dehydrogenase with NADH, which possessed an absorption maximum at 325 nm, led to appearance of the spectrum of free NADH, with a maximum at 340 nm, and this indicated dissociation of the complex. Barron and Levine (1952) found that NAD^+ and ethanol not only protect yeast alcohol dehydrogenase from inactivation by iodoacetate and iodosobenzoate, but also diminish the number of its titratable SH groups. Later

*Hoch et al. (1960) have described an instructive case where a fall in the number of titratable SH groups of yeast alcohol dehydrogenase on adding a coenzyme analog, N-methylnicotinamide, was due to the presence of impurities of heavy metal ions in the preparations of this analog.

Li and Vallee (1963, 1965) found that selective carboxymethylation by iodoacetate of two of the 24 SH groups (of one per active site) of liver enzyme led to complete inactivation; NAD^+ and NADH protect the enzyme from inactivation and the SH groups from alkylation. The inactive carboxymethylated apoenzyme unexpectedly preserved its capability to bind coenzyme specifically, albeit less firmly as indicated by the appearance of the characteristic Cotton effect. These findings apparently mean that the SH group does not participate directly in binding the coenzyme, but

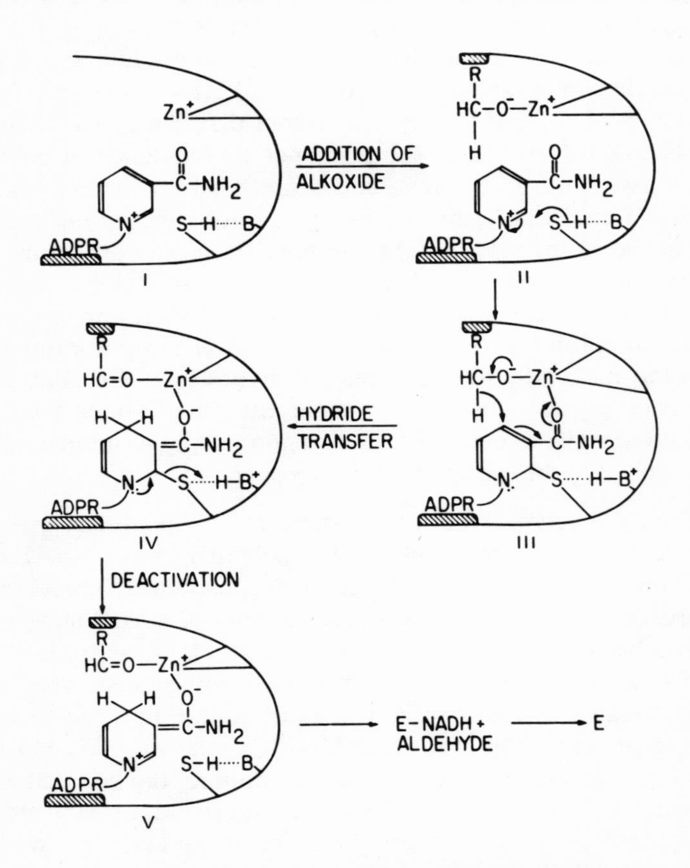

Fig. 21. Mechanism of action of alcohol dehydrogenase (Evans and Rabin, 1968). I) Complex of the enzyme (E) with NAD^+; II) ternary complex E$-NAD^+-$alcohol; III) activated complex E$-NAD^+-$alcohol; IV) activated complex E$-$NADH$-$aldehyde; V) complex E$-$NADH$-$aldehyde (ADPR = ADP-ribose).

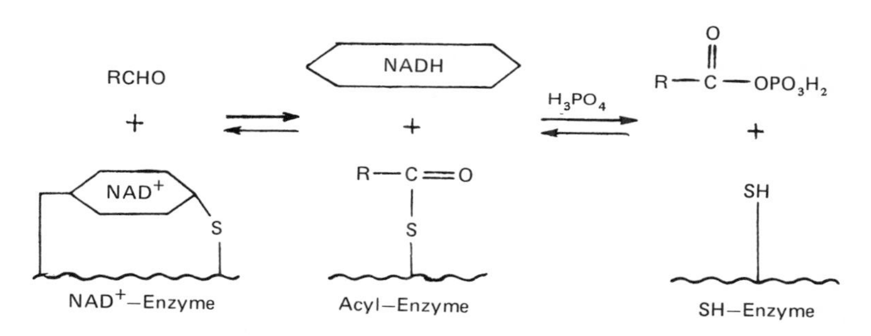

Fig. 22. "Aldehydolysis" of the complex of 3-phosphoglyceraldehyde dehydrogenase with NAD^+ (Racker and Krimsky, 1952).

that its protection by NADH is probably due to steric hindrance produced either directly by the bound coenzyme molecule, or by a change due to coenzyme binding in the structure of the apoenzyme. Auricchio and Bruni (1969) came to analogous conclusions in studying the SH groups of the yeast enzyme.

Evans and Rabin (1968) consider, however, that the data of Li and Vallee do not exclude combination of the SH group with the coenzyme in the course of catalysis. They found that protection of the SH groups of liver alcohol dehydrogenase could be achieved not only with coenzyme, but also with o-phenanthroline (which reacts with protein-bound Zn^{2+}) and also ADP and ADP-ribose (ADPR). On this basis they postulated that the SH group is located in the active site between the Zn^{2+} ion and the region that binds the adenine part of the coenzyme molecule. In their opinion the SH group combines with position 2 of the nicotinamide ring of the coenzyme at an intermediate step in the catalytic action of the enzyme (Fig. 21); the activation of the ternary complex (enzyme–NAD^+–substrate) and the facilitation of hydride transfer are due to this.

The question of the participation of the reactive SH group of glyceraldehyde-3-phosphate (Gra-3-P) dehydrogenase in coenzyme binding has a long history. As in the case of alcohol dehydrogenase, this question is not yet fully clarified. Rapkine (1938; Rapkin et al., 1939) had already noted that NAD^+ partly protects Gra-3-P dehydrogenase from inactivation by oxidized glutathione. Racker and Krimsky (1952, 1958; Racker, 1954, 1955) found that the broad absorption band at 360 nm, characteristic of Gra-3-P dehydrogenase that contains tightly bound NAD^+ is greatly diminished by p-mercuribenzoate, iodoacetate, 1,3-diphosphoglycerate, and acetyl phosphate. They proposed that in the enzyme the SH group forms a covalent bond with the nicotinamide ring of NAD^+. According to their hypothesis, glyceraldehyde-3-phosphate combines not with a free SH group of the dehydrogenase, as Boyer and Segal (1954) (see p. 156) had proposed, but with the protein–S–NAD^+ linkage by an "aldehydolysis"; in this the hydrogen of the aldehyde group of the substrate is transferred to NAD^+ with the formation of NADH and 3-phosphoglyceryl-enzyme (Fig. 22). The findings of Velick (1953) that p-mercuribenzoate evokes splitting of the coenzyme from the protein also point to participation of SH groups in the binding of NAD^+.

The hypothesis of Racker and Krimsky has, however, encountered a number of objections (Kosower, 1956; Boyer, 1959; Colowick et al., 1966). Kosower expressed the opinion that the spectral properties of the enzyme–NAD^+ complex were more reminiscent of those of a charge-transfer complex than of a covalent compound. It was soon shown that NAD^+ can in fact form charge-transfer complexes with tryptophan and other indole derivatives (Cilento and Giusti, 1959; Cilento and Tedeschi, 1961; Shifrin, 1964). On the other hand, NAD^+ can also form complexes with mercaptans (Van Eys and Kaplan, 1957). Boross and Cseke (1967) found that the position of the absorption maximum (360 nm) and the pH dependence of the spectrum of the complex of Gra-3-P dehydrogenase with NAD^+ were more reminiscent of the properties of the complex of NAD^+ with mercaptans (λ_{max} = 330 nm) than with tryptophan (λ_{max} = 300-305 nm). Boross and Cseke proposed that the interaction of a mercaptide ion with the pyridine ring of NAD^+ was the basis of the appearance of the characteristic spectrum with a maximum at 360-365 nm. Alkylating agents destroy (iodoacetate) or modify (iodoacetamide, bromoacetone) this spectrum, but in contrast with p-mercuribenzoate they do not cause splitting of NAD^+ from the protein (Park et al., 1960; Friedrich, 1965). The blocking of the most reactive SH groups of Gra-3-P dehydrogenase with Ag^+ ions similarly does not cause splitting off of NAD^+ (Boross, 1965). Inactive enzyme with oxidized SH groups preserves its ability to bind NAD^+, but the character of the binding appears to differ somewhat between active and inactive enzymes (Astrachan et al., 1957). Finally, it has been shown that combination of NAD^+ greatly increases the rate of alkylation by

TABLE 14. Influence of the Coenzyme on the Rate Constants for Alkylation of the Catalytically Active SH Group of Muscle Glyceraldehyde-3-Phosphate Dehydrogenase (5°C, tris-HCl buffer, ionic strength 0.05) (Boross et al., 1969)

Coenzyme, 4 equiv.	pH	k, $M^{-1} \cdot min^{-1}$
	Iodoacetate	
NADH	7.4	180
None	8.5	250
NAD$^+$	7.4 or 8.5	4800
	Iodoacetamide	
NADH	7.4	35
None	7.4	540
None	8.5	3700
NAD$^+$	7.4	140
NAD$^+$	8.5	350

iodoacetate and of oxidation of the SH group of the active site of Gra-3-P dehydrogenase* (in contrast with the action of NAD$^+$ on alcohol dehydrogenase) (Table 14) (Trentham, 1968; Boross et al., 1969; Little and O'Brien, 1969; Fenselau, 1970). These findings rule out the formation of a covalent bond between the SH group and NAD$^+$. The dissociation of NAD$^+$ under the action of p-mercuribenzoate, observed by Velick (1953, 1958) is probably due to conformational changes in the protein (see Chapter 6, Section 3). The acyl enzyme, like other forms whose SH is blocked, has lowered affinity for coenzyme (Trentham, 1968, 1971; Harrigan and Trentham, 1973); this may facilitate replacement of NADH by NAD$^+$ in the course of the catalyzed reaction (step 3 on p. 157).

* The increase in the rate of alkylation of the SH group by iodoacetate in the presence of bound NAD$^+$ could be explained by an influence of the NAD$^+$ on the reactivity or accessibility of the catalytically active SH group. But the finding that NAD$^+$ greatly inhibits the reaction of this group with iodoacetamide (Fenselau, 1970) and fluoro-2,4-dinitrobenzene (Shaltiel and Soria, 1969) contradicts this explanation. So it is more probable that the pyridinium cation of NAD$^+$ attracts and orientates the iodoacetate anion, and thus facilitates its alkylation of the SH group of the active site (Fenselau, 1970). The proximity of the catalytically active SH group to the nicotinamide of the bound NAD$^+$ is confirmed by x-ray analysis (Buehner et al., 1973).

A curious fact was discovered in a study of ferredoxin-NADP reductase of chloroplasts: NADP protects this enzyme from inactivation by polar organomercurial reagents, but accelerates its inactivation by the nonpolar phenylmercury acetate (Zanetti and Forti, 1969). Apparently in this case too the coenzyme does not form a direct bond with the SH group, but only changes the character of its environment.

In studying the SH groups of pig heart aspartate aminotransferase (EC 2.6.1.1) we obtained data that indicated that one SH group was located near the substrate-binding site of the enzyme (Torchinsky, 1964; Torchinskii and Sinitsyna, 1970). Selective blocking of this group with p-mercuribenzoate (after prior blocking of two nonessential SH groups) inhibits the activity of the enzyme by 95% and greatly diminishes its affinity for substrate analogs (and apparently for substrates themselves). The results of experiments with one of the pseudosubstrates, erythro-β-hydroxyaspartate, which forms a complex with the enzyme with an absorption maximum at 492 nm, were particularly clear. As can be seen from Fig. 23, blocking of this SH group of the enzyme

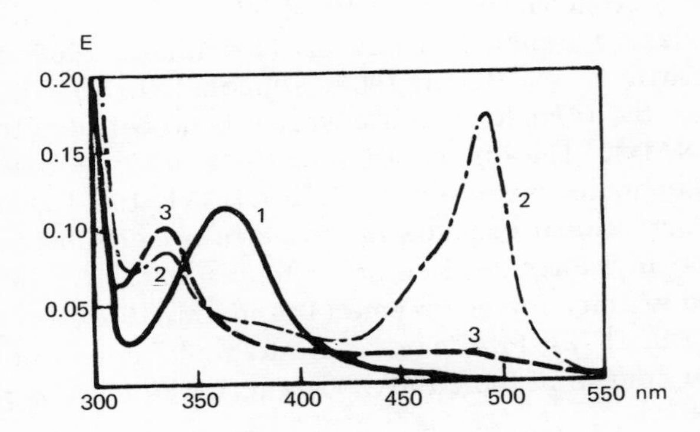

Fig. 23. The influence of p-mercuribenzoate on the interaction of partially alkylated aspartate aminotransferase with erythro-β-hydroxyaspartate. 1) Absorption spectrum of the enzyme in which two SH groups are blocked with N-ethylmaleimide without fall in activity (pH 7.8); 2) the same as 1, +0.01 M erythro-β-hydroxy-DL-aspartate; 3) spectrum of the enzyme in which two SH groups are alkylated and a third blocked with p-mercuribenzoate, in the presence of 0.01 M erythro-β-hydroxy-DL-aspartate.

with p-mercuribenzoate completely prevented formation of this complex; no maximum at 492 nm arises. Birchmeier and Christen (1971) found that this SH group of the aminotransferase was alkylated by N-ethylmaleimide only in the presence of substrates, with an accompanying large fall in activity. They suggested that the induced accessibility of the group reflected a transient conformational change in the enzyme in the course of the catalytic cycle.

The presence of an SH group in the active site region of aspartate aminotransferase from pig heart cytosol was convincingly confirmed by Okamoto and Morino (1973) by affinity labeling. They found that the substrate analog bromo[^{14}C]pyruvate enters into a transamination reaction with the pyridoxamine form of the enzyme. In the course of this reaction the enzyme became inactivated and an SH group alkylated. Torchinskii et al. (1972) and Birchmeier et al., (1972) established that the "syncatalytically" reacting SH group (i.e., the group reactive only in the course of the enzyme-catalyzed reaction) belongs to residue 390 in the primary structure of the aminotransferase. Birchmeier et al. (1973) found that syncatalytic modification of this residue by N-ethylmaleimide, 5,5'-dithio-bis-(2-nitrobenzoate), or tetranitromethane diminished the enzymic activity to less than 5% of the original. Replacement, however, of the nitrothiobenzoate group by cyanide, with formation of S-cyanocysteine, led to regeneration of 60% of the activity. It follows that the syncatalytically reactive SH group (which is in the active site or close to it) plays no essential role in catalysis or in the maintenance of the active conformation of the aminotransferase. Karpeiskii, Polyanovskii, and colleagues (Demidkina et al., 1973; Bocharov et al., 1973) have shown that enzyme in which the cysteine residue has been syncatalytically oxidized by tetranitromethane (with simultaneous nitration of tyrosine) is incapable of transamination, although it forms an imine between the ketoacid substrate and its own pyridoxamine phosphate group. Transamination does not occur even at temperatures (40-45°C) at which ketoacids and pyridoxamine transaminate nonenzymically. This too is consistent with the idea that the oxidized cysteine residue prevents the relative positioning of groups that is necessary for the catalytic act.

In several enzymes a bond between SH groups and substrates or coenzymes is apparently mediated by a metal ion. This hypothesis was given convincing backing in a study of prolidase (EC 3.4.3.7), an enzyme that hydrolyzes peptide bonds formed by the imino group of proline or hydroxyproline. Glycyl-L-proline is a typical substrate for this enzyme. Smith et al. (1954) showed that purified preparations of the enzyme lost almost all their activity in the absence of manganese ions; addition of these converted the enzyme from the inactive into the active form. p-Mercuribenzoate strongly inhibited the activity of prolidase, which could be restored by adding cysteine or glutathione. It is interesting that prolidase activity is inhibited by iodoacetamide only if it is added before Mn^{2+} ions; in their presence it has hardly any effect on the activity. These findings suggest that the Mn^{2+} ion may combine with the SH group and protect it from iodoacetamide (but not from p-mercuribenzoate).

On the basis of these experiments, and also of a study of the specificity of prolidase, Smith and colleagues proposed a scheme with an intermediate enzyme–substrate complex, shown here:

CH_2
OC NH ? P
CH_2 R
H_2C N——Mn^{++}—S—O
T
? E
H_2C——CH O^- I
CO N

As shown in the scheme, the function of the SH group of prolidase consists in forming a mercaptide "bridge" which plays an essential part in binding and activating the substrate. Bryce and Rabin (1964) proposed a similar role for the SH group in the active site of leucine aminopeptidase (Fig. 24).

According to the findings of Toda et al. (1968), the SH group is apparently one of the ligands that binds Ca^{2+} in the active sites of taka-amylase A from *Aspergillus oryzae* and α-amylase from *Bacillus subtilis*, removal of Ca^{2+} with EDTA leads to "unmasking" of the SH group of the amylase, and its subsequent modification greatly decreases the affinity of the enzyme for Ca^{2+}. There are indications in the literature that SH groups are also involved in binding Cu^{2+} in galactose oxidase (Kelly-Falcoz et al., 1965) and in azurin (Finazzi-Agró et al., 1970), iron in hemerythrin (Klotz and Klotz, 1959), in homogentisate oxidase (Flamm and Crandall,

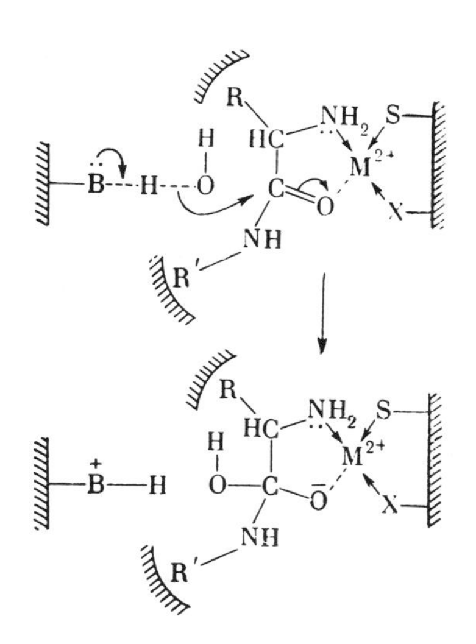

Fig. 24. Scheme of the enzyme–substrate interactions in the active site of leucine aminopeptidase (Bryce and Rabin, 1964).

1963), and in the nonheme electron-carrier proteins: ferredoxin, adrenodoxin, and rubredoxin (Lovenberg et al., 1963; Blomstrom et al., 1964; Lovenberg and Williams, 1969; Tanaka et al., 1966; Suzuki, 1967; Bachmayer et al., 1967, 1968; Poe et al., 1970). The metalloproteins in which the metal ion is apparently bound by an SH group are listed below:

Protein	Metal ion	Protein	Metal ion
Alcohol dehydrogenase	Zn^{2+}	Rubredoxin	Fe^{3+}
Taka-amylase A	Ca^{2+}	Adrenodoxin	Fe^{2+}
α-Amylase	Ca^{2+}	Hemerythrin	Fe^{2+}
Azurin	Cu^{2+}	Ferritin	Fe^{2+}
Galactose oxidase	Cu^{2+}	Histidine ammonia-lyase	Cd^{2+}, Mn^{2+}, or Zn^{2+}

Klee and Gladner (1972) have shown that only four SH groups per tetramer of histidine ammonia-lyase are reactive in the absence of denaturants. Since a single labeled peptide is obtained on digestion after treatment with radioactive iodoacetate, there is probably one reactive SH group per subunit. Oxidation greatly lowers the affinity of the enzyme for Cd^{2+}, Mn^{2+}, and Zn^{2+}, which activate it, and oxidation, treatment with Ellman's reagent, and alkylation all greatly diminish its activity (but do not destroy it completely) and this fall can be reversed in the case of the first two treatments. Klee and Gladner cite earlier evidence that the substrate may be bound through the metal ion, and propose that the SH group in turn assists in binding this ion.

Earlier suggestions that SH groups were involved in the binding of zinc ions in carboxypeptidase A (Vallee et al., 1960; Vallee, 1963) and in carbonic anhydrase (Rickli and Edsall, 1962) seem to have been incorrect, as found for carboxypeptidase by x-ray crystallography (Lipscomb et al., 1968) and chemical studies (Walsh et al., 1970), and for carbonic anhydrase by x-ray analysis (Liljas et al., 1972) which indicated that the zinc ion of this enzyme was bound by three histidine residues.

The nature of the so-called acid labile or inorganic sulfide found in ferredoxin and some other nonheme iron-containing proteins has aroused lively discussion. While some authors considered that this sulfide is a component of ferredoxin that is distinct from cysteine SH groups (Malkin and Rabinowitz, 1966; Hong and Rabinowitz, 1967; Hong et al., 1969; Jeng and Mortenson, 1968; Lovenberg and McCarthy, 1968) others suggested that it was formed on acid degradation of cysteine residues (β-elimination) (Bayer et al., 1965; Bayer and Parr, 1966; Gersonde and Druskeit, 1968).

Bayer et al. described the release of H_2S on heating cysteine methyl ester in the presence of Mohr's salt, ferrous ammonium sulfate, but Malkin and Rabinowitz (1966) could not repeat their findings; further, they could find no dehydroalanine residues in apoferredoxin, and these should have been formed from cysteine residues by β-elimination. Later Bayer et al. (1969) also came to the conclusion that the labile sulfur of ferredoxin does not belong to cysteine residues; they postulated that plant ferredoxin has a

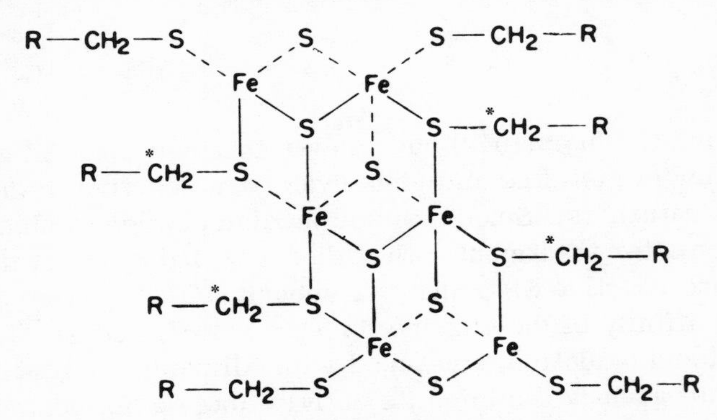

Fig. 25. Model of the active center of ferredoxin from *Clostridium pasteurianum* (Poe et al., 1970). Analysis of the PMR spectrum showed that half the cysteine methylene groups experienced twice as much spin density as the other half. They were assigned to residues (marked with an asterisk) whose sulfur atoms were bonded to two iron atoms instead of one.

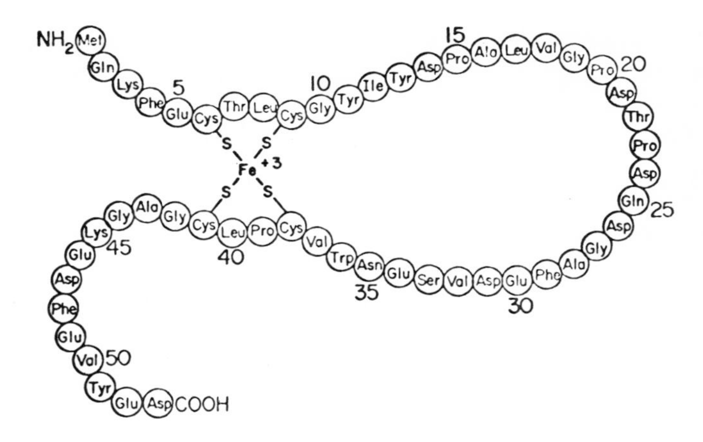

Fig. 26. Diagram of the structure of rubredoxin from *Micrococcus aerogenes* (Bachmayer et al., 1967) as drawn by Ali et al. (1972).

nonplanar structure of the Fe–S–S–Fe type in its active site. Figure 25 shows a model of the complex formed by iron ions with cysteine residues and inorganic sulfide in ferredoxin from *Clostridium pasteurianum.*

Mayerle et al. (1973) have described the preparation and properties of the compound $(Et_4N_2)[FeS(SCH_2)_2C_6H_4]_2$, which they regarded as an analog of the active center of two-iron ferredoxins. The similarity suggests that the minimal structure of this center may be represented as

Cys – S   S   S – Cys
        Fe   Fe
Cys – S   S   S – Cys

In contrast with ferredoxin, rubredoxin contains no acid-labile sulfide. The ferric iron is bound by the SH-groups of four cysteine residues, forming, according to x-ray structural analysis, a distorted tetrahedral complex (Fig. 26) (Bachmayer et al., 1967; Herriott et al., 1970; Watenpaugh et al., 1972).

It should be noted that a metal ion bound by a mercaptide bond to a sulfur atom can form further co-ordinate bonds with various functional groups of the protein (nitrogen, carboxyl, etc.). Chelates are thus formed, and these may play a direct catalytic role, or may assist in establishing bonds between various parts of the protein molecule or between separate subunits, bonds that ensure maintenance of the unique three-dimensional conformation of the enzymes.

3. The Role of SH Groups in the Maintenance of the Native Conformation of Enzymes

In 1949 Desnuelle and Rovery expressed the opinion that the inactivation of urease (EC 3.5.1.5) that occurred on acylation of its SH groups with phenyl isocyanate was a consequence of reversible changes in the structure of the enzyme. Swenson and Boyer (1957) came to a similar conclusion when they studied the reaction of p-mercuribenzoate with muscle aldolase (EC 4.1.2.13). In neither case did the inactivation occur on blocking the most reactive SH groups, but it accompanied blocking of the more slowly reacting ones. Szabolcsi and colleagues (Szabolcsi and Biszku, 1961; Szabolcsi et al., 1964, 1970; Závodszky et al., 1972) confirmed that p-mercuribenzoate evoked conformational changes in aldolase, which they followed by measuring the increased susceptibility of its mercaptide derivatives to tryptic hydrolysis (Fig. 27), and also the changes in its ORD and intrinsic viscosity. They found that although blocking of the first 8-12 SH groups (2-3 per subunit) with p-mercuribenzoate had no effect on the enzymic activity, it nevertheless evoked small reversible changes in the conformation of the enzyme, as shown by its increased lability to trypsin. Addition of the substrate, fructose diphosphate, to the partially substituted aldolase apparently restored the original conformation of the enzyme, since it restored its resistance to tryptic attack. When more than about 8-12 SH groups were blocked, inactivation of the aldolase was observed, presumably due to a change of conformation.

To probe such phenomena more deeply, it is necessary to find which thiol groups are modified. Perham and Anderson (1970) have made some progress with this, showing that two residues per subunit react rapidly with iodoacetamide, but only one of these with Ellman's reagent. A total of three residues react rapidly or at a moderate speed with each reagent, the same three in each case. Szajàni et al. (1973) identified these three residues after reaction with bromo[^{14}C]acetate. The remaining five groups per subunit do not react in the native enzyme. Comparison of these

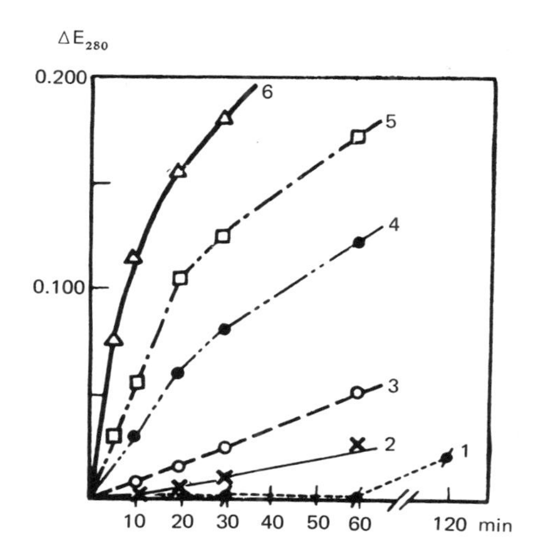

Fig. 27. The influence of p-mercuribenzoate on the rate of tryptic hydrolysis of aldolase (Szabolcsi and Biszku, 1961). 1) Native aldolase; 2-6) aldolase containing 3, 5, 8, 9.5, and 14 mercaptide groups respectively. These numbers of Szabolcsi and Biszku should not be taken as exact, since they assumed a molecular weight of 149,000, and it is now known to be about 160,000 (Kawahara and Tanford, 1966).

results with those of Szabolcsi and colleagues again emphasizes the ability of p-mercuribenzoate to penetrate, albeit slowly, into masked regions of protein molecules.

Szabolcsi (1958; Szabolcsi et al., 1959) and Elödi (1960) also observed changes in the secondary and tertiary structure of muscle glyceraldehyde-3-phosphate dehydrogenase on blocking its SH groups with p-mercuribenzoate. These changes were manifested in increased viscosity and negative optical rotation of solutions of the enzyme, and in its increased sensitivity to proteases. The blocking of the SH groups of aldolase and glyceraldehyde-3-phosphate dehydrogenase thus leads to changes in their structure like the changes seen during denaturation. They can be reversed by removing the p-mercuribenzoate with cysteine or other thiols, but at late stages of the denaturation the removal of the reagent does not lead to complete restoration of the initial conformation and activity of the enzyme molecules.

The kinetics of denaturation of the protein on blocking the SH groups of glyceraldehyde-3-phosphate dehydrogenase was studied by Friedrich and Szabolcsi (1967). They found that eight SH groups of the dehydrogenase react instantaneously with p-mercuribenzoate, and that the blocking of the remaining eight requires a relatively long incubation and follows first-order kinetics with a rate constant of 0.131 min^{-1}. The changes in protein conformation that occur when mercaptide bonds are formed were followed by measuring the hypsochromic shift in the absorption band of the enzyme at 295 nm. The conformational changes also followed first-order kinetics with a rate constant of 0.136. The authors concluded that the rate of mercaptide formation was limited by the rate of unfolding (denaturation) of the protein which led to the "unmasking" of the SH groups; as soon as an SH group was "unmasked" it reacted rapidly with p-mercuribenzoate, and this in turn led to further unfolding of the globule (see also Vas and Boross, 1970).

Guha et al. (1968) studied the changes in the structure of ox heart cytoplasmic malate dehydrogenase that arise on blocking its buried SH groups. They found that three of the six groups of the native enzyme react with p-mercuribenzoate without any loss of activity or change in optical rotatory dispersion. The remaining three SH groups are blocked only in 2.6 M urea, and this is accompanied by loss of activity and a marked change in rotatory dispersion (a fall in the values of $[m']_{233}$, $-b_0$, a_0, and λ_c). The enzyme is inactivated more slowly than mercaptides are formed, but at the same rate as rotatory dispersion changes. This indicates that the loss of activity is a consequence of conformational changes that set in after blocking of the SH groups.

The investigations of Madsen and Cori (1956; Madsen, 1956) on muscle phosphorylase (EC 2.4.1.1) have contributed considerably to the development of ideas on the participation of SH groups in maintaining the quaternary structure of enzymes and other proteins. They observed that blocking the SH groups of phosphorylase *a* with p-mercuribenzoate, methylmercury nitrate, or iodoacetamide not only greatly diminished the enzymic activity, but also dissociated the enzyme into its four subunits. Subsequent addition of cysteine to the mercaptide of the enzyme restored its original tetrameric structure. Madsen and Cori compared the rates of inactivation of the phosphorylase and of its dissociation into subunits with the rate of combination with p-mercuribenzoate (followed by increase of absorption at 250 nm) and found that the inactivation was only slightly slower than mercaptide formation (rate constants of 43 and 51 $M^{-1} \cdot sec^{-1}$ respectively) and preceded dissociation of the tetramer (Fig. 28). These authors proposed that the cause of the inactivation is a rapid change in the protein structure which follows blocking of the SH groups, and that this finally leads to dissociation into subunits.

After publication of the work of Madsen and Cori, a large number of proteins were found to dissociate into subunits under the influence of thiol reagents. They include liver glutamate dehydrogenase (Rogers et al., 1962), formyltetrahydrofolate synthetase (Himes and Rabinowitz, 1962, Nowak and Himes, 1971), hemerythrin (Keresztes-Nagy and Klotz, 1963; Duke et al. 1971), fumarase (Hill and Kanarek, 1964), the protein of turnip yellow mosaic virus (Kaper and Houwing, 1962; Kaper and Jemfer, 1967), potato virus X (Reichmann and Hatt, 1961), liver fatty acid synthetase (Butter-

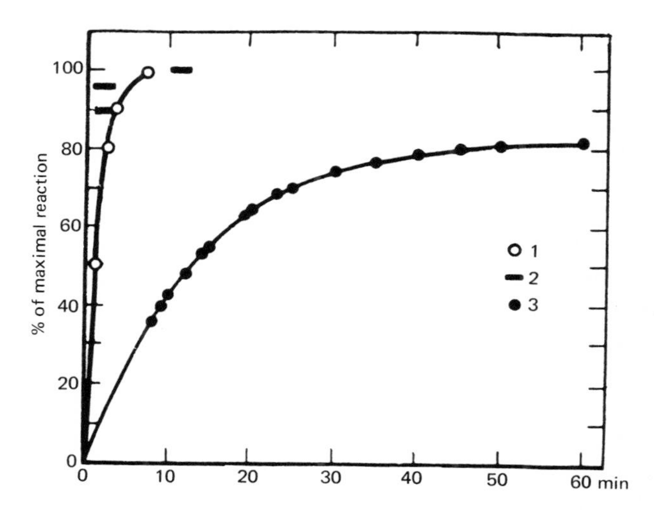

Fig. 28. Comparison of the rate of inactivation of phosphorylase a with its rate of dissociation into subunits under the influence of p-mercuribenzoate (Madsen, 1956). 1) Inhibition of enzymic activity, calculated from the rate constant for the reaction of p-mercuribenzoate with the enzyme (43 $M^{-1} \cdot sec^{-1}$); 2) inhibition of enzymic activity, measured experimentally under the conditions in which the light scattering was measured; 3) dissociation into subunits as followed by light scattering.

worth et al., 1967), aspartate transcarbamylase from *Escherichia coli* (Gerhart and Schachman, 1965), yeast hexokinase (Lazarus et al., 1968), liver pyruvate carboxylase (Palacian and Neet, 1970), succinyl-CoA synthetase of *E. coli* (Grinnell and Nishimura, 1970; Leitzmann et al., 1970), human hemoglobin (Chiancone et al., 1970), glyceraldehyde-3-phosphate dehydrogenase (Smith and Schachman, 1971), etc. The dissociation of aspartate transcarbamylase by p-mercuribenzoate was used to discover and separate the catalytic and regulatory subunits of this allosteric enzyme. With some enzymes, ribulose-1,5-diphosphate carboxykinase and 17β-hydroxysteroid dehydrogenase for example, the disorganization of structure produced by thiol reagents is accompanied not by dissociation, but the reverse — aggregation of the protein molecules (Akazawa et al., 1968).

The reactions of phosphorylase and aspartate transcarbamylase molecules with p-mercuribenzoate follow the "all-or-none" principle, i.e., either all or none of the SH groups of a particular enzyme molecule react (Madsen and Cori, 1956; Gerhart and Schachman, 1968). This follows from the finding that in partially reacted mixtures the undissociated enzyme contains no p-mercuribenzoate whereas the dissociated product has fully reacted. One of the possible explanations of this is that the SH groups of these enzymes are poorly accessible to the reagent and that combination of the inhibitor with the first SH group of the native enzyme is the rate-determining step which is followed by a rapid conformational change that improves the accessibility of the remaining SH groups of the molecule (the so-called "zipper" mechanism). Another explanation is that the SH groups of the undissociated molecule do not react with p-mercuribenzoate and that they become accessible only after a dissociation, which may in this case be the rate-limiting step of the reaction. This explanation presumes the pre-existence of a monomer—oligomer equilibrium before reaction of the SH groups. Gerhart and Schachman (1968) found that the addition to aspartate transcarbamylase of both a substrate analog (succinate) and carbamoyl phosphate, which are bound by the catalytic subunits, leads to a sixfold increase in the rate of reaction of p-mercuribenzoate with the SH groups of the regulatory subunits. They consider that in the presence of succinate and carbamoyl phosphate the enzyme goes into a "swollen" conformation, in which it more easily dissociates into subunits.

This "all-or-none" situation provides a warning against facile interpretations based on no more than measurement of the total number of groups per molecule modified. Identification of the actual residues, e.g., by peptide mapping, is important in elucidating the process of modification.

It should be noted that changes in the conformation of enzymes brought about by mercaptide-forming reagents can lead not only to falls, but also to rises of enzymic activity. Kielley and Bradley (1956) noted that the addition of a small quantity of p-mercuribenzoate to myosin in the presence of Ca^{2+} ions increased its ATPase activity 3-4 times; increase in reagent concentration, however, led to complete inhibition of the enzymic activity (see also Petrushkova and Bocharnikova, 1968). Similarly aspartate transcarbamylase can be stimulated with Hg^{2+} ions (Gerhart, and Pardee, 1962), malate dehydrogenase with Hg^{2+} and p-mercuribenzoate (Kuramitsu, 1968; Silverstein and Sulebele, 1970), and glutamate dehydrogenase (Rogers et al., 1962, 1963) and dihydrofolate reductase (Kaufman, 1964) by organic mercury compounds. It is interesting that the activation of dihydrofolate reductase can be achieved either with urea (4 M) or by treatment with low concentration of methylmercury bromide (20 μM), p-mercuribenzoate, or mersalyl. The similarity of the effects of urea and of organomercurials suggests that in both cases the cause of the activation is a change in the conformation of the molecule. Iodine and N-bromosuccinimide, which oxidize the SH group, also activate dihydrofolate reductase (Kaufman, 1966; Freisheim and Huennekens, 1969). Conceivably some allosteric enzymes appear on this list because they have evolved so that they are not normally, in the absence of effector, in their most active conformation.

Little et al. (1969) found that liver fructose-1,6-diphosphatase is activated when 5-6 of the 20 SH groups of its molecule are modified with p-mercuribenzoate, iodoacetamide, N-ethylmaleimide, o-iodosobenzoate, and a number of disulfides (cystamine, Ellman's reagent, diethyldisulfide, etc.). The increase in activity was 220-450% on reaction with disulfides, and about 100% on reaction with the other reagents listed. The activated enzyme differed from the native enzyme in solubility and in sensitivity to the action of the allosteric inhibitor AMP. It is highly likely that the cause of the activation is here too conformational changes whose degree and character depends on the nature of the bound thiol reagent (see also Pontremoli et al., 1967).

What are the causes of the changes in the macromolecular structure of enzymes and other protein that occur when their SH groups are blocked? One of them could be the breaking of intramolecular bonds formed by SH groups. In Chapter 4, Section 2b, we reviewed the kinds of bonds in which SH groups could participate, and rejected the possibilities of thioester bonds and thiazoline and thiazolidine structures, and we came to the conclusion that SH groups might take part in hydrophobic interactions. Evidently blocking SH groups that had been involved in such interactions could destroy the packing of the nonpolar amino acid side chains within the protein molecule and destabilize its conformation. It has often been noted that poorly accessible or "buried" SH groups are more important in maintaining protein structure than readily reacting ones. This observation is probably explained by the fact that the readily reacting groups are located on the surface of the globule, while the slowly reacting or "buried" groups are inside the molecule and so may be involved in intramolecular interactions which are destroyed when they are blocked.

Usually SH groups do not form covalent bonds that could be involved in stabilizing the conformation of protein molecules. An exception is the bonds that are mediated by a metal ion, i.e., mercaptide (semipolar) bonds that participate in forming chelate complexes. The significance of such chelates in maintaining the macromolecular structure in a number of metalloenzymes (glutamate dehydrogenase, alcohol dehydrogenase, etc.) is experimentally well established. Thus, for example, Kägi and Vallee (1960) showed with yeast alcohol dehydrogenase that the Zn^{2+} ions which seem to form mercaptide bonds apparently participate in binding the subunits together to form the catalytically active oligomer.

Another cause of the structural changes that occur on blocking of the SH groups may be the direct deformatory influence of the bound molecules of inhibitor (their hydrophilic or hydrophobic radicals or charged groups) on the nearby parts of the protein globule and its effect on the interactions between subunits. The often noted dependence of the effect of blocking an SH group on the nature of the inhibitor molecule (in particular on its size and charge) argues in favor of this point of view. Thus, for example, formyltetrahydrofolate dehydrogenase readily dissociates under the influence of p-mercuribenzoate, but mercuric ions, which also inactivate the enzyme, do not cause it to dissociate (Himes and Rabinowitz, 1962; Nowak and Himes, 1971). Similar results were obtained in a study of the dissociation of human hemoglobin into its α- and β-chains; p-mercuribenzoate and chloromercurinitrophenols but not $HgCl_2$ evoked dissociation. The last even reversed the dissociation that had been produced by the organomercurials (Stefanini et al., 1972).

In the opinion of Boyer (1959, 1960) the changes in conformation of proteins that occur on blocking of their SH groups should not be regarded as evidence of direct participation of such groups in forming the structure of the proteins. He emphasized two points that have already been noted above: firstly that the major changes in protein structure occur usually only as a result of blocking slowly reacting or poorly accessible SH groups; secondly, these changes do not occur instantaneously after the SH groups are blocked, but a little time later. Starting from the fact that the approach of the inhibitor to the SH groups is sterically hindered, Boyer suggests that this steric hindrance may be periodically removed during small fluctuations in the conformation of the protein. The temporary removal of the barrier makes possible the blocking of the SH groups, and this in turn prevents return of the protein into its original, energetically more favorable, and consequently more stable, conformation.* Repetition of this process can lead to destabilization of the protein molecule and, finally, to its denaturation.

There are, therefore, various explanations of the conformational changes that occur in proteins on blocking their SH groups. Whether the SH group is important *per se* in maintaining the conformation or whether it is merely the point of combination of a deforming molecule is a question that must be solved in each individual case by careful studies with a number of different thiol reagents.

---

* On the basis of a similar idea Vas and Boross (1970) were able to explain peculiarities of the kinetics of the reaction of p-mercuribenzoate with one of the less reactive SH groups of glyceraldehyde-3-phosphate dehydrogenase.

4. Conclusion

On the basis of the findings presented in this chapter one can come to the following conclusion. The functions of the SH groups of proteins are extremely varied. In some enzymes they play a catalytic role, i.e., they take a direct part in forming intermediates (e.g., thioesters) in the course of the reaction catalyzed by the enzyme. Disulfide–dithiol interconversions, found in the active sites of some oxidative enzymes, apparently play a part in intermediate stages of electron (plus proton) transport from substrates to acceptors. The SH groups can also participate in binding substrates and cofactors (metal ions and prosthetic groups) to the enzymes, i.e., they can form part of the binding sites of apoenzymes. Finally, SH groups can in some cases contribute to the stabilization of the catalytically active conformation of the protein molecules of the enzymes.

Inhibition of the activity of a number of enzymes by reagents for SH groups is due to destruction of the three-dimensional structure (conformation) of these enzymes. This destruction may occur either as a result of breakage of intramolecular bonds in which SH groups were involved, or as a result of the deforming (or destabilizing) action of an SH-bound inhibitor on the nearby parts of the protein molecule.

Chapter 7

The Role of S–S Groups in Proteins

1. The Role of S – S Groups in Stabilizing the Macromolecular Structure of Proteins

The disulfide group is the most widespread of the covalent bonds in proteins that can link different parts of a polypeptide chain or different chains, and can thus play a part in maintaining the secondary, tertiary, and quaternary structures of proteins. The restriction of rotation around S–S bonds gives to the proteins that contain them an added rigidity in their macrostructure. Disulfide bonds are mainly found in extracellular proteins.

The disulfide bonds of proteins can be divided into two types:

a. Intrachain bonds, which unite separate parts of a single polypeptide chain and form loops within it. Such bonds participate in forming and stabilizing the secondary and tertiary structure of proteins. Examples of intrachain S–S bonds include those of pancreatic ribonuclease, egg lysozyme, pepsin, serum albumin, the bond within the A-chain of insulin, etc.
b. Interchain S–S bonds, which, since they unite separate polypeptide chains, take part in forming the quaternary structure of the protein. Examples include the bonds between the A- and B-chains of insulin, those between the light and heavy chains of immunoglobulins, and the bonds in the molecules of enzymes such as chymotrypsin, tryp-

tophenase (Morino and Snell, 1967), and in polymeric forms of thetin-homocysteine methyltransferase (Klee and Cantoni, 1960) and histidine-ammonia lyase (Klee, 1970).

Complete splitting of S—S bonds usually leads to destruction of the unique three-dimensional structure of a protein macromolecule and to loss of its biological activity. Inactivation on scission of S—S bonds has been found with many enzymes (ribonuclease, desoxyribonuclease, lysozyme, pepsin, trypsin, chymotrypsin, etc.), and also of the protein inhibitors of trypsin, of cobra toxin, and of peptide hormones. The changes in secondary and tertiary structure of proteins that follow scission of their S—S bonds can be followed by measuring optical rotatory dispersion, circular dichroism, hydrogen—deuterium exchange, and the hydrodynamic properties of the molecules. Harrington and Sela (1959) found that scission of the four S—S bonds of ribonuclease by oxidation or reduction led to destruction of the secondary structure of this protein and its conversion to a random conformation, as indicated by changes in its specific rotation $[\alpha]_D$ from −74° to −91° and in its parameter λ_c (in the Drude equation) from 233 to 223 nm. Data from polarization of fluorescence and circular dichroism also showed that reduction of the four S—S bonds in 8 M urea was accompanied by marked destruction of secondary and tertiary structure of the protein; nevertheless fully reduced ribonuclease retains residual noncovalent structure which makes a contribution to the optical activity and only disappears in the presence of 6 M guanidine hydrochloride (Fig. 29) (Young and Potts, 1963; Tamburro et al., 1970).

It should be noted that there are two rather different types of method for studying conformational changes in proteins. Methods such as optical rotatory dispersion and viscosity measure properties of the molecule as a whole. Effects from many parts of the molecule are summed in the parameter measured. Quite large conformational changes could occur without being observed if changes in the contributions from some parts of the molecule cancelled changes in the contributions from others. The other kind of method observes the influence of the rest of the molecule on a particular probe, e.g., in the study of the circular dichroism of a prosthetic group, or in a study of the NMR signal of a proton distinguishable from the rest of the protons of the molecule. More precise information is available from this type of method, but it refers only to one particular local environment and the method therefore has the disadvantage that quite large changes in the conformation of distant parts of the molecule may not be detected. Even such measurements of the local environment of a group may be subject to cancellation effects, e.g., if a negative charge approaches the probe and

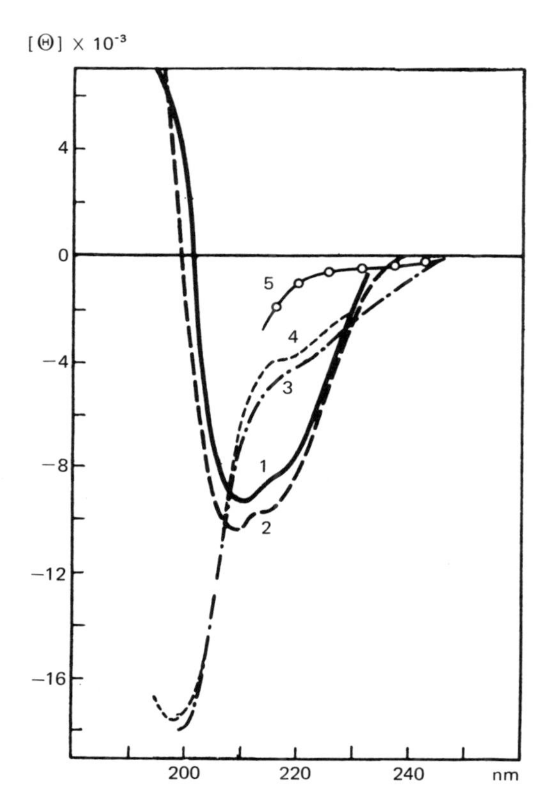

Fig. 29. Influence of the reduction of the S—S bonds on the circular dichroism of ribonuclease (Tamburro et al., 1970). 1) Ribonuclease in 0.1 M acetic acid; 2) as 1, after reduction of two S—S bonds by phosphorothioate; 3) as 1, after reduction of four S—S bonds by mercaptoethanol in 8 M urea and gel filtration; 4) as 3, after alkylation of the SH groups with iodoacetate; 5) completely reduced ribonuclease in 6 M guanidine hydrochloride.

another recedes, the net effect on the parameter examined may be zero. Most methods lie between the two extremes, because they entail summation of effects on a number of probes. Thus polarization of fluorescence examines the environment of the groups to which the fluorescence is due, and it will usually be difficult to limit these to a single group at a unique position in the protein. The value of introducing a "reporter group" at a specific site in a molecule becomes clear, as does the importance of using complementary methods.

Several studies have shown that reduction of the S–S bonds of serum albumin is accompanied by significant changes in its conformation, which are reflected in an increased viscosity and changes in the parameters of optical rotatory dispersion; in particular the parameter b_0 of the Moffit equation is changed from −330 to −90* (Harrap and Woods, 1965; Markus and Karush, 1957; Jirgensons and Ikenaka, 1959; Urnes and Doty, 1961).

Hen egg lysozyme contains four intrachain S–S bonds. The reduction of these and conversion of the SH groups formed into mixed disulfides by reaction with cystine leads to loss of enzymic activity, as well as to an unfolding of the peptide chain as indicated by optical rotatory dispersion and ultraviolet absorption spectrum (Bradshaw et al., 1967).

Confirmation of the important role of disulfide groups in forming the protein macrostructure can be found in the fact that the half-cystine residues involved in S–S links are the least subject to evolutionary changes. Examples of how conservative half-cystine residues are in the evolution of related proteins are the homologies in the positions of the S–S bridges and in the amino acid sequences adjacent to them found in the structures of the serine proteases trypsin, chymotrypsins A and B, elastase, and thrombin (Brown et al., 1967; Smillie and Hartley, 1967; Hartley, 1970). The disulfide bridges are homologous also in two other related proteases: pig pepsin and calf rennin (Foltmann and Hartley, 1967).

Proteins that contain intrachain S–S bond are, in general, more resistant to various denaturing influences than those that do

* In studying the influence of scission of S–S bonds on the optical rotatory dispersion of proteins two kinds of effect should be distinguished: firstly, the effect of conformational change, and, secondly, the effect of the disappearance of the asymmetrical disulfide group, which contributes to the rotatory dispersion (see Chapter 2, Section 1).

not. Naturally various noncovalent interactions, such as hydrophobic and hydrogen bonds, make a large contribution to the stabilization of the macromolecular structure of proteins too, and many proteins with relatively stable macrostructure contain no S–S bonds (e.g., muscle aldolase, pig heart aspartate aminotransferase).

Not all the S–S bonds in a particular protein are equally important for the maintenance of structure and function. In some proteins a number of the S–S bonds can be selectively broken with retention of biological activity (see Table 15).

According to Nagy and Straub (1969) the maximal number of S–S bonds that can be electrolytically reduced with retention of at least 90% of the biological activity is 0 in lysozyme, 2 in ribonuclease, 2 in chymotrypsin, 3 in trypsin, and 2 in soybean trypsin inhibitor. Partially reduced proteins that retain biological activity apparently have a lowered stability; thus in the case of re-

TABLE 15. Selective Scission of the S–S Bonds of Proteins with Complete or Partial Retention of Biological Activity

Protein	Total number of S–S bonds	Number of bonds split with retention of activity	Splitting agent	Reference
Ribonuclease	4	2	Monothiophosphate	Neumann et al., 1967
Ribonuclease	4	~2-3	UV irradiation	Risi et al., 1967
Deoxyribonuclease	2	1	β-Mercaptoethanol	Price et al., 1969
α-Chymotrypsin	5	2	Monothiophosphate	Berezin et al., 1972
Trypsinogen	6	2	Dithioerythritol	Sondack and Light, 1971
β-Trypsin	6	1	$NaBH_4$	Liu et al., 1971
Basic pancreatic trypsin inhibitor	3	1	Dithiothreitol or $NaBH_4$	Liu and Meienhofer, 1968; Kress et al., 1968; Liu et al., 1971
Soya bean trypsin inhibitor	2	1	$NaBH_4$	DiBella and Liener, 1969
Papain	3	1	β-Mercaptoehtanol in 8 M urea	Shapira and Arnon, 1969; Arnon and Shapira, 1969
Lysozyme	4	~2	Sulfite	Azari, 1966
Pituitary growth hormone	2	2	Dithiothreitol	Bewley et al., 1969

duced ribonuclease and growth hormone a lowered resistance to tryptic attack was noted (Neumann et al., 1967; Bewley et al., 1969).

Sperling et al. (1969) found that the S–S bond between the fourth and fifth half-cystine residues (numbering from the N-terminus after Spackman et al., 1960) of pancreatic ribonuclease is reduced by dithiothreitol or dithioerythritol much faster than the other three bonds, and this allows its selective scission. The partially reduced ribonuclease reacts with one or two equivalents of mercuric ions to form derivatives of the types –S–Hg–S– and –S–Hg^{+} respectively. The mono- and dimercury derivatives, and also the carboxymethyl derivative, obtained after scission of this S–S bond do not differ from the native ribonuclease in catalytic or physicochemical properties, or in resistance to tryptic digestion. Hence this bond is evidently dispensable in maintaining the macromolecular structure of this protein and its catalytic action. The ease of splitting of this bond is explained by the fact that, as may be seen from the three-dimensional structure of ribonuclease, it is to a high degree in contact with solvent.

In contrast with dithiothreitol, monothiophosphate readily reduces two S–S bonds in native ribonuclease, that between the 4th and 5th residues and that between the 3rd and 8th (Neumann et al., 1967). The derivative so obtained is similar in a number of ways to native ribonuclease, e.g., it has equal activity with respect to RNA, but it differs in that it is more sensitive to trypsin and is more active with respect to cytidine-2',3'-cyclic phosphate. Thus the breakage of two disulfide bonds is accompanied by some conformational change, which somewhat affects the active site. The splitting of all four S–S bonds with dithiothreitol or monothiophosphate (in 8 M urea) leads to complete inactivation of ribonuclease. But the derivative in which all four S–S bonds are replaced with –S–Hg–S– retains 5% and 25% of enzymic activity with respect to RNA and to cytidine-2',3'-cyclic phosphate respectively (Sperling and Steinberg, 1971).

It is interesting that many proteins that contain S–S groups do not contain SH groups and vice versa; the number of proteins that contain both types of group is comparatively small and includes serum albumin, papain, ficin, ovalbumin, and β-lactoglobulin. One reason for this may be that the coexistence of SH and S–S groups in one macromolecule could create potential instability, since thiol–disulfide exchange could occur between the groups. Such an intramolecular reaction is prevented in the proteins that do contain both SH and S–S groups by their spatial separation or by their special chemical environments. Nevertheless a few proteins are known in which such a reaction nevertheless appears to take place under mild conditions (in aqueous solution of low ionic strength and weakly alkaline pH). One such protein is serum albumin, whose molecule consists of a single polypeptide chain (of molecular weight 66,000) containing 17 S–S bonds and one SH group (which is partly in the form of a mixed disulfide, see Chapter 1, Section 10). Foster and colleagues (Sogami et al., 1969; Wong and Foster, 1969; Nikkel and Foster, 1971) obtained data indicating that the microheterogeneity of ox serum albumin was due, at least in part, to an intramolecular thiol–disulfide exchange which led to altered pairing of the half-cystine residues. Blocking this reaction with iodoacetamide or N-ethylmaleimide prevented the development of microheterogeneity when albumin solutions were stored. The reaction mechanism for the "aging" of ox mercaptalbumin, as proposed by Nikkel and Foster (1971), is shown in Fig. 30. They believe that at pH 7.5–9.0 the native mercaptalbumin

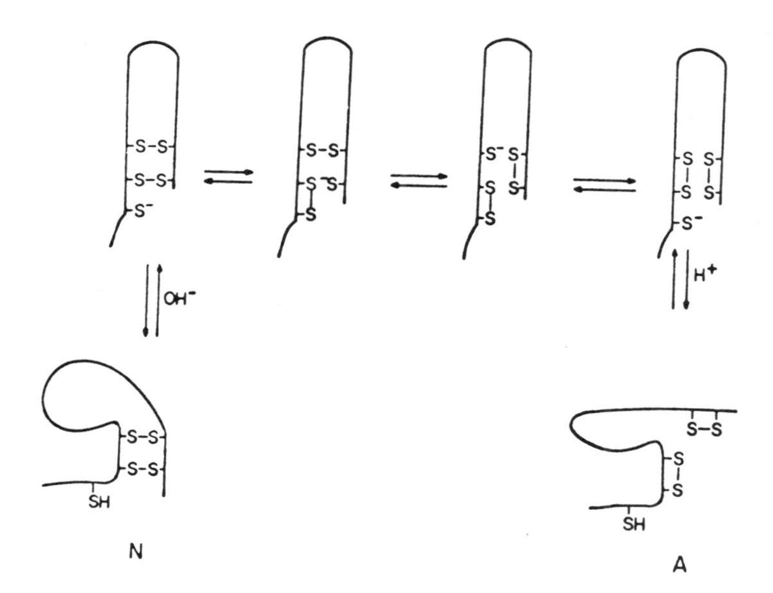

Fig. 30

molecule (the N form in the figure) undergoes a conformational change which brings the SH group into juxtaposition with some S–S bonds. The ionized SH group can then initiate thiol–disulfide exchange. This results in the formation of a new component (the A form in Fig. 30) with modified physical properties. McKenzie and Shaw (1972) suggest that ox β-lactoglobulin exists in two forms in which either residue 119 or 121 bears a free SH group and the other residue forms an S–S bridge with residue 106; these two forms appear to be in equilibrium.

Brocklehurst et al. (1972b) showed that bromelain could exist in a form of low reactivity, which could be transformed into the active form in various ways. This led Brocklehurst and Kierstan (1973) to discover in *Papaya* latex an inactive form of papain, which they called propapain. In it the SH group of residue 25, which is essential in the enzymically active form (see Chapter 6, Section 1a), is combined in disulfide linkage, probably with either residue 22 or 63, the other of which bears a free SH group. Addition of a thiol or of cyanide liberates the SH group of residue 25, but the nucleophile thus added to the former partner of this residue is then expelled by the previously free SH group, thus forming the disulfide bond between residues 22 and 63. Activation therefore requires a thiol or cyanide, but produces no increase in the SH content of the protein.

Another reason why proteins do not usually contain both SH and S–S groups may be that the two forms are usually found in different environments. Extracellular proteins tend to contain disulfide groups and intracellular proteins sulfhydryl groups. There are clearly exceptions to this, some of them linked with special functions, such as the disulfide groups of flavoproteins (Chapter 6, Section 1b) and the SH groups of plant proteases (Chapter 6, Section 1a).

The S–S bond is not the only cysteine-based crosslink found in peptides. The antibiotics nisin and subtilin contain sulfide (thioether) crosslinks, in the form of the amino acids lanthionine (Ala–S–Ala) and β-methyllanthionine (Abu–S–Ala). The structure of nisin is (Gross and Morell, 1971):

```
                ┌──────── S ────────┐      ┌──── S ────┐
Ile—Thr*—Ala—Ile—Ser*—Leu—Ala—Abu—Pro—Gly—Ala—
 1                                              11
```

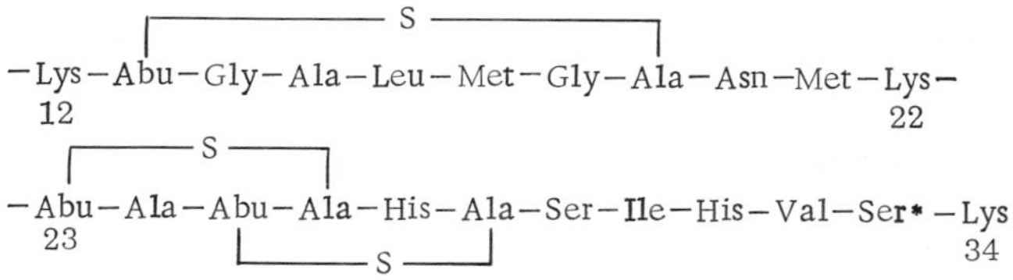

```
     ┌──────────── S ────────────┐
–Lys–Abu–Gly–Ala–Leu–Met–Gly–Ala–Asn–Met–Lys–
 12                                      22
  ┌──── S ────┐
–Abu–Ala–Abu–Ala–His–Ala–Ser–Ile–His–Val–Ser*–Lys
 23        └──── S ────┘                      34
```

where Ser* and Thr* represent residues of the dehydration products of serine and threonine, i.e., 2-amino- acrylic and crotonic acids. Subtilin has a very similar structure (Gross et al., 1973) with crosslinks identical in nature and position. It seems likely that the crosslinks are formed by addition of cysteine to Ser* and Thr*, as in the alkaline formation of lanthionine (Chapter 2, Section 6). This is consistent with the observations (Ingram, 1969) that cysteine and serine are biological precursors of lanthionine in nisin, and that cysteine and threonine are precursors of β-methyllanthionine. It is also consistent with the D-configuration in the Abu moieties of methyllanthionine and in one Ala moiety of lanthionine (Gross et al., 1973). The thioether links and D-configurations may enable these antibiotics to resist biological degradation.

2. The Formation of S–S Groups in the Renaturation and Biosynthesis of Proteins

Anfinsen and colleagues (Anfinsen and Haber, 1961; Anfinsen et al., 1961) first discovered that pancreatic ribonuclease, denatured with reduction of its S–S bonds by β-mercaptoethanol in 8 M urea, underwent spontaneous reactivation when a dilute solution from which the reagents had been removed stood in air at pH 8.2 and 24°C; the SH groups were simultaneously reoxidized to S–S groups. The reactivated ribonuclease did not differ from the original native enzyme in catalytic, immunochemical, or physical properties, or in the position of its S–S bridges (White, 1961; Bello, et al., 1961). Although Pflumm and Beychok (1969) found a difference in the circular dichroism spectrum at 240 nm between native ribonuclease A and the product obtained by reoxidizing reduced ribonuclease by Anfinsen's method, this difference was not found when the material was reoxidized in the presence of 10 μM β-mercaptoethanol. When the reduced ribonuclease was reoxidized in the presence of urea an inactive form of the enzyme with wrongly paired half-cystine residues was formed; after removal of the urea this inactive form transformed itself spontaneously into the active one in the presence of low concentrations of β-mercaptoethanol (Haber and Anfinsen, 1962).

On the basis of these experiments the so-called "thermodynamic" hypothesis was formulated, that the secondary and tertiary structures of proteins and the position of the S–S bonds are completely determined by the primary structure, i.e., correspond with the most stable, energetically most favored conformation of the polypeptide chain, and that to achieve this conformation no extra genetic information is required beyond that expressed in the sequential arrangement of the amino acid residues (Anfinsen, 1962; Epstein et al., 1963). It should be noted, however, that some proteins can achieve the correct refolding during the reoxidation of reduced polypeptide chains only in the presence of specific cofactors. Thus the complete renaturation of desoxyribonuclease and of taka-amylase A requires the addition of Ca^{2+} ions (Price et al., 1969; Takagi and Isemura, 1965; Friedmann and Epstein, 1967). Serum albumin can apparently be renatured successfully only in the presence of fatty acids (Andersson, 1969).

Experiments on the reoxidation and spontaneous renaturation of reduced proteins have been successful with egg lysozyme (Epstein and Goldberger, 1963; Goldberger and Epstein, 1963; Imai et al., 1963; Yutani et al., 1968), taka-amylase A (Isemura et al., 1963; Yutani et al., 1969), *E. coli* alkaline phosphatase (Levinthal et al., 1962), pepsinogen (Frattali et al., 1963; Nakagawa and Perlman, 1970), soybean trypsin inhibitor (Steiner et al., 1965), cobra toxin (Yang, 1967), proinsulin (Steiner and Clark, 1968), and polyalanyl-immunoglobulin (Freedman and Sela, 1966). In the case of chymotrypsinogen A similar experiments were successful only after immobilization of the zymogen by covalent attachment to a solid support (Brown et al., 1972).

An interesting question is to what extent the protein renaturation just described, in which an unfolded polypeptide chain reproducibly regains its complicated three-dimensional structure with striking precision, can be regarded as an adequate model for the folding of polypeptide chains that are nascent in biosynthesis *in vivo*. To answer this question one should compare the rates of of the two processes. Under optimal conditions (at protein concentrations of 10–20 μg/ml, pH 8, and 24°C) complete reactivation of reduced ribonuclease is achieved in 1–3 h (Epstein et al., 1962; White, 1967); under more physiological conditions (pH 7.4 and 37°C) the rate of spontaneous reactivation of ribonuclease is still lower and only 10% of the activity is regenerated in an hour. The half-time for reactivation of reduced lysozyme in aerobic oxidation was about 20 min at pH 8.4 and 37°C in the presence of 2.6 mM β-mercaptoethanol (Epstein and Goldberger, 1963). These rates are much smaller than the rates of synthesis of protein molecules *in vivo* which vary from several seconds to a few minutes (see, for example, Wilhelm and Haselkorn, 1970).

One of the reasons for the comparatively low rate of the renaturation of these proteins is apparently the "mismatching" of half-cystine residues. Kauzmann (1959) pointed out that in pairing the eight SH groups formed by the reduction of the four S–S bonds of ribonuclease, 105 isomers can be formed. This is because there are seven possible partners for any single residue to form the first S–S bond, and when it has been formed the next residue has five possible partners, and so on; $7 \times 5 \times 3 \times 1 = 105$. The mis-

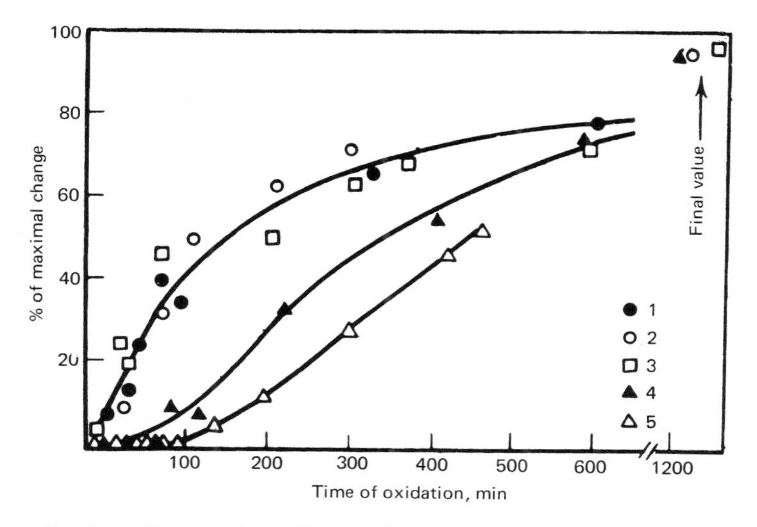

Fig. 31. Kinetics of reoxidation of reduced pancreatic ribonuclease (Anfinsen et al., 1961). 1 and 2) Disappearance of SH groups, determined by titration with p-mercuribenzoate (1) or by reaction with iodo[^{14}C]acetate (2); 3) change in optical rotation; 4 and 5) changes in enzymic activity, measured with ribonucleic acid (4) or uridine-2'3'-cyclicphosphate (5).

matched bridges are then converted by thiol–disulfide exchange reactions into the correct ones, corresponding to the most stable conformation of the protein. Even in the first experiments on the reoxidation of reduced ribonuclease it was noticed that the disappearance of titratable SH groups is faster than the appearance of enzymic activity (Fig. 31). It was natural to imagine that there may be factors in the cell that regulate and accelerate the folding of polypeptide chains in the final stage of protein biosynthesis. In fact Goldberger et al. (1963, 1964), and also Venetianer and Straub (1963a,b) found in pancreatic and liver microsomes an "activating" enzyme, which greatly increased the rate of reactivation of the reduced forms of ribonuclease, egg lysozyme, pepsinogen, and soybean trypsin inhibitor. It was shown that this enzyme does not catalyze the direct oxidation of SH groups, but accelerates the thiol–disulfide exchange reaction. This conclusion was reached, in particular, from experiments with ribonuclease that had been reoxidized in 8 M urea and consequently contained mismatched S–S bonds. In the absence of the activating enzyme the conversion of this wrongly oxidized ribonuclease into the active form took about 24 h; in the presence of the microsomal enzyme and of 1 mM β-mercaptoethanol this conversion took place in a few minutes (Givol et al., 1964, 1965). Later the activating enzyme was isolated in a pure form from ox liver microsomes and studied in detail (De Lorenzo et al., 1966a,b). It proved to contain three half-cystine residues, of which one was essential for activity. The enzyme was active only in the reduced form; it could be activated either by prior incubation with β-mercaptoethanol, or by including this substance in the incubation medium with the enzyme and protein substrate (Fuchs et al., 1967). The localization of the enzyme from many tissues in the microsomes and its broad substrate specificity support the idea that its role *in vivo* is to catalyze thiol–disulfide exchange in the process by which newly synthesized polypeptide chains acquire their unique conformation which is stabilized by S–S bridges. In other words the enzyme rapidly "corrects" the "mistakes" made in pairing half-cystine residues. It may act as the proteins pass through the endoplasmic reticulum on their way out of the cell (Sunshine et al., 1971).

Bradshaw et al. (1967) studied the kinetics of reactivation of the mixed-disulfide derivative of egg lysozyme with cystine. This derivative is catalytically inactive; on adding β-mercaptoethanol or cysteine it rapidly reactivates (the half-time of reactivation is about 6 min) through a series of thiol–disulfide exchange reactions. The initial rate of reactivation is about three times greater than the rate of reactivation of reduced lysozyme in air.

Saxena and Wetlaufer (1970) found that the regeneration of enzymic activity on reoxidation of reduced egg lysozyme is accelerated in the presence of a mixture of oxidized and reduced glutathione of ratio 1:10 and concentration of reduced glutathione of 5 mM (half-time of reactivation about 5 min). Other thiol–disulfide pairs (cystine–cysteine and cystamine–cysteamine) were as effective as glutathione. Saxena and Wetlaufer found that the rate of regeneration of activity of lysozyme in their system did not depend on the concentration of mercaptide ions or of the glutathione. Hence the rate-determining step in the process of renaturation is apparently not the thiol–disulfide exchange itself but the accompanying conformational changes in the protein.

Successful renaturation was at first obtained only with proteins that contained intrachain S–S bonds. There were considerable difficulties in attempting to conduct such experiments with proteins that contained interchain S–S bonds (e.g., insulin) (Dixon and Wardlaw, 1960; Meienhofer et al., 1963). Considering the relatively low yields of insulin (1-2%) obtained on oxidizing a mixture of its component (reduced) chains A and B, Givol et al. (1964, 1965) suggested that insulin is synthesized *in vivo* in the form of a single-chain protein precursor, which, after formation of its disulfide bridges, undergoes limited proteolysis like that which occurs in the conversion of the single-chain chymotrypsinogen into the triple-chain chymotrypsin. This suggestion has been fully confirmed. Several authors have shown that insulin is biosynthesized through the stage of proinsulin, which consists of a single, long, polypeptide chain. The N-terminal sequence of proinsulin consists of the B-chain of insulin, and the C-terminal sequence of proinsulin consists of the A-chain; the two sequences are joined by a peptide which is split out by proteases to form the active hormone (Fig. 32) (Steiner et al., 1967; Steiner and Oyer, 1967; Steiner and Clark, 1968; Clark and Steiner, 1969; Schmidt and Arens, 1968; Chance and Ellis, 1969; Tung and Yip, 1969; Trakatellis and Schwartz, 1970; Grant and Coombs, 1970; Nolan et al., 1971). Unlike insulin, proinsulin can be reactivated in good yield after reduction by reoxidation (Steiner and Clark, 1968). It is interesting

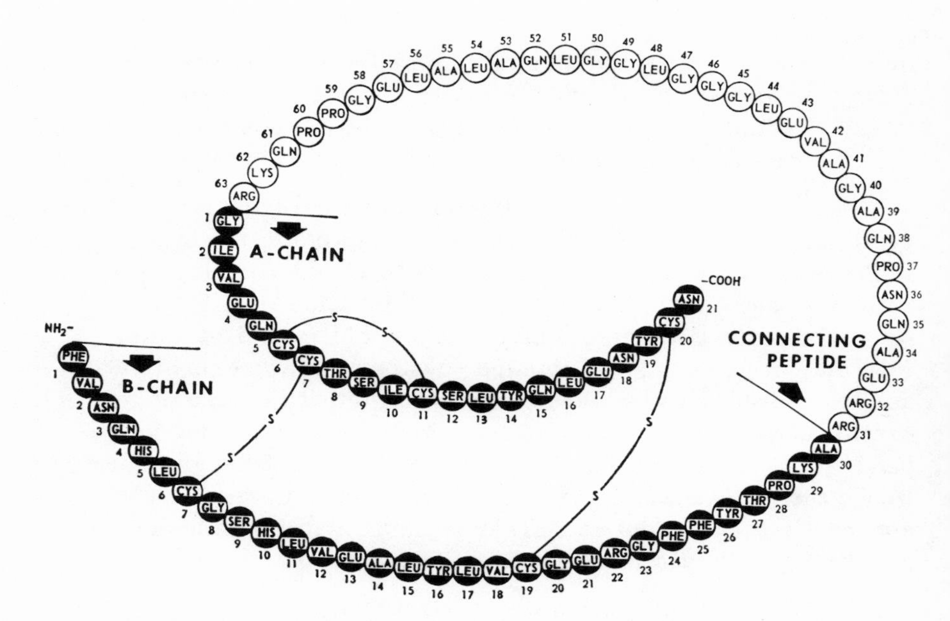

Fig. 32. Primary structure of pig proinsulin (adapted from Chance et al., 1968, by Grant and Cooms, 1970).

that species differences in the structure of the connecting peptide are much greater than in insulin itself, so that several different sequences of approximately the same length allow correct refolding. Brandenburg and Wollmer (1973) have described successful reoxidation by air (with a yield of up to 75%) of reduced adipoyl insulin, the insulin derivative in which the α-amino group of Gly-1 of chain A is combined with the ε-amino group of Lys-29 of chain B through a residue of adipic acid. From this it follows that the role of the connecting peptide with regard to the correct pairing of half-cystines can be played by a short, nonbiological crosslink.

Thus the formation of insulin *in vivo* is not complicated by the need to combine reduced A- and B-chains (contrary to the findings of Humbel, 1965). Special measures have to be taken to obtain a good yield for such a combination *in vitro* . Zahn et al. (1965, 1966) raised the yield of active insulin on recombination of the chains to 20–30% by prior partial reoxidation of the isolated A-chain; there was an optimal yield of insulin if the B-chain was added and the oxidation completed at a certain time after oxidation of the A-chain began, presumably along enough for its one intrachain S—S bond to be formed. Later Zahn et al. (1968) found that the best results were obtained on reaction of the S-sulfo form of B-chain [$B(SSO_3H)_2$] with the partially reduced A-chain [$A(SS)SH_2$] at pH 11.5. Jiang et al. (1963; Du et al., 1965) also worked out an improved procedure for combining A- and B-chains; they recommended oxidizing a mixture of reduced A- and B-chains in a molar ratio of 1.5:1 in air at pH 10.6 and 3-5°C; the yield of active insulin was about 50% in their experiments, so that the presence of excess of A-chain improved the regeneration of activity. Later Katsoyannis and Tometsko (1966; Katsoyannis, 1967) worked out a method for recombining chains A and B by direct reaction of excess of the SH form of chain A with the S-sulfo form of chain B, as shown:

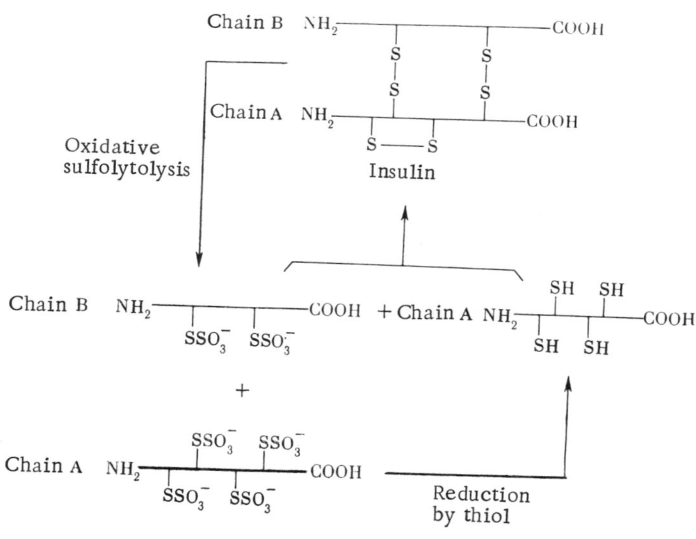

This method gives a yield of active insulin of 60-80%. The successes in the resynthesis of insulin suggest that the conformations of chains A and B are favorable for formation of the "correct" S—S bridges.

3. S—S Groups in Immunoglobulins

S—S groups play an important part in maintaining the macromolecular structure of the immunoglobulins. Edelman (1959; Edelman and Poulik, 1961) first observed that treatment of human γ-globulin G (IgG) with β-mercaptoethanol in 6 M urea led to a splitting of its molecule into subunits. Similar results were obtained by Franěk (1961), who observed dissociation of human and animal IgG into subunits after scission of their S—S bonds with sulfite in the presence of 8 M urea. These experiments served as the start of a series of investigations in which it was found that the molecule of IgG consists of four polypeptide chains: two identical light chains of molecular weight 20,000-25,000 and two identical heavy chains of molecular weight 50,000-55,000 (Porter, 1962; Fleischman et al., 1962, 1963; Edelman and Gally, 1964; Marler et al., 1964). According to Porter's scheme, each light chain is combined with a heavy chain by a single S—S bond; the heavy chains are in turn bound together by S—S bonds. Porter's scheme was put forward first for rabbit IgG, but it proved correct in its main features also for various other animal species and man and has been the basis of all subsequent investigations into antibody structure (Fig. 33). In accord with the scheme a single S—S bond has been found joining each light chain (at or very close to its C-terminus) to one heavy chain (i.e., two of such bonds per molecule) in all IgG molecules studied. Nevertheless the number and positions of the S—S bonds that join the two heavy chains together vary not only between immunoglobulins G from different species, but between different subtypes of protein from a single species. Thus the number of S—S bonds between the heavy chains of human IgG varies from two to five (Frangione et al., 1969). Edelman et

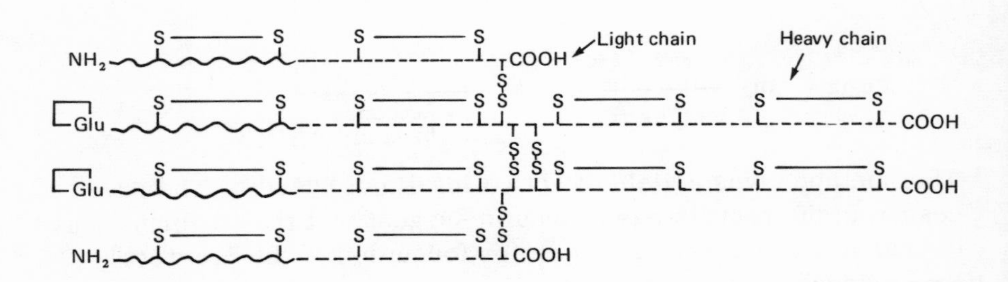

Fig. 33. Scheme of immunoglobulin G structure (modified from Porter, 1962, and Putnam, 1969; see also Edelman et al., 1969; Milstein and Pink, 1970). The variable regions are marked with wavy lines and the constant regions with straight lines. Glu- represents pyrrolidone carboxylic acid, but some IgG heavy chains have a different terminal residue. The intrachain disulfide links are highly conserved. The interchain link between the heavy and light chains always arises at or near the C-terminus of the light chains, but is variable in its point of attachment to the heavy chain. The heavy—heavy interchain links are variable in number and position.

al. (1969) established the complete primary structure and the positions of the S–S bonds in the molecule of a human IgG, so it may be compared with the many partial structures now available.

Besides the interchain bonds, IgG molecules contain several intrachain ones. Thus rabbit IgG contains 3 interchain S–S bonds (two heavy-light and one heavy-heavy) and 14 or 16 intrachain bonds (Rejnek et al., 1969; O'Donnell et al., 1970). The intrachain bonds are often much harder to reduce than interchains ones; complete reduction of all the intrachain bonds can be achieved only in comparatively harsh conditions, such as treating the globulin with 0.54 M β-mercaptoethanol in 12 M urea at 40°C (Utsumi and Karush, 1964; Markus et al., 1962). The intrachain S–S bonds are apparently the most conservative, least variable part of the structure of IgG. Each light chain consists of two regions of about 110 residues each, and the heavy gamma chains consist of four such regions. Each of these regions, whether in a light or heavy chain, contains a disulfide bridge which forms a loop of about 60 residues, from about the 30th residue to about the 90th of the region.

Of these regions of about 110 residues, the N-terminal one of both heavy and light chains is "variable"; immense variation is shown in sequence between different molecules, and it probably reflects the immense diversity in the specificities of the antibodies of a single animal, since the antigen-binding site is made up of this region of a heavy and a light chain together. The constancy of the S–S loop of this otherwise variable part of the chains is therefore particularly notable. The rest of the molecule, i.e., one region of about 110 residues per light chain and three such regions per heavy chain, makes up the "constant" region. In fact it shows some diversity within a species as several subtypes of IgG occur, and there is allelic variation as well. But the four subtypes of human IgG heavy chains, in spite of differences of sequence, have the same number of intrachain disulfide bonds and they are in almost identical positions (Frangione et al., 1969). Nevertheless there is in rabbit IgG an extra disulfide bridge in the second region (from the N-terminus) of the heavy chains (O'Donnell et al., 1970).

The function of the highly conserved loops of about sixty residues is not understood. The repeating nature of the loops, like that of the sequences of the regions, suggests that they have arisen by gene duplications, but this does not explain their conservation. The need for the interchain links is more obvious, since heavy–light combination is needed to form the complete antigen-binding site, and heavy–heavy combination makes the antibody bivalent; this in turn assists precipitation of multivalent antigens.

Human and animal blood contains not only immunoglobulins G, but also immunoglobulins A, M, D, and E. IgM (macroglobulin) proteins have molecular weights of about 900,000 and consist of five subunits linked by S–S bridges; the subunits are apparently similar to IgG in that each consists of two light and two heavy chains (Miller and Metzger, 1965; Morris and Inman, 1968; Beale and Feinstein, 1969).

4. S – S Groups in Peptide Hormones

Peptide hormones that contain S–S bonds include insulin, which contains two interchain and one intrachain bond (Fig. 32), and also the hormones of the neurohypophysis, vasopressin and oxytocin:

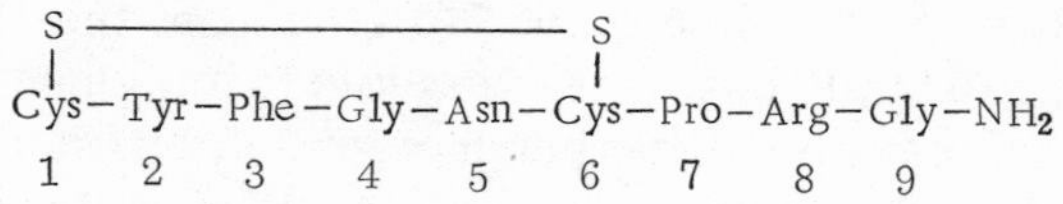

```
              S ———————————————— S
              |                  |
Vasopressin:  Cys–Tyr–Phe–Gly–Asn–Cys–Pro–Arg–Gly–NH2
              1   2   3   4   5   6   7   8   9

              S ———————————————— S
              |                  |
Oxytocin:     Cys–Tyr–Ile–Gly–Asn–Cys–Pro–Leu–Gly–NH2
              1   2   3   4   5   6   7   8   9
```

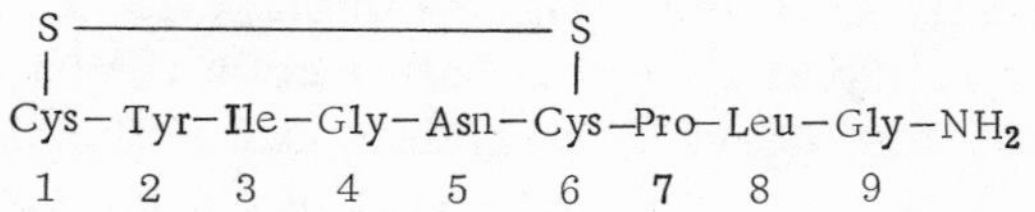

The molecules of all three of these hormones contain hexapeptide disulfide rings. Opening of the rings by reduction of the S–S bonds with subsequent alkylation leads to loss of biological activity (see review by Walter et al., 1967). Acyclic analogs of vasopressin in which the half-cystine residues are replaced by alanine residues are also completely devoid of activity (Huguenin and Guttmann, 1965). With oxytocein (the product of reduction of oxytocin) a weak prolonged hypotensive action was observed, but this could be explained by a slow conversion into the cyclic disulfide *in vivo* (Yamashiro et al., 1966).

Fong, Schwartz, and co-workers (Fong et al., 1959, 1960, 1962; Schwartz et al., 1960), studying the action of vasopressin on the permeability of rat kidney tubules and toad bladder, found that (1) the action of vasopressin is inhibited by substances that block tissue SH groups, and (2) tritium-labeled hormone was bound by the tissues and liberated when the tissues were treated with thiols (cysteine, β-mercaptoethanol, etc.). On the basis of these findings Schwartz and colleagues put forward the hypothesis that vasopressin and oxytocin bind to tissue receptor proteins by a thiol–disulfide exchange reaction between an SH group of the receptor and the S–S group of the hormone, which forms a mixed disulfide. To test this hypothesis analogs of the hormones were synthesized without the S–S bridge but with the cyclic structure of the natural hormones, such as that in which the –S–S– grouping of oxytocin was replaced by $-CH_2-S-$, and that in which the –S–S– grouping of lysine vasopressin (the natural hormone of the pig) was replaced with $-S-CH_2-S-$ (Rudinger and Jǒst, 1964; Schwartz et al., 1964; Jǒst and Rudinger, 1967; Sakakibara and Hase, 1968). It proved that these analogs possessed biological activity, although much less than the natural hormones (analogs of oxytocin in which the S–S bond is replaced with –S–Se– or –Se–Se– have high biological activity according to Walter and du Vigneand, 1965, 1966, and Walter and Chan, 1967). Hence formation of a mixed disulfide with a receptor cannot be a necessary condition for the action of vasopressin and oxytocin. The destruction of the binding properties of tissues on blocking of their SH groups could be due to changed conformation of the receptor proteins.

There has been discussion in several papers on the role of the S–S bond between residues 6 and 11 of the A chain of insulin in the mechanism of action of this hormone. Cadenas et al. (1961) found that prior perfusion of isolated rat heart with 1 mM N-ethylmaleimide greatly diminished the binding of ^{131}I-labeled insulin by the tissue and its action in stimulating transport of sugars into the cells. N-Ethylmaleimide did not exhibit a similar action in experiments in which the insulin was administered to rats by intravenous injection before the hearts were removed. Carlin and Hechter (1962) studied the influence of insulin and N-ethylmaleimide on the transport of xylose and α-aminoisobutyric acid and on glycogen synthesis in isolated rat diaphragm. Contrary to Cadenas et al., they found that incubation of the diaphragm with insulin before addition of N-ethylmaleimide not only did not prevent development of the inhibitory influence of the N-ethylmaleimide on the action of insulin, but in many cases increased this influence. This fact suggests that insulin and N-ethylmaleimide interact with different functional groups of the tissue proteins. Carlin and Hechter suggested that N-ethylmaleimide does not affect the primary interaction of insulin with its receptor but can block reactions that follow the binding of the hormone. Clauser et al. (1965) and Vasil'eva and Il'in (1970) have also noted the inhibitory action of N-ethylmaleimide on the stimulatory effect of insulin on sugar uptake by cells, but the mechanism of this effect remains unsolved. Vasil'eva and Il'in (1970) found that N-ethylmaleimide increases(!) the stimulatory action of insulin on glycine uptake by heart tissue (in contrast with its effect in the case of sugars) but completely abolishes the influence of the hormone on glycine incorporation into proteins.

In order to elucidate the role of the intrachain S–S bond of insulin, Jŏst et al. (1968) synthesized an analog of the hormone in which the cystine residue of positions 6 and 11 of chain A was replaced by cystathionine; they found that this analog possessed hypoglycemic action, although less than that of the natural hormone. Katsoyannis et al. (1973) synthesized the analog of insulin in which the half-cystine residues in positions 6 and 11 of the A-chain were replaced by alanine. This analog possessed about 10% of the activity of the natural hormone. Thus the intrachain S–S bond does not take direct part in the action of insulin; the function of this bond in insulin as in vasopressin and oxytocin is apparently only in the maintenance of the biologically active conformation of the hormone molecule.

In contrast with the S–S bonds of insulin, vasopressin, and oxytocin, those of growth hormone can be reduced, and the SH groups thereby formed alkylated, with retention of biological activity. This retention can be explained by the fact that scission of the S–S bonds does not lead to appreciable changes in the conformation of the hormone molecule, as judged by physicochemical findings (circular dichroism, viscosity, etc.) (Dixon and Li, 1966; Bewley et al., 1969).

Another hormone is the hypothalamic peptide that inhibits pituitary secretion of growth hormone. It also contains an S–S bridge, between residues 3 and 14 of its sequence of 14 (Brazeau et al., 1973). The fungal product malformin A is somewhat to a hormone in its action of causing distorted growth of higher plants. It is a cyclic pentapeptide, with a disulfide bond between two half-cystine residues of D-configuration; they are separated on one side of the ring by Val and on the other by Leu–Ile (Anzai and Curtis, 1965).

5. S – S Groups in Fibrous Proteins

S–S groups play an important part in the structure of certain fibrous proteins, namely; keratins, silk fibroin, and invertebrate collagens. McBride and Harrington (1965, 1967) found that treatment with β-mercaptoehtanol of the collagen from the cuticle of the earthworm *Ascaris lumbricoides* leads to a reduction in its molecular weight from 900,000 to 62,000 and to the appearance of titratable SH groups, which are absent in the native collagen; they concluded that the molecule of this protein consists of subunits joined by S–S bridges. Disulfide bonds have also been found in collagens from the muscular layer of *Ascaris lumbricoides* and in some other invertebrate collagens (Blanquet and Lenhoff, 1966; Pikkarainen et al., 1968; Fujimoto et al., 1969). In this respect invertebrate collagens differ from vertebrate collagens, which normally contain no S–S bonds (Harding, 1965; Piez, 1968). An exception to this rule is apparently collagen from dog kidney glomeruli, in which cystine residues were found (Kefalides, 1968).

Contrary to earlier findings, cystine residues have also been found in silk fibroin (Schroeder and Kay, 1955; Earland and Raven, 1961; Gustus, 1964; Lucas, 1966). Tashiro and Otsuki (1970) showed that treatment with dithiothreitol led to a fall in the molecular weight of silk fibroin, isolated from the glands of the silkworm *Bombyx mori* from about 3.7×10^5 to $1.7 \pm 0.3 \times 10^5$; they suggested that the molecule of silk fibroin consists of two subunits joined by a single S–S bond.

Many papers are devoted to the study of the properties and role of the S–S bonds in keratins. To the group of keratins belong the structural proteins that form part of hair, wool, horns, hooves, and nails; the proteins of bird feathers and fish scales are very similar. Keratins are characterized by an unusually high sulfur content (about 3.6%) most of which occurs in the S–S groups of cystine (wool contains only 0.12% of methionine-sulfur) (Cuthbertson and Phillips, 1945; Fletcher et al., 1963; Reis et al., 1967). The chemistry of wool keratin has been best studied, and its development has been stimulated by the interests of the textile industry (see reviews by Zahn, 1951; Alexander and Hudson, 1954; Ward and Lundgren, 1954; Speakman, 1963; Crewther et al., 1965).

Some of the reactions of S–S groups (with alkali, cyanide, sulfite, and some other reagents) were first studied in detail in connection with investigations on wool keratin. Thus the formation of lanthionine on treating proteins with alkali was first discovered in studying keratins (Horn et al., 1941, 1942).

Phillips and co-workers (Elsworth and Phillips, 1938, 1941; Middlebrook and Phillips, 1942; Cuthbertson and Phillips, 1945) found that only 50% of the S–S groups of wool keratin react with bisulfite and thioglycolate at pH 5.0. On the basis of these and other data they put forward a division of the S–S groups of keratin into two main fractions, each of which was divided into two subfractions (A + B) and (C + D) which differed in reactivity to bisulfite, thiols, and alkali (see also Lindley, 1959). It was shown later, however, that the incompleteness of reaction with sulfite was due to the reversible nature of this reaction and that displacement of the equilibrium by blocking the SH groups formed with Hg^{2+} allowed carrying it to completion (Leach, 1959, 1960). Almost complete reduction of the S–S bonds of keratin could also be achieved by using high concentrations of thiols (of the order of 1-4 M) (Gillespie and Springell, 1961; Leach and O'Donnell, 1961; Thompson and O'Donnell, 1961; 1962). Thus the firm division of the S–S groups of kertain into types has hardly been justified (Crewther et al., 1965). Leach et al. (1965) developed an electrolytic method for reduction of the S–S bonds in the presence of 0.07 M β-mercaptoethanol (see Chapter 2, Section 10). With this method it was possible to reduce 100% of the S–S bonds of keratin at pH 9.0 and more than 90% at pH 7.0. Weigman (1968) described a reduction of 85% of the S–S bonds of keratin under mild conditions using dithiothreitol. Benzyl mercaptan and tributylphosphine have also been used successfully to reduce the S–S bonds of keratin (Maclaren, 1962; Maclaren et al., 1968).

Keratins are insoluble proteins; their conversion into a soluble form is possible only after splitting of the S–S bonds by reduction, oxidation, or sulfitolysis. Scission of the S–S bonds alone, however, is not enough to dissolve keratins; the further influence of an alkaline medium or addition of urea is necessary. After treating wool with alkaline thioglycolate and alkylation with iodoacetate two protein fractions (kerateines) were isolated: one with lower sulfur content (1-2%) and one with higher (6-7%) than the wool itself. The proteins of lower sulfur content constituted 50-60% of the weight of the solubilized wool and were derived from the microfibrils. The proteins with higher sulfur content constitute 20-30% of the weight and came from the matrix; half-cystine is up to 20% of their amino acid residues. Both these fractions are inhomogeneous, and have been separated into a number of subfractions (Gillespie, 1959, 1962, 1963; Gillespie and Simmonds, 1960; Springell et al., 1964; Crewther et al., 1965; Thompson and O'Donnell, 1967; O'Donnell and Thompson, 1968; Frater, 1969). The main groups of high-sulfur proteins appear to have molecular weights of 11,000, 17,000, 19,000, and 23,000 (see Swart and Haylett, 1973). The sequences of several of those of weight 11,000 have been found, and they prove to have closely related molecules of 97 or 98 residues with Ac–Ala–Cys–Cys– at the N-terminus, and a total of 16-18 residues of half-cystine (Swart and Haylett, 1971). These largely occur in regions extremely rich in this residue. The group of molecular weight 17,000 is devoid of lysine but has about 10% of arginine residues, together with about 25% of half-cystine residues. The one whose sequence has been found (Swart and Haylett, 1973) has N-terminal threonine (with a free amino group) and a fairly even distribution of residues of different kinds, in marked contrast with the proteins of the 11,000 and 19,000 groups. The sequences of three protein of the 19,000 group have been found (Elleman, 1972; Elleman and Dopheide, 1972). Like those of the 11,000 group they begin with Ac–Ala–Cys–Cys–, but differ thereafter. Within the group they are closely similar, and they also show much internal homology. In one of them, for example, half-cystine comprises 32 out of 151 residues, and a 10-residue sequence is repeated 5 consecutive times with little variation. The regions near the N- and C-termini contain much of the half-cystine of the molecule, and between them there is a region of 19 residues devoid of cystine and proline. The overall sequence suggests a structure of two compact globular units which are separated by a flexible chain and might move fairly freely relative to each other. This agrees with some models of the wool fiber that are based on its mechanical and physical properties.

The application of mild reducing agents to hair breaks some of its disulfide bridges and makes the fibers pliable. This may be used in "permanent" waving, since the hair can then be "set" in the desired position by reforming the bridges with mild oxidizing agents. Such a "cold" process is an alternative to heat treatments.

The ability of wool fibers and human hairs to stretch under a load, especially in a wet state at raised temperatures, is well known (it is used in the textile industry and in waving hair). The final length of the fibers after removal of the load may, according to the conditions, be greater than the original (a "set"), or shorter (a "supercontraction"). Although the chemical mechanism behind these changes of length is not fully clear, some authors consider that the stretching of keratin fibers involves the breaking of S–S bonds and that the setting of the fibers after stretching involves the formation of new S–S– or other covalent cross-linking (Speakman, 1933, 1936, 1963; Zahn et al., 1961; Ziegler, 1964; Crewther et al., 1965). It has been established that the mechanical properties of keratin fibers are changed on scission of the S–S bonds and restored on their regeneration. Burley (1956) made the interesting suggestion that the conformational changes in keratin fibers when they are stretched and shorten can be facilitated by thiol–disulfide exchange:

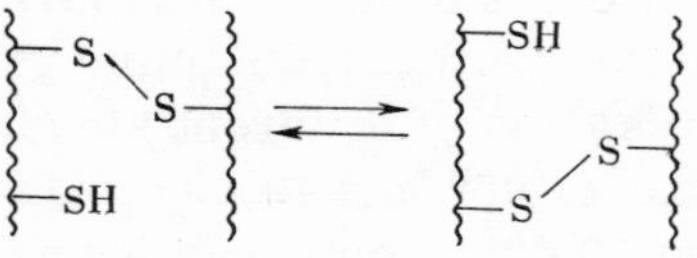

He showed that blocking the SH groups of keratin with N-ethylmaleimide lowers the rate of lengthening of the fibers under a load and the degree of supercontraction. The findings of Crewther et al. (Crewther and Dowling, 1961; Crewther et al., 1967) that the rate of supercontraction of wool fibers in LiBr solution is diminished after treatment of the fibers with N-ethylmaleimide, and also after conversion of the S–S groups into the thioether groups of lanthionine or of dimethylene-S,S'-dicysteine, also support the significance of thiol–disulfide exchange. Caldwell et al. (1964) found that the rate of setting of wool fibers after stretching depends on its content of SH groups (see also Crewther, 1966). Weigman and Rebenfeld (1968) came to the conclusion, on the basis of experiments in which fibers were treated with N-ethylmaleimide, that the stabilization of the structure of the fibers after stretching is not solely the "crystallization" of the extended β-form, but involves the formation of chemical bonds with the participation of SH groups.

6. Concluding Remarks

This chapter was mainly devoted to description of the static function of S–S groups; their participation in the formation and stabilization of the macromolecular structure of proteins. The contribution of S–S bonds to such stabilization varies from one protein to another. In many cases this contribution is decisive; breaking of the S–S bonds leads to a loss by the protein of its unique conformation and biological activity. In other, less common cases, the protein wholly or partly retains its original conformation and biological activity despite scission of the S–S bonds. We have reviewed the role of S–S bonds in the examples of several enzymes, immunoglobulins, hormones, and fibrous proteins. Naturally these examples do not exhaust the extensive literature on the role of S–S bonds in various proteins.

It should be noted that in addition to their static function, reactive S–S bonds can also play a dynamic part in some proteins. The most soundly based example is that of the dynamic (catalytic) function of reactive S–S groups that form part of the active sites of some oxidative enzymes, in particular lipoamide dehydrogenase. The disulfide groups of these enzymes take part in the transfer of electrons and protons from substrates to acceptors, undergoing in the process reversible conversion into dithiol groups (see Chapter 6, Section 1b).

The hypothesis that S–S groups participate in the mechanism of action of peptide hormones (oxytocin, vasopressin, and insulin) was not confirmed in experiments with analogs of these hormones and has now been abandoned. The combination of peptide hormones to tissue receptor proteins apparently occurs by noncovalent interactions, and not by thiol–disulfide exchange. This reaction, however, plays an important role in the process of renaturing reduced protein molecules, possibly in their biosynthesis, and also in the aggregation and polymerization of certain proteins, and in the activation of some plant and bacterial proteases.

References

Abbott, E. H., and Martell, A. E. (1970), J. Amer. Chem. Soc. 92:1754.
Adler, E., Euler, H. von, and Günther, G. (1938), Scand. Arch. Physiol. 80:1.
Akazawa, T., Sugiyama, T., and Nakayama, N. (1968), Arch. Biochem. Biophys. 128:646.
Akhtar, M., and Wilton, D. C. (1972), Annu. Rep. Progr. Chem. 69B:140.
Albers, R. W., and Koval, G. K. (1961), Biochem. Biophys. Acta 52:29.
Alberts, A. W., and Vagelos, P. R. (1966), J. Biol. Chem. 241:5201.
Alberts, A. W. Majerus, P. W., Talamo, B., and Vagelos, P. R. (1964), Biochemistry 3:1563.
Alberts, A. W., Majerus, P. W., and Vagelos, P. R. (1965), Biochemistry 4:2265.
Alexander, N. M. (1958), Anal. Chem. 30:1292.
Alexander, P., and Hudson, R. F. (1954), "Wool – Its Chemistry and Physics," Chapman & Hall, London.
Ali, A., Fahrenholz, F., Garing, J. C., and Weinstein, B. (1972), J. Amer. Chem. Soc. 94:2556.
Allison, A. C., and Cecil, R. (1958), Biochem. J. 69:27.
Allison, W. S., and Connors, M. J. (1970), Arch. Biochem. Biophys. 136:383.
Allison, W. S., and Harris, J. H. (1965), Abstracts 2nd FEBS Meeting (Vienna), A205, p. 140.
Allison, W. S., and Swain, L. C. (1973), Arch. Biochem. Biophys. 155:405.
Amiconi, G., Antonini, E., Brunori, M., Nason, A., and Wyman, J. (1971), Eur. J. Biochem. 22:321.
Ampulski, R. S., Ayers, V. E., and Morell, S. A. (1969), Anal. Biochem. 32:163.
Anderson, P. J. (1972), Can. J. Biochem. 50:111.
Anderson, P. J., and Perham, R. N. (1971), Biochem. J. 117:291.
Andersson, L.-O. (1966), Biochim. Biophys. Acta 117:115.
Andersson, L.-O. (1969), Arch. Biochem. Biophys. 133:277.
Andersson, L.-O. (1970), Biochim. Biophys. Acta 200:363.
Andersson, L.-O., and Berg, G. (1969), Biochim. Biophys. Acta 192:534.
Anfinsen, C. B. (1958), In "Symposium on Protein Structure" (Neuberger A., ed.), p. 223, Methuen, London.
Anfinsen, C. B. (1962), Int. Cong. Biochem. 5th, 4:66.

Anfinsen, C. B., and Haber, E. (1961), J. Biol. Chem. 236:1361.

Anfinsen, C. B., Haber, E., Sela, M., and White, F. H. (1961), Proc. Nat. Acad. Sci. U.S. 47:1309.

Anson, M. L. (1945), Advan. Protein Chem. 2:361.

Anzai, K., and Curtis, R. W. (1965), Phytochemistry 4:263.

Arnon, R., and Shapira, E. (1969), J. Biol. Chem. 244:1033.

Asquith, R. S., and Garcia-Dominguez, J. J. (1968), J. Soc. Dyers Colour. 84:211.

Asquith, R. S., and Hirst, L. (1969), Biochim. Biophys. Acta 184:345.

Asquith, R. S. and Shah, A. V. (1971), Biochim. Biophys. Acta 244:547.

Astrachan, L., Colowick, S.P., and Kaplan, N.O. (1957), Biochim. Biophys. Acta 24:141.

Augenstine, L. G., and Ghiron, C. A. (1961), Proc. Nat. Acad. Sci. U.S. 47:1530.

Auricchio, F., and Bruni, C. B. (1969), Biochim. Biophys. Acta 185:461.

Autor, A. P., and Fridovich, I. (1970), J. Biol. Chem. 245:5214.

Ayling, J., Pirson, R., and Lynen, F. (1972), Biochemistry 11:526.

Azari, P. (1966), Arch. Biochem. Biophys. 115:230.

Bachmayer, H., Piette, L. H., Yasunobu, K. T., and Thiteley, H. R. (1967), Proc. Nat. Acad. Sci. U.S. 57:122.

Bachmayer, H., Benson, A. M., Yasunobu, K. T., Garrard, W. T., and Whiteley, H. R. (1968), Biochemistry, 7:986.

Bacq, Z. M., and Alexander, P. (1961), "Fundamentals of Radiobiology," 2nd ed., pp. 457-483, Pergamon, Oxford.

Baddeley, G. (1950), J. Chem. Soc. 633.

Bailey, J. L. (1962), "Techniques in Protein Chemistry," Elsevier, Amsterdam.

Bailey, J. L., and Cole, R. D. (1959), J. Biol. Chem. 234:1733.

Baker, A. W., and Harris, G. H. (1960), J. Amer. Chem. Soc. 82:1923.

Baker, B. R. (1967), "Design of Active Site Directed Irreversible Enzyme Inhibitors," Wiley, New York.

Banks, T. E., and Shafer, J. A. (1970), Biochemistry 9:3343.

Banks, T. E., and Shafer, J. A. (1972), Biochemistry 11:110.

Barnard, E. A., and Stein, W. D. (1959), J. Mol. Biol. 1:339.

Barns, R. J., and Keech, D. B. (1968), Biochim. Biophys. Acta 159:514.

Barnett, R. E., and Jencks, W. P. (1969), J. Amer. Chem. Soc. 91:6758.

Barron, E. J., and Mooney, L. A. (1968), Anal. Chem. 40:1742.

Barron, E. S. G. (1951), Advan. Enzymol. 11:201.

Barron, E. S. G. (1953), Tex. Rep. Biol. Med. 11:653.

Barron, E. S. G., and Dickman, S. (1949), J. Gen. Physiol. 32:595.

Barron, E. S. G., and Levine, S. (1952), Arch. Biochem. Biophys. 41:175.

Barron, E. S. G., Miller, Z. B., and Kalnitsky, G. (1947), Biochem. J. 41:62.

Basford, R. E., and Huennekens, F. M. (1955), J. Amer. Chem. Soc. 77:3873.

Battell, M. L., Smilie, L. B., and Madsen, N. B. (1968), Can. J. Biochem. 46:609.

Bayer, E., Eckstein, H., Hagenmaier, H., Josef, D., Koch, J., Krauss, P., Röder, A., and Schretzmann, P. (1969), Eur. J. Biochem. 8:33.

Bayer, E., and Parr, W. (1966), Angew. Chem., Int. Ed. Engl. 5:840.

Bayer, E., Parr, W., and Kazmaier, B. (1965), Arch. Pharm. 298:196.

Beale, D., and Feinstein, A. (1969), Biochem. J. 112:187.

Behme, M. T. A., and Cordes, E. H. (1967), J. Biol. Chem. 242:5500.

Belitser, V. A., and Lobachevskaya, O. V. (1961), Dokl. Akad. Nauk SSSR 137:1226.

Bello, J., Harker, D., and De Jarnette E. (1961), J. Biol. Chem. 236:1358.
Bender, M. L. (1960), Chem. Rev. 60:53.
Bender, M. L., and Brubacher, L. J. (1964), J. Amer. Chem. Soc. 86:5333.
Bender, M. L., and Brubacher, L. J. (1966), J. Amer. Chem. Soc. 88:5880.
Benesch, R., and Benesch, R. E. (1948), Arch. Biochem. 19:35.
Benesch, R. E., and Benesch, R. (1950), Arch. Biochem. 28:43.
Benesch, R. E., and Benesch, R. (1953), J. Amer. Chem. Soc. 75:4367.
Benesch, R. E., and Benesch, R. (1955), J. Amer. Chem. Soc. 77:5877.
Benesch, R. E., and Benesch, R. (1958), J. Amer. Chem. Soc. 80:1666.
Benesch, R., and Benesch, R. E. (1962), Methods Biochem. Anal. 10:43.
Benesch, R., Benesch, R. E., and Rogers, W. J. (1954), In "Glutathione" (Colowick, S., et al., ed.), p. 31, Academic, New York.
Benesch, R. E., Lardy, H. A., and Benesch, R. (1955), J. Biol. Chem. 216:663.
Benisek, W. F. (1971), J. Biol. Chem. 246:3151.
Bennett, H. S., and Watts, R. M. (1958), In "General Cytochemical Methods" (Danielli, J. F., ed), Vol. 1, p. 317, Academic, New York.
Berezin, I. V., Kazanskaya, N. F., Khludova, M. S., and Khusainova, R. B. (1968), Biokhimiya 33:644 (Engl. transl., p. 526).
Berezin, I. V., Kazanskaya, N. F., and Khludova, M. S. (1972), Dokl. Akad. Nauk SSSR 202:463.
Bergel, F., and Harrap, K. R. (1961), J. Chem. Soc. 4051.
Bergson, G. (1962), Arkiv Kemi 18:409.
Bergson, G., and Schotte, L. (1958), Arkiv Kemi 13:43.
Bersin, I. (1935), Ergebn. Enzymforsch. 4:68.
Bersin, I., and Logemann, W. (1933), Hoppe-Seyler's Z. Physiol. Chem. 220:209.
Bersin, T. (1933), Hoppe-Seyler's Z. Physiol. Chem. 222:177.
Bersin, T., and Steudel, J. (1938), Chem. Ber. 71:1015.
Bewley, T. A., Brovetto-Cruz, J., and Li, C. H. (1969), Biochemistry 8:4701.
Bewley, T. A., Dixon, J. S., and Li, C. H. (1968), Biochim. Biophys. Acta 154:420.
Beychok, S. (1965), Proc. Nat. Acad. Sci. U. S. 53:999.
Beychok, S., and Breslow, E. (1968), J. Biol. Chem. 243:151.
Birchmeier, W., and Christen, P. (1971), FEBS Lett. 18:209.
Birchmeier, W., Wilson, K. J., and Christen, P. (1972), FEBS Lett. 26:113.
Birchmeier, W., Wilson, K. J., and Christen, P. (1973), J. Biol. Chem. 248:1751.
Birkett, D. J., Price. N. C., Radda, G. K., and Salmon, A. G. (1970), FEBS Lett. 18:209.
Birktoft, J. J., and Blow, D. M. (1972), J. Mol. Biol. 68:187.
Bitny-Szlachto, S. (1965), Progr. Biochem. Pharmacol. 1:112.
Blake, C. C. F., Mair, G. A., North, A. C. T., Phillips, D. C., and Sarma, V. R. (1967), Proc. Roy. Soc. London, Ser. B 167:365.
Blanquet, R., and Lenhoff, H. M. (1966), Science 154:152.
Bloch, K., and Clarke, H. T. (1938), J. Biol. Chem. 125:275.
Blomstrom, D. C., Knight, E., Phillips, W. D., and Weiher, J. F. (1964), Proc. Nat. Acad. Sci. U.S. 51:1085.
Bloxham, D. P., Clark, M. G., and Holland, P. C. (1973), Biochem. Soc. Trans. 1:1272.
Blumenfeld, O. O., and Perlmann, G. E. (1961), J. Biol. Chem. 236:2472.
Blundell, T. L., Dodson, G. G., Dodson, E., and Hodgkin, D. C., Vijayan, M. (1971), Recent Progr. Horm. Res. 27:1.

Blundell, T., Dodson, G., Hodgkin, D., and Mercola, D. (1972), Advan. Protein Chem. 26:279.
Bocchini, V., Alioto, M. R., and Najjar, V. A. (1967), Biochemistry 6:313.
Bocharov, A. L., Demidkina, T. V., Polyanovskii, O. L., and Karpeiskii, M. Ya. (1973), Mol. Biol. 7:620.
Bogle, G. S., Burgess, V. R., Forbes, W. F., and Savige, W. E. (1962), Photochem. Photobiol. 1:277.
Bohak, Z. (1964), J. Biol. Chem. 239:2878.
Bond, J. S., Francis, S. H., and Park, J. H. (1970), J. Biol. Chem. 245:1041.
Bongartz, J. (1888), Ber. 21:483.
Boross, L. (1965), Biochim. Biophys. Acta 96:52.
Boross, L. (1969), Acta Biochim. Biophys. 4:57.
Boross, L., and Cseke, E. (1967), Acta Biochim. Biophys. 2:47.
Boross, L., Cseke, E., and Vas, M. (1969), Acta Biochim. Biophys. 4(3):301.
Börresen, H. C. (1963), Anal. Chem. 35:1096.
Borsook, H., Ellis, E. L., and Huffman, H. M. (1937), J. Biol. Chem. 117:281.
Boyer, P. D. (1954), J. Amer. Chem. Soc. 76:4331.
Boyer, P. D. (1959), In "The Ebzymes" (Boyer, P., et al. eds.), Vol. 1, p. 511, Academic, New York.
Boyer, P. D. (1960), Brookhaven Symp. Biol. 13:1.
Boyer, P. D., and Segal, H. L. 1954), In "A Symposium on the Mechanism of Enzyme Action (McElroy, W. D., and Glass, B., eds.), p. 520, Johns Hopkins, Baltimore.
Boyko, J., and Fraser, M. J. (1964), Can. J. Biochem. 42:1677.
Boyland, E., and Chasseaud, L. F. (1969), Advan. Enzymol. 32:173.
Boyland, E., and Speyer, B. E. (1970), Biochem. J. 119:463.
Bradbury, A. F., and Smyth, D. G. (1973), Biochem. J. 131:637.
Bradley, D. F., and Calvin, M. (1955), Proc. Nat. Acad. Sci. U.S. 41:563.
Bradshaw, R. A., Kanarek, L., and Hill, R. L. (1967), J. Biol. Chem. 242:3789.
Brandenburg, D., and Wollmer, A. (1973), Hoppe-Seyler's Z. Physiol. Chem. 354:613.
Brazeau, P., Vale, W., Burgus, R., Ling, N., Rivier, J., and Guillemin, R. (1973), Science 179:77.
Bressler, R., and Wakil, S. J. (1961), J. Biol. Chem. 236:1643.
Brewer, C. F., and Riehm, J. P. (1967), Anal. Biochem. 18:248.
Bright, H. (1964), J. Biol. Chem. 239:2307.
Brocklehurst, K., and Kierstan, M. P. J. (1973), Nature (London) New Biol. 242:167.
Brocklehurst, K., and Little, G. (1972), Biochem. J. 128:471.
Brocklehurst, K., and Little, G. (1973), Biochem. J. 133:67.
Brocklehurst, K., Kierstan, M., and Little, G. (1972a), Biochem. J. 128:811.
Brocklehurst, K., Crook, E. M., and Kierstan, M. (1972b), Biochem. J. 128:979.
Brocklehurst, K., Carlsson, J., Kierstan, M. P. J., and Crook. E. M. (1973), Biochem. J. 133:573.
Brockman, J. A. Stokstand, E. L. R., Patterson, E. L., Pierce, J. V., and Macchi, M. E. (1954), J. Amer. Chem. Soc. 76:1827.
Brois, S. J., Pilot, J. F., and Parnum, B. W. (1970), J. Amer. Chem. Soc. 92:7629.
Brown, C. S., and Cunningham, L. W. (1970), Biochemistry 9:3878.
Brown, J. C., Swaisgood, H. E., and Horton, H. R. (1972), Biochem. Biophys. Res. Commun. 48:1068.

Brown, J. P., and Perham, R. N. (1974), Biochem. J. 137:505.
Brown, J. P., and Perham, R. N. (1972), FEBS Lett. 26:221.
Brown, J. R., and Hartley, B. S. (1963), Biochem. J. 89:59P.
Brown, J. R., and Hartley, B. S. (1966), Biochem. J. 101:214.
Brown, J. R., Kauffman, D. L., and Hartley, B. S. (1967), Biochem. J. 103:497.
Brown, P. E. (1967), Nature (London) 213:363.
Brown, P. R., and Edwards, J. O. (1969), Biochemistry 8:1200.
Brown, W. D. (1960), Biochim. Biophys. Acta 44:365.
Brox, L. W., and Hampton, A. (1968), Biochemistry 7:2589.
Brubacher, L. J., and Bender, M. L. (1966), J. Amer. Chem. Soc. 88:5871.
Brubacher, L. J., and Glick, B. R. (1974), Biochemistry 13:915.
Bruice, T. C. (1961), In "Organic Sulfur Compounds" (Kharasch, N., ed), p. 431, Pergamon, New York.
Bruice, T. C., and Benkovic, S. (1966), "Bioorganic Mechanisms," Vol. 1, Benjamin, New York.
Bruice, T. C., and Fedor, L. R. (1964), J. Amer. Chem. Soc. 86:738, 739.
Brumby, P. E., Miller, R. W., and Massey, V. (1965), J. Biol. Chem. 240:2222.
Bryant, R. G., Yeh, H. J. C., and Strengle, T. R. (1969), Biochem. Biophys. Res. Commun. 37:603.
Bryce, G. F., and Rabin, B. R. (1964), Biochem. J. 90:513.
Buchanan, J. M. (1973), Advanc. Enzymol. 39:91.
Buehner, M., Ford, G. C., Moras, D., Olsen, K. W., and Rossman, M. G. (1973), Proc. Nat. Acad. Sci. U. S. 70:3052.
Buell, M. V., and Hansen, R. E. (1960), J. Amer. Chem. Soc. 82:6042.
Bühner, M., and Sund, H. (1969), Eur. J. Biochem. 11:73.
Bullock, M. W., Brockman, J. A., Patterson, E. L., Pierce, J. V., and Stokstad, E. L. R. (1952), J. Amer. Chem. Soc. 74:3455.
Bullock, M. W., Brockman, J. A., Patterson, E. L., Pierce, J. V., Von Saltza M. H., Sanders, F., and Stokstad, E. L. R. (1954), J. Amer. Chem. Soc. 76:1828.
Burleigh, B. D., and Williams, C. H. (1972), J. Biol. Chem. 247:2077.
Burley, R. W. (1956), Proc. Int. Wool. Textile Res. Conf. Australia, 1955, Vol. D, p. 88.
Burton, H. (1958), Biochim. Biophys. Acta 29:193.
Bustin, M., Lin, M. C., Stein, W. H., and Moore, S. (1970), J. Biol. Chem. 245:846.
Butterworth, P. H. W., Yang, P. C., Bock, J. W. (1967a), Arch. Biochem. Biophys. 118:716.
Butterworth, P. H. W., Yang, P. C., Bock, R. M., and Porter, J. W. (1967b), J. Biol. Chem. 242:3508.
Cadenas, E., Kaji, H., Park, C. R., and Rasmussen, H. (1961), J. Biol. Chem. 236:PC63.
Caldwell, J. B., Leach, S. J., Meschers, A., and Milligan, B. (1964), Text. Res. J. 34:627.
Calvin, M. (1954), In "Glutathione" (Colowick, S., et al., eds.), p. 3, Academic, New York.
Campaigne, E. (1961), in "Organic Sulfur Compounds" (Kharasch, N., ed.), p. 134, Pergamon, Oxford.
Cantau B., Brunel, C., and Pudles, J. (1968), Biochim. Biophys. Acta 167:511.
Carlin, H., and Hechter, O. (1962), J. Biol. Chem. 237:PC 1371.

Carmack, M., and Neubert, L. A. (1967), J. Amer. Chem. Soc. 89:7134.

Carter, J. R. (1959), J. Biol. Chem. 234:1705.

Cecil, R. (1951), Biochem. J. 49:183.

Cecil, R. (1963), In "The Proteins" (Neurath, H., ed.), p. 460, Academic, New York.

Casola, L., and Massey, V. (1966), J. Biol. Chem. 241:4985.

Casola, L., Brumby, P. E., and Massey, V. (1966), J. Biol. Chem. 241:4977.

Catsimpoolas, N., and Wood, J. L. (1964), J. Biol. Chem. 239:4132.

Catsimpoolas, N., and Wood, J. L. (1966), J. Biol. Chem. 241:1790.

Cavallini, D., De Marco, C., and Dupré, S. (1968), Arch. Biochem. Biophys. 124:18.

Cavins, J. F., and Friedman, M. (1968), J. Biol. Chem. 243:3357.

Cecil, R. (1950), Biochem. J. 47:572.

Cecil, R., and Loening, U. E. (1960), Biochem. J. 76:146.

Cecil, R., and McPhee, J. R. (1955), Biochem. J. 60:496.

Cecil, R., and McPhee, J. R. (1957), Biochem. J. 66:538.

Cecil, R., and McPhee, J. R. (1959), Advan. Protein Chem. 14:255.

Cecil, R., and Snow, N. S. (1962), Biochem. J. 82:247, 255.

Cecil, R., and Thomas, M. A. W. (1965), Nature (London) 206:1317.

Cecil, R., and Wake, R. G. (1962), Biochem. J. 82:401.

Chaiken, I. M., and Smith, E. L. (1969), J. Biol. Chem. 244:5087, 5095.

Challenger, F. (1959), "Aspects of the Organic Chemistry of Sulfur," Butterworths, London.

Chan, W. W.-C. (1968), Biochemistry 7:4247.

Chan, W., Morse, D. E., and Horecker, B. L. (1967), Proc. Nat. Acad. Sci. U. S. 67:1013.

Chance, R. E., and Ellis, R. M. (1969), Arch. Intern. Med. 123:229.

Chance, R. E., Ellis, R. M., and Bromer, W. W. (1968), Science 161:165.

Chang, S. H., and Wilken, D.-R. (1966), J. Biol. Chem. 241:4251.

Chapman, A., Sanner, T., and Pihl, A. (1969), Biochim. Biophys. Acta 178:74.

Chiancone, E., Currell, D. L., Vecchini, P., Antonini, E., and Wyman, J. (1970), J. Biol. Chem. 245:4105.

Chibnall, A. C. (1942), Proc. Roy. Soc. London Ser. B. 131:136.

Chinard, F. P., and Hellerman, L. (1954), Methods Biochem. Anal. 1:1.

Chung, A. E., Franzen, J. S., and Braginski, J. E. (1971), Biochemistry 10:2872.

Cilento, G., and Giusti, P. (1959), J. Amer. Chem. Soc. 81:3801.

Cilento, G., and Tedeschi, P. (1961), J. Biol. Chem. 236:907.

Claeson, G. (1968), Acta Chem. Scand. 22:2429.

Clark, W. M. (1960), "Oxidation Reduction Potentials of Organic Systems," pp. 471-487, Williams and Willkins, Baltimore.

Clark, J. L., and Steiner, D. F. (1969), Proc. Nat. Acad. Sci. U. S. 62:278.

Clarke, H. T. (1932), J. Biol. Chem. 97:235.

Clark-Walker, G. D., and Robinson, H. C. (1961), J. Chem. Soc. 2810.

Clauser, H., Chambaut, A. M., Eboué-Bonis, D., and Volfin, P. (1965), Ann. Endocrinol. 26:714.

Cleland, W. W. (1964), Biochemistry 3:480.

Cliffe, E. E., and Waley, S. G. (1961), Biochem. J. 79:475.

Coates, E., Marsden, C., and Rigg, B. (1969), Trans. Faraday Soc. 65:863.

Cohen, W., and Petra, P. H. (1967), Biochemistry 6:1047.

Cohn, E. J., and Edsall, J. T. (1943), In "Proteins, Amino Acids and Peptides," p. 84, Reinhold, New York.
Cole, R. D. (1967), In "Methods in Enzymology," Vol. 11 (Hirs, C. H. W., ed.), p. 206, Academic, New York.
Cole, R. D., Stein, W. H., and Moore, S. (1958), J. Biol. Chem. 233:1359.
Coleman, D. L., and Blout, E. R. (1968), J. Amer. Chem. Soc. 90:2405.
Coleman, J. E., and Vallee, B. L. (1961), J. Biol. Chem. 236:2244.
Colman, R. F. (1968), J. Biol. Chem. 243:2454.
Colman, R. F., and Black, S. (1965), J. Biol. Chem. 240:1796.
Colowick, S., Lazarow, A., Racker, E., Schwarz, D. R., Stadtman, E., and Waelsch, H. (eds.) (1954), "Glutathione," Academic, New York.
Colowick, S. P., van Eys, J., and Park, J. H. (1966), In "Comprehensive Biochemistry," Vol. 14 (Florkin, M., and Stotz, E. H., eds.), p. 1, Elsevier, Amsterdam.
Connellan, J. M., and Folk, J. E. (1969), J. Biol. Chem. 244:3173.
Connors, K. A., and Bender, M. L. (1961), J. Org. Chem. 26:2498.
Conway, A., and Koshland, D. E. (1967), Biochim. Biophys. Acta 133:593.
Cooley, S. L., and Wood, J. L. (1951), Arch. Biochem. Biophys. 34:372.
Coombs, T. L., Omote, Y., and Vallee, B. L. (1964), Biochemistry 3:653.
Cori, G. T., Slein, M. W., and Cori, C. F. (1948), J. Biol. Chem. 173:605.
Cornell, N. W., and Crivaro, K. E. (1972), Anal. Biochem. 47:203.
Cotner, R. C., and Clagett, C. O. (1973), Anal. Biochem. 54:170.
Cremona, T., Kowal, J., and Horecker, B. L. (1965), Proc. Nat. Acad. Sci. U. S. 53:1395.
Crestifield, A. M., Skupin, J., Moore, S., and Stein, W. H. (1960), Fed. Proc. 19:341.
Crestfield, A. M., Moore, S., and Stein, W. H. (1963), J. Biol. Chem., 238:622.
Crewther, W. G. (1966), J. Soc. Dyers Colour. 82:54.
Crewther, W. G., and Dowling, L. M. (1961), Textile Res. J. 31:31.
Crewther, W. G., Fraser, R. D. B., Lennox, F. G., and Lindley, H. (1965), Advan. Protein Chem. 20:191.
Crewther, W. G., Dowling, L. M., Inglis, A. S., and McLaren, J. A. (1967), Textile Res. J. 37:736.
Crook, E. M. (ed.) (1959), Biochem. Soc. Sympos. No. 17.
Cseke, E., and Boross, L. (1970), Acta Biochim. Biophys. 5:385.
Cullis, A. F., Muirhead, H., Perutz, M. F., Rossman, M. G., and North, A. C. T. (1962), Proc. Roy. Soc. (Lodnon) A265:161.
Cunningham, L. W. (1964), Biochemistry 3:1629.
Cunningham, L. W., and Nuenke, B. J. (1959), J. Biol. Chem. 234:1447.
Cunningham, L. W., and Nuenke, B. J. (1960), J. Biol. Chem. 235:1711.
Cunningham, L. W., and Nuenke, B. J. (1961), J. Biol. Chem. 236:1716.
Cunningham, L. W., and Schepman, A. M. (1963), Biochim. Biophys. Acta 73:406.
Cunningham, L. W., Nuenke, B. J., and Strayhorn, W. D. (1957), J. Biol. Chem. 228:835.
Cuthbertson, W. R., and Phillips, H. (1945), Biochem. J. 39:7.
Daigo, K., and Reed, L. J. (1962), J. Amer. Chem. Soc. 84:666.
Danehy, J. P. (1966), In "The Chemistry of Organic Sulfur Compounds," Vol. 2. (Kharasch, N., and Meyers, C. Y., eds.), p. 337, Pergamon, Oxford.

Danehy, J. P., and Noel, C. J. (1960), J. Amer. Chem. Soc. 82:2511.
Davidson, B. E., and Hird, F. J. R. (1967), Biochem. J. 104:473.
Davidson, B. E., Sajgò, M., Noller, H. F., and Harris, J. I. (1967), Nature (London), 216:1181.
De Barreiro, O. L. C. (1969), Biochim. Biophys. Acta 178:412.
De Deken, R. H., Broekhuysen, J., Bechet, J., and Mortier, A. (1956), Biochim. Biophys. Acta 19:45.
De Lorenzo, F., Fuchs, S., and Anfinsen, C. B. (1966a), Biochemistry 5:3961.
De Lorenzo, F., Goldberger, R. F. Steers, E., Givol, D., and Anfinsen, C. B. (1966b), J. Biol. Chem. 241:1562.
De Luca, M., and McElroy, W. D. (1966), Arch. Biochem. Biophys. 116:103.
De Marco, C., Graziani, M. T., and Mosti, R. (1966), Anal. Biochem. 15:40.
Demidkina, T. V., Bocharov, A. L., Polyanovskii, O. L. Karpeiskii, M. Ya. (1973), Mol. Biol. 7:461.
De Rey-Pailhade, J. (1888), C. R. Acad. Sci. 106:1683; 107:43.
Desnuelle, P., and Rovery, M. (1949), Biochim. Biophys. Acta 3:26.
Di Bella, F. P., and Liener, I. E. (1969), J. Biol. Chem. 244:2824.
Dimroth, P., Dittmar, W., Walther, G., and Eggerer, H. (1973), Eur. J. Biochem. 37:305.
Diopon, J., and Olomucki, M. (1972), Biochim. Biophys. Acta 263:220.
Dixon, C. J., and Grant, D. W. (1970), J. Phys. Chem. 74:941.
Dixon, G. A., and Wardlaw, A. C. (1960), Nature (London) 188:721.
Dixon, J. S., and Li, C. H. (1966), Science 154:785.
Dixon, H. B. F., and Tipton, K. F. (1973), Biochem. J. 133:837.
Dixon, M., and Tunnicliffe, H. E. (1923), Proc. Roy. Soc. London, Ser. B, 94:266.
Dodson, R. M., and Nelson, V. C. (1968), J. Org. Chem. 33:3966.
Dohan, J. S., and Woodward, G. E. (1939), J. Biol. Chem. 129:393.
Donehue, J. (1969), J. Mol. Biol. 45:231.
Donovan, J. W. (1964), Biochemistry 3:67.
Donovan, J. W. (1967), Biochem. Biophys. Res. Commun. 29:734.
Donovan, J. W., and White, T. M. (1971), Biochemistry 10:32.
Dose, K. (1966), Biophysik 3:259.
Dose, K. (1968), Photochem. Photobiol. 8:331.
Dose, K., and Risi, S. (1972), Photochem. Photobiol. 15:43.
Drenth, J., Jansonius, J. N., Koekoek, R., Swen, H. M., and Wolthers, B. G. (1968), Nature (London) 218:929.
Drenth, J., Jansonius, J. N., Koekoek, R., Sluyterman, L. A. A., and Wolthers, B. G. (1970), Phil. Trans. Roy. Soc. B 257:231.
Du, Y.-C., Jiang, R.-Q., and Tsou, C.-L. (1965), Sci. Sinica (Peking) 14:229.
Duke, J., Barlow, G. H., and Klapper, M. H. (1971), Biochim. Biophys. Acta 229:155.
Dunnill, P., and Green, D. W. (1966), J. Mol. Biol. 15:147.
Durell, J., and Fruton, J. S. (1954), J. Biol. Chem. 207:487.
Eager, J. E., and Savige, W. E. (1963), Photochem. Photobiol. 2:25.
Earland, C., and Raven, D. J. (1961), Nature (London) 192:1185.
Edelhoch, H., Katchalski, E., Maybury, R. H., Hughes, W. L., and Edsall, J. T. (1953), J. Amer. Chem. Soc. 75:5058.

Edelman, G. M. (1959), J. Amer. Chem. Soc. 81:3155.
Edelman, G. M., and Gally, J. A. (1964), Proc. Nat. Acad. Sci. U.S. 51:846.
Edelman, G. M., and Poulik, M. D. (1961), J. Exp. Med. 113:861.
Edelman, G. M., Cunningham, B. A., Gall, W. E., Gottlieb, P. D., Rutishauser, U., and Waxdal, M. J. (1969), Proc. Nat. Acad. Sci. U. S. 63:78.
Edmondson, D., Massey, V., Palmer, G., Beacham, L. M., and Elion, G. B. (1972), J. Biol. Chem. 247:1597.
Edsall, J. T. (1965), Biochemistry, 4:28.
Edwards, D. J., Heinrikson, R. L., and Chung, A. E. (1974), Biochemistry 13:677.
Edwards, J. B., and Keech, D. B. (1967), Biochim. Biophys. Acta 146:576.
Edwards, S. W., and Knox, W. E. (1956), J. Biol. Chem. 220:79.
Elcombe, M. M., and Taylor, J. C. (1968), Acta Crystallogr. A24:410.
Eldjarn, L., and Pihl, A. (1956), Proc. Int. Conf. Radiobiol. 4th, p. 249.
Eldjarn, L., and Pihl, A. (1957a), J. Amer. Chem. Soc. 79:4589.
Eldjarn, L., and Pihl, A. (1957b), J. Biol. Chem. 225:499.
Eldjarn, L., and Pihl, A. (1960), In "Errera and Forssberg's Mechanisms in Radiobiology," Vol. 2, Academic, New York.
Eldjarn, L., Pihl, A., and Shapiro, B. (1956), Proc. Int. Conf. on Peaceful Uses Atom Energy, Geneva, Vol. 11, p. 335.
Elleman, T. C. (1972), Biochem. J. 128:1229; 130:833.
Elleman, T. C., and Dopheide, T. A. (1972), J. Biol. Chem. 247:3900.
Ellman, G. L. (1959), Arch. Biochem. Biophys. 82:70.
Elödi, P. (1960), Biochim. Biophys. Acta 40:272.
Elovson, J., and Vagelos, P. R. (1968), J. Biol. Chem. 243:3603.
Elson, E. L., and Edsall, J. T. (1962), Biochemistry 1:1.
Elsworth, F. F., and Phillips, H. (1938), Biochem. J. 32:837.
Elsworth, F. F., and Phillips, H. (1941), Biochem. J. 35:135.
Epstein, C. J., and Goldberger, R. F. (1963), J. Biol. Chem. 238:1380.
Epstein, C. J., Goldberger, R. F., Young, D. M., and Anfinsen, C. B. (1962), Arch. Biochem. Biophys. Suppl. 1, p. 223.
Epstein, C. J., Goldberger, R. F., and Anfinsen, C. B. (1963), Cold Spring Harbor Sympos. Quant. Biol. 28:439.
Eriksson, A. S., and Mannervik, B. (1970), FEBS Lett. 7:26.
Evans, N., and Rabin, B. R. (1968), Eur. J. Biochem. 4:548.
Fava, A., and Iliceto, A. (1953), Ricerca Sci. 23:839.
Fava, A., and Iliceto, A. (1958), J. Amer. Chem. Soc. 80:3478.
Fava, A., Iliceto, A., and Camera, E. (1957), J. Amer. Chem. Soc. 79:833.
Fawcett, D. M. (1968), Can. J. Biochem. 46:1433.
Feher, F., and Schulze-Rettmer, R. (1958), Z. Anorg. Allg. Chem. 295:262.
Fenselau, A. (1970), J. Biol. Chem. 245:1239.
Ferdinand, W., Stein, W. H., and Moore, S. (1965), J. Biol. Chem. 240:1150.
Fernández-Morán, H., Reed, L. J., Koike, M., and Willms, C. R. (1964), Science 145:930.
Field, L., and White, J. E. (1973), Proc. Nat. Acad. Sci. U. S. 70:328.
Fields, R. (1971), Biochem. J. 124:581.
Fieser, L. F., and Turner, R. B. (1947), J. Amer. Chem. Soc. 69:2335.
Finazzi-Agrò, A., Rotilio, G., Avigliano, L., Guerrieri, P., Boffi, V., and Mondovì, B. (1970), Biochemistry 9:2009.

Flamm, W. G., and Crandall, D. I. (1963), J. Biol. Chem. 238:389.
Flashner, M., Hollenberg, P. F., and Coon, M. J. (1972), J. Biol. Chem. 247:8114.
Fleischman, J. B., Pain, R. H., and Porter, R. R. (1962), Arch. Biochem. Biophys. Suppl. 1, p. 174.
Fleischman, J. B., Porter, R. R., and Press, E. M. (1963), Biochem. J. 88:220.
Fletcher, J. C., and Robson, A. (1963), Biochem. J. 87:553.
Fletcher, J. C., Robson, A., and Todd, J. (1963), Biochem. J. 87:560.
Folk, J. E., and Cole, P. W. (1966), J. Biol. Chem. 241:5518.
Folk, J. E., Cole, P. W., and Mullooly, J. P. (1967), J. Biol. Chem. 242:4329.
Foltmann, B., and Hartley, B. S. (1967), Biochem. J. 104:1064.
Fonda, M. L., and Anderson, B. M. (1969), J. Biol. Chem. 244:666.
Fondy, T. P., and Everse, J. (1964), Fed. Proc. 23:424.
Fondy, T. P., Everse, J., Driscoll, G. A., Castillo, F., Stolzenbach, F. E., and Kaplan, N. O. (1965), J. Biol. Chem. 240:4219.
Fong, C. T. O., Schwartz, I. L., Popenoe, E. A., Silver, L., and Schoessler, M. A. (1959), J. Amer. Chem. Soc. 81:2592.
Fong, C. T. O., Silver, L., Christman, D., and Schwartz, I. L. (1960), Proc. Nat. Acad. Sci. U. S. 46:1273.
Fong, C. T. O., Silver, L., Popenoe, E. A., and Debons, A. F. (1962), Biochim. Biophys. Acta 56:190.
Fontana, A., Scoffone, E., and Benassi, C. A. (1968a), Biochemistry 7:980.
Fontana, A., Veronese, F. M., and Scoffone, E. (1968b), Biochemistry 7:3901.
Footner, G. B., and Smiles, S. (1925), J. Chem. Soc. 127:2887.
Forbes, W. F., and Hamlin, C. R. (1968), Can. J. Chem. 46:3033.
Forbes, W. F., and Hamlin, C. R. (1969), Exp. Gerontol. 4:151.
Forbes, W. F., and Savige, W. E. (1962), Photochem. Photobiol. 1:1.
Forest, P. B., and Kemp, R. G. (1968), Biochem. Biophys. Res. Commun. 33:763.
Foss, O. (1961), In "Organic Sulfur Compounds" (Kharasch, N., ed.), Pergamon, Oxford.
Fraenkel-Conrat, H. (1950), Arch. Biochem. Biophys. 27:109.
Fraenkel-Conrat, H. (1955), J. Biol. Chem. 217:373.
Fraenkel-Conrat, H. (1959), In "Sulfur in Proteins," Proc. Symp. Falmouth (1958), p. 339.
Francis, S. H., Meriwether, B. P., and Park, J. H. (1973), Biochemistry 12:346.
Franěk, F. (1961), Biochem. Biophys. Res. Commun. 4:28.
Franěk, F., and Nezlin, R. S. (1963), Biokhimiya 28:193 (Engl. transl. p. 153).
Frangione, B., Milstein, C., and Franklin, E. C. (1969), Nature (London) 221:145.
Franklin, J. L., and Lumpkin, H. E. (1952), J. Amer. Chem. Soc. 74:1023.
Franzen, V. (1957), Chem. Ber. 90:623.
Frater, R. (1969), Aust. J. Biol. Sci. 22:1086.
Frattali, V., Steiner, R. F., Millar, D. B. S., and Edelhoch, H. (1963), Science 199:1186.
Fredga, A. (1937), Arkiv Kemi, Mineral. Geol. 12A, No. 13.
Freedberg, W. B., and Hardman, J. K. (1971), J. Biol. Chem. 246:1439.
Freedman, M. H., and Sela, M. (1966), J. Biol. Chem. 241:2383.
Freedman, M. H., and Sela, M. (1966), J. Biol. Chem. 241:2383.
Freedman, R. B., and Radda, G. K. (1968), Biochem. J. 108:383.
Freisheim, J. H., and Huennekens, F. M. (1969), Biochemistry 8:2271.

French, T. C., Dawid, I. B., and Buchanan, J. M. (1963), J. Biol. Chem. 238:2186.
Friedman, M., Cavins, J. F., and Wall, J. S. (1965), Amer. Chem. Soc. 87:3672.
Friedman, M., Krull, L. H., and Cavins, J. F. (1970), J. Biol. Chem. 245:3868.
Friedmann, E. (1903), Beitr. Chem. Physiol. Pahtol. 3:1.
Friedmann, T., and Epstein, C. J. (1967), J. Biol. Chem. 242:5131.
Friedrich, P. (1965), Biochim. Biophys. Acta 99:371.
Friedrich, P., and Szabolcsi, G. (1967), Acta Biochim. Biophys. 2(1):59.
Friedrich, P., Polgar, L., and Szabolcsi, G. (1964), Nature (London) 202:1214.
Fruchter, R. G., and Crestfield, A. M. (1965), J. Biol. Chem. 240:3868.
Fruton, J. S. (1950), Yale J. Biol. Med. 22:263.
Fruton, J. S., and Clarke, H. T. (1934), J. Biol. Chem. 106:667.
Fuchs, S., De Lorenzo, F., and Anfinsen, C. B. (1967), J. Biol. Chem. 242:398.
Fujimoto, D., Ikeuchi, T., and Nozawa, S. (1969), Biochim. Biophys. Acta 188:295.
Garfinkel, D. (1958), J. Amer. Chem. Soc. 80:4833.
Garfinkel, D., and Edsall, J. T. (1958), J. Amer. Chem. Soc. 80:3823.
Gawron, O. (1966), In "The Chemistry of Organic Sulfur Compounds," Vol. 2 (Kharasch, N., and Meyers, C. Y., eds.), p. 351, Pergamon, Oxford.
Gawron, O., and Fernando, J. (1961), J. Amer. Chem. Soc. 83:2906.
Gawron, O., and Odstrchel, G. (1967), J. Amer. Chem. Soc. 89:3263.
Gawron, O., Mahboob, S., and Fernando, J. (1964), J. Amer. Chem. Soc. 86:2283.
Gehring, U., and Harris, J. I. (1968), FEBS Lett. 1:150.
Gehring, U., Riepertinger, C., and Lynen, F. (1968), Eur. J. Biochem. 6:264.
Gehring, U., Riepertinger, C., and Lynen, F. (1968), Eur. J. Biochem. 6:264.
Gehart, J. C., and Pardee, A. B. (1962), J. Biol. Chem. 237:891.
Gerhart, J. C., and Schachman, H. K. (1965), Biochemistry 4:1054.
Gerhart, J. C., and Schachman, H. K. (1968), Biochemistry 7:538.
Gersonde, K., and Druskeit, W. (1968), Eur. J. Biochem. 4:391.
Gerwin, B. I. (1967), J. Biol. Chem. 242:451.
Gerwin, B. I., Stein, W. H., and Moore, S. (1966), J. Biol. Chem. 241:3331.
Gevers, W., Kleinkauf, H., and Lipmann, F. (1969), Proc. Nat. Acad. Sci. U. S. 63:1335.
Ghiron, C. A., Volkert, W. A., and Lahmeyer, H. (1971), Photochem. Photobiol. 13:431.
Ghosh, J. C., and Gangili, S. C. (1935), Biochem. Z. 279:296.
Ghosh, J. C., Raychaudhuri, S. N., and Ganguli, S. C. (1932), J. Indian Chem. Soc. 9:43, 53.
Giardina, B., Binotti, I., Amiconi, G., Antonini, E., Brunori, M., and McMurray, C. H. (1971), Eur. J. Biochem. 22:327.
Gilhuus-Moe, C. C., Kristensen, T., Bredesen, J. E., Zimmer, T. L., and Laland, S. G. (1970), FEBS Lett. 7:287.
Gillespie, J. M. (1959), In "Sulfur in Proteins," Proc. Symp. Falmouth (1958), p. 51.
Gillespie, J. M. (1962), Aust. J. Biol. Sci. 15:564, 572.
Gillespie, J. M. (1963), Aust. J. Biol. Sci. 16:241, 259.
Gillespie, J. M., and Simmonds, D. H. (1960), Biochim. Biophys. Acta 39:538.
Gillespie, J. M., and Springell, P. H. (1961), Biochem. J. 79:280.
Givol, D., Goldberger, R. F., and Anfinsen, C. B. (1964), J. Biol. Chem. 239:PC 3114.
Givol, D., De Lorenzo, F., Goldberger, R. F., and Anfinsen, C. B. (1965), Proc. Nat. Acad. Sci. U. S. 53:676.
Glaser, C. B., Maeda, H., and Meienhofer, J. (1970), J. Chromatogr. 50:151.

Glazer, A. N., and Smith, E. L. (1961), J. Biol. Chem. 236:416.
Gleisner, J. M., and Liener, I. E. (1973), Biochim. Biophys. Acta 317:482.
Goddard, D. R., and Michaelis, L. (1935), J. Biol. Chem. 112:361.
Gold, A. M. (1968), Biochemistry 7:2106.
Gold, A. H., and Segal, H. L. (1964), Biochemistry 3:778.
Gold, A. H., and Segal, H. L. (1965), Biochemistry 4:1506.
Goldberger, R. F., and Epstein, C. J. (1963), J. Biol. Chem. 238:2988.
Goldberger, R. F., Epstein, C. J., and Anfinsen, C. B. (1963), J. Biol. Chem. 238:628.
Goldberger, R. F., Epstein, C. J., and Anfinsen, C. B. (1964), J. Biol. Chem. 239:1406.
Goldfine, H., and Ailhaud, G. P. (1971), Biochem. Biophys. Res. Commun. 45:1127.
Goldfine, H., Ailhaud, G. P., and Vagelos, P. R. (1967), J. Biol. Chem. 242:4466.
Goldman, P., and Vagelos, P. R. (1964), In "Comprehensive Biochemistry," Vol. 15 (Florkin, M., and Stotz, E. H., eds.), p. 71, Elsevier, Amsterdam.
Gonzalez Porqué, P., Baldesten, A., and Richard, P. (1970), J. Biol. Chem. 245:2363, 2371.
Goodman, I., and Salce, L. (1965), Biochim. Biophys. Acta 100:283.
Gordy, W., and Stanford, S. C. (1940), J. Amer. Chem. Soc. 62:497.
Goren, H. J., and Barnard, E. A. (1970), Biochemistry 9:959, 974.
Goren, H. J., Glick, D. M., and Barnard, E. A. (1968), Arch. Biochem. Biophys. 126:607.
Gorin, G. (1965), Progr. Biochem. Pharmacol. 1:142.
Gorin, G., and Doughty, G. (1968), Arch. Biochem. Biophys. 126:547.
Gorin, G., Fulford, R., and Deonier, R. C. (1968), Experientia 24:26.
Grafius, M. A., and Neilands, J. B. (1955), J. Amer. Chem. Soc. 77:3389.
Graevskii, É. Ya. (1969), "Sulfhydryl Groups and Radiosensitivity," Atomizdat, Moscow.
Grant, P. T., and Coombs, T. L. (1970), Essays Biochem. 6:69.
Grassetti, D. R., and Murray, J. F. (1967), Arch. Biochem. Biophys. 119:41.
Grassetti, D. R., and Murray, J. F. (1969), Anal. Chim. Acta 46:139; J. Chromatogr. 41:121.
Grassetti, D. R., Murray, J. F., and Ruan, H. T. (1969), Biochem. Pharmacol. 18:603.
Grazi, E., and Rossi, N. (1968), J. Biol. Chem. 243:538.
Grazi, E., Conconi, F., and Vigi, V. (1965), J. Biol. Chem. 240:2465.
Green, D. E., Needham, D. M., and Dewan, J. G. (1937), Biochem. J. 31:2327.
Gregory, J. D. (1955), J. Amer. Chem. Soc. 77:3922.
Grinnell, F., and Nishimura, J. S. (1970), Biochim. Biophys. Acta 212:150.
Grisaro, V., and Sharon, N. (1964), Biochim. Biophys. Acta 89:152.
Gross, E., and Morell, J. L. (1971), J. Amer. Chem. Soc. 93:4634.
Gross, E., Kiltz, H. H., and Nebelin, E. (1973), Hoppe-Seyler's Z. Physiol. Chem. 354:810.
Gruber, W., Warzecha, K., Pfleiderer, G., and Wieland, T. (1962), Biochem. Z. 336:107.
Gruen, L. C., and Harrap, B. S. (1971), Anal. Biochem. 42:377.
Guest, J. R., and Yanofsky, C. (1966), J. Biol. Chem. 241:1.
Guha, A., Englard, S., and Listowskii, I. (1968), J. Biol. Chem. 243:609.
Guidotti, G. (1967), J. Biol. Chem. 242:3673.
Guidotti, G., and Konigsberg, W. (1964), J. Biol. Chem. 239:1474.
Gundlach, H. G., Moore, S., and Stein, W. H. (1959a), J. Biol. Chem. 234:1761.

Gundlach, H. G., Stein, W. H., and Moore, S. (1959b), J. Biol. Chem. 234:1754.
Gurd, F. R. N. (1967), In "Methods in Enzymology," Vol. 11 (Hirs, C. H. W., ed.), p. 532, Academic, New York.
Gurd, F. R. N., and Goodman, D. S. (1952), J. Amer. Chem. Soc. 74:670.
Gurd, F. R. N., and Murray, G. R. (1954), J. Amer. Chem. Soc. 76:187.
Gurd, F. R. N., and Wilcox, P. E. (1956), Advan. Protein Chem. 11:311.
Gustus, E. L. (1964), J. Biol. Chem. 239:115.
Gutfreund, H., and McMurray, C. H. (1970), Biochem. Soc. Symp. No. 31, p. 39.
Gutmann, A. (1908), Ber. 41:1650.
Guttman, S. (1966), Helv. Chim. Acta 49:83.
Haarman, W. (1943), Biochem. Z. 314:1, 18.
Habeeb, A. F. S. (1966), Biochim. Biophys. Acta 115:440.
Haber, E., and Anfinsen, C. B. (1962), J. Biol. Chem. 237:1839.
Hagen, U. (1956), Arzneimittel-Forschung 7:384.
Halcomb, S., Bond, J. S., Kloepper, R., and Park, J. H. (1968), Fed. Proc. 27:292.
Hammond, B. R., and Gutfreund, H. (1959), Biochem. J. 72:349.
Haraldson, L., Olander, C. J., Sunner, S., and Varde, E. (1960), Acta Chem. Scand. 14:1509.
Harding, J. J. (1965), Advan. Protein Chem. 20:109.
Harding, J. S., and Owen, L. N. (1954), J. Chem. Soc. 1528:1536.
Harding, M. M., and Long, H. A. (1968), Acta Crystallogr. B24:1096.
Harington, C. R., and Mead, T. H. (1935), Biochem. J. 29:1602.
Harrap, B. S., and Gruen, L. C. (1971), Anal. Biochem. 42:398.
Harrap, B. S., and Woods, E. F. (1965), Biopolymers 3:595.
Harrigan, P. J., and Trentham, D. R. (1973), Biochem. J. 135:695.
Harrington, W. F., and Sela, M. (1959), Biochim. Biophys. Acta 31:427.
Harris, J. I. (1964), Nature (London) 203:30.
Harris, J. I., and Perham, R. N. (1968), Nature (London) 219:1025.
Harris, J. I., Meriwether, B. P., and Park, J. H. (1963), Nature (London) 197:154.
Harting, J., and Velick, S. (1952), Fed. Proc. 11:226.
Harting, J., and Velick, S. F. (1954), J. Biol. Chem. 207:857, 867.
Hartley, B. S. (1970), Biochem. J. 119:805.
Hartman, F. C. (1970), Biochemistry, 9:1783.
Hasselbach, W., and Taugner, G. (1970), Biochem. J. 119:265.
Hayakawa, T., Kanzaki, T., Kitamura, T., Fukuyoshi, Y., Sakurai, Y., Koike, K., Suematsu, T., and Koike, M. (1969), J. Biol. Chem. 244:3660.
Heaton, G. S., Rydon, H. N., and Schofield, J. A. (1956), J. Chem. Soc. 3157.
Heffter, A. (1907), Med. Naturwiss. Arch. 1:81.
Heinrikson, R. L. (1970), Biochem. Biophys. Res. Commun. 41:967.
Heinrikson, R. L. (1971), J. Biol. Chem. 246:4090.
Heitmann, P. (1968a), Eur. J. Biochem. 3:346.
Heitmann, P. (1968b), Eur. J. Biochem. 5:305.
Heitz, J. R., and Anderson, B. M. (1968), Arch. Biochem. Biophys. 127:637.
Heitz, J. R., Anderson, C. D., and Anderson, B. M. (1968), Arch. Biochem. Biophys. 127:627.
Hellerman, L. (1937), Physiol. Rev. 17:454.

Hellerman, L., and Perkins, M. E. (1934), J. Biol. Chem. 107:241.
Hellerman, L., Chinard, F. P., and Ramsdell, P. A. (1941), J. Amer. Chem. Soc. 63:2551.
Hellerman, L., Chinard, F. P., and Deitz, V. R. (1943), J. Biol. Chem. 147:443.
Hellerman, L., Schellenberg, K. A., and Reiss, O. K. (1958), J. Biol. Chem. 233:1468.
Hellerman, L., Coffey, D. S., and Neims, A. H. (1965), J. Biol. Chem. 240:290.
Hellström, N. (1931), Z. Phys. Chem. A157:242.
Hellström, N. (1933), Z. Phys. Chem. A163:33.
Herriott, J. R., Sieker, L. C., Jensen, L. H., and Lovenberg, W. (1970), J. Mol. Biol. 50:391.
Hexter, C. S., and Westheimer, F. H. (1971), J. Biol. Chem. 246:3934.
Heyl, D., Harris, S. A., and Folkers, K. (1948), J. Amer. Chem. Soc. 70:3429.
Hill, R. L., and Kanarek, L. (1964), Brookhaven Symp. Biol. No. 17, p. 80.
Himes, R. H., and Rabinowitz, J. C. (1962), J. Biol. Chem. 237:2903.
Hird, F. J. R. (1962), Biochem. J. 85:320.
Hird, F. J. R., and Yater, J. R. (1961), Biochem. J. 80:612.
Hirs, C. H. W. (1967a), In "Methods in Enzymology," Vol. 11 (Hirs, C. H. W., ed), p. 59, Academic, New York.
Hirs, C. H. W. (1967b), In "Methods in Enzymology," Vol. 11 (Hirs, C. H. W. ed.), p. 201, Academic, New York.
Hoch, F. L., Martin, R. G., Wacker, W. E. C., and Vallee, B. L. (1960), Arch. Biochem. Biophys. 91:166.
Holbrook, J. J., and Jeckel, R. (1969), Biochem. J. 111:689.
Holbrook, J. J., Cooke, R. D., and Kingston, I. B. (1973), Biochem. J. 135:901.
Holbrook, J. J., Pfleiderer, G., Mella, K., Volz, M., Leskowac, W., and Jeckel, R. (1967), Eur. J. Biochem. 1:476.
Hollaway, M. R., Mathias, A. P., and Rabin, B. R. (1964), Biochim. Biophys. Acta 92:111.
Hollaway, M. R., Antonini, E., and Brunori, M. (1971), Eur. J. Biochem. 24:332.
Holzer, H., and Holzer, E. (1952), Hoppe-Seyler's Z. Physiol. Chem. 291:67.
Hong, J. S., and Rabinowitz, J. C. (1967), Biochem. Biophys. Res. Commun. 29:246.
Hong, J. S., Champion, A. B., and Rabinowitz, J. C. (1969), Eur. J. Biochem. 8:307.
Hong, R., and Nisonoff, A. (1965), J. Biol. Chem. 240:3883.
Hooton, B. T. (1968), Biochemistry 7:2063.
Hopkins, F. G. (1921), Biochem. J. 15:286.
Hopkins, F. G. (1929), J. Biol. Chem. 84:269.
Hopkins, F. G., and Morgan, E. J. (1938), Biochem. J. 32:611.
Hopkins, F. G., and Morgan, E. J. (1948), Biochem. J. 42:23.
Hopkins, G., and Hunter, L. (1942), J. Chem. Soc. 638.
Hordvik, A. (1963), Acta Chem. Scand. 17:2575.
Horn, M. J., Jones, D. B., and Ringel, S. J. (1941), J. Biol. Chem. 138:141.
Horn, M. J., Jones, D. B., and Ringel, S. J. (1942), J. Biol. Chem. 144:87, 93.
Huggins, C., Tapley, D. F., and Jensen, E. V. (1951), Nature (London) 167:592.
Huggins, M. L. (1953), J. Amer. Chem. Soc. 75:4123.
Hughes, W. L. (1950), Cold Spring Harbor Sympos. Quant. Biol. 14:79.
Hughes, W. L. (1954), In "The Proteins," Vol. 2, Pt. B (Neurath, H., and Bailey, K., eds.), p. 663, Academic, New York.

Hughes, W. L., and Straessle, R. (1950), J. Amer. Chem. Soc. 72:452.
Huguenin, R. L., and Guttmann, St. (1965), Helv. Chim. Acta 48:1885.
Huisman, T. H. J., and Dozy, A. M. (1962), J. Lab. Clin. Med. 60:302.
Humbel, R. E. (1965), Proc. Nat. Acad. Sci. U. S. 53:853.
Humbel, R. E., Derron, R., and Neumann, P. (1968), Biochemistry 7:621.
Humphrey, R. E., and Hawkins, J. M. (1964), Anal. Chem. 36:1812.
Humphrey, R. E., and Potter, J. L. (1965), Anal. Chem. 37:164.
Husain, S. S., and Lowe, G. (1965), Chem. Commun. 345.
Husain, S. S., and Lowe, G. (1968a), Biochem. J. 108:855.
Husain, S. S., and Lowe, G. (1968b), Biochem. J. 108:861.
Husain, S. S., and Lowe, G. (1970a), Biochem. J. 116:689.
Husain, S. S., and Lowe, G. (1970b), Biochem. J. 117:333, 341.
Imai, K., Takagi, T., and Isemura, T. (1963), J. Biochem. Tokyo 53:1.
Il'in, V. S., and Timova, G. V. (1969), In "Chemical Factors for Regulating the Activity and Biosynthesis of Proteins" (Orekhovich, V. N. ed.), p. 360, Meditsina, Moscow.
Inglis, A. S., and Liu, T.-Y. (1970), J. Biol. Chem. 245:112.
Ingram, L. C. (1969), Biochim. Biophys. Acta 184:216.
Isemura, T., Takagi, T., Maeda, V., and Yutani, K. (1963), J. Biochem. Tokyo 53:155.
Ishikawa, E., Olwer, R. M., and Reed, L. J. (1966), Proc. Nat. Acad. Sci. U. S. 56:534.
Isles, T. E., and Jocelyn, P. C. (1963), Biochem. J. 88:84.
Itano, H. A., and Robinson, E. A. (1972), J. Biol. Chem. 247:4819.
Ivanov, Ch., Alexiev, B., and Krsteva, M. (1963), Biochim. Biophys. Acta 69:195.
Ivanov, Ch., Alexiev, B., and Nishanyan, P. (1967), In "Abstracts of the Symposium on Structure and Function of Peptides and Proteins," Riga, p. 74.
Iyer, K. S., and Klee, W. A. (1973), J. Biol. Chem. 248:707.
Jackson, R. C., Harrap, K. R., and Smith, C. A. (1968), Biochem. J. 110:37P.
Jacob, E. J., Butterworth, P. H. W., and Porter, J. W. (1968), Arch. Biochem. Biophys. 124:392.
Jacobson, G. R., and Stark, G. R. (1973), J. Biol. Chem. 248:8003.
Jacobson, G. R., Schaffer, M. H., Stark, G. R., and Vanaman, T. C. (1973), J. Biol. Chem. 248:6583.
Jacoby, W. B. (1958), J. Biol. Chem. 232:89.
Jaenicke, L., and Lynen, F. (1960), In "The Enzymes," Vol. 3 (Boyer, P., et al., eds.), p. 3, Academic, New York.
Jencks, W. P. (1964), Progr. Phys. Org. Chem. 2:63.
Jencks, W. P., and Salvesen, K. (1971), J. Amer. Chem. Soc. 93:4433.
Jencks, W. P., Cordes, S., and Carriuolo, J. (1960), J. Biol. Chem. 235:3608.
Jeng, D.-Y., and Mortenson, L. E. (1968), Biochem. Biophys. Res. Commun. 32:984.
Jiang, R. Q., Du, Y. C., and Tsou, C. L. (1963), Sci. Sinica (Peking) 12:452.
Jirgensons, B., and Ikenaka, T. (1959), Makromol. Chem. 31:112.
Jirousek, L., and Pritchard, E. T. (1971a), Biochim. Biophys. Acta 229:618.
Jirousek, L., and Pritchard, E. T. (1971b), Biochim. Biophys. Acta 243:230.
Jocelyn, P. C. (1962), Biochem. J. 85:480.
Jocelyn, P. C. (1967), Eur. J. Biochem. 2:327.
Jocelyn, P. C. (1972), "Biochemistry ofthe SH Group," Academic, London.
Jones, D. H., and Nelson, W. L. (1969), Biochemistry, 8:2622.

Josien, M. L., Dizabo, P., and Saumagne, P. (1957), Bull. Soc. Chim. Fr. 423.
Jőst, K., and Rudinger, J. (1967), Collect. Czech. Chem. Commun. 32:1229.
Jőst, K., Rudinger, J., Klostermeyer, H., and Zahn, H. (1968), Z. Naturforsch. B. 23:1059.
Jung, G., Breitmaier, E., and Voelter, W. (1972), Eur. J. Biochem. 24:438.
Kägi, J. H. R., and Valle, B. L. (1960), J. Biol. Chem. 235:3188.
Kalan, S. B., Neistadt, A., and Weil, L. (1965), Fed. Proc. 24:225.
Kanaoka, Y., Machida, M., Kokubun, H., and Sekine, T. (1968), Chem. Pharm. Bull. Tokyo 16:1747.
Kanarek, L., Bradshaw, R. A., and Hill, R. L. (1965), J. Biol. Chem. 240:2755.
Kaper, J. M., and Houwing, C. (1962), Arch. Biochem. Biophys. 97:449.
Kaper, J. M., and Jemfer, F. G. (1967), Biochemistry 6:440.
Karush, F., Klinman, N. R., and Marks, R. (1964), Anal. Biochem. 9:100.
Katsoyannis, P. G., Okada, Y., and Zalut, C. (1973), Biochemistry 12:2516.
Katsoyannis, P. G. (1967), Recent Progr. Horm. Res. 23:505.
Katsoyannis, P. G., and Tometsko, A. (1966), Proc. Nat. Acad. Sci. U. S. 55:1554.
Katsoyannis, P. G., Trakatellis, A. C., Johnson, S., Zalut, C., and Schwartz, G. (1967), Biochemistry 6:2642.
Katzen, H. M., Tietze, F., and Stetten, D. (1963), J. Biol. Chem. 238:1006.
Kaufman, B. T. (1964), J. Biol. Chem. 239:PC 669.
Kaufman, B. T. (1966), Proc. Nat. Acad. Sci. U. S. 56:695.
Kauzmann, W. (1959), In "Sulfur in Proteins," Proc. Symp. Falmouth 1958, p. 93.
Kawahara, K., and Tanford, C. (1966), Biochemistry 5:1578.
Ke, B. (1957), Biochim. Biophys. Acta 25:650.
Kefalides, N. A. (1968), Biochemistry 7:3103.
Kelly-Falcoz, F., Greenberg, H., and Horecker, B. L. (1965), J. Biol. Chem. 240:2966.
Kemp, R. G., and Forest, P. B. (1968), Biochemistry 7:2596.
Kendall, E. C. McKenzie, B. F., and Mason, H. L. (1929), J. Biol. Chem. 84:657.
Keresztes-Nagy, S., and Klotz, I. M. (1963), Biochemistry 2:923.
Kharasch, N. (ed.) (1961), "Organic Sulfur Compounds," Pergamon, Oxford.
Kice, K. L. (1968), Accounts Chem. Res. 1:58.
Kidder, G. W., and Dewey, V. C. (1949), Arch. Biochem. 20:433.
Kielley, W. W., and Bradley, L. B. (1956), J. Biol. Chem. 218:653.
Kilmartin, J. V., and Rossi-Bernardi, L. (1971), Biochem. J. 124:31.
Kimmel, J. R., and Parcells, A. J. (1960), Fed. Proc. 19:341.
Kimmel, J. R., and Smith, E. L. (1954), J. Biol. Chem. 207:515.
Kimmel, J. R., and Smith, E. L. (1957), Advan. Enzymol. 19:267.
King, T. P. (1961), J. Biol. Chem. 236:PC5.
Kirsch, J. F., and Igelström, M. (1966), Biochemistry 5:783.
Kirsch, J. F., and Katchalski, E. (1965), Biochemistry 4:884.
Kirtley, M. E., and Rudney, H. (1967), Biochemistry 6:230.
Klapper, M. H. (1970), Biochem. Biophys. Res. Commun. 38:172.
Klee, C. B. (1970), J. Biol. Chem. 245:3143.
Klee, C. B., and Gladner, J. A. (1972), J. Biol. Chem. 247:8051.
Klee, W. A., and Cantoni, G. L. (1960), Biochim. Biophys. Acta 45:545.
Klein, I. B., and Kirsch, J. F. (1969a), Biochem. Biophys. Res. Commun. 34:575.
Klein, I. B., and Kirsch, J. F. (1969b), J. Biol. Chem. 244:5928.

Kline, L., and Barker, H. A. (1950), J. Bacteriol. 60:349.
Klotz, I. M. (1960), Brookhaven Symp. Biol. No. 13, p. 25.
Klotz, I. M., and Campbell, B. J. (1962), Arch. Biochem. Biophys. 96:92.
Klotz, I. M., and Carver, B. R. (1961), Arch. Biochem. Biophys. 95:540.
Klotz, I. M., and Fiess, H. A. (1951), J. Phys. Colloid. Chem. 55:101.
Klotz, I. M., and Klotz, T. A. (1959), In "Sulfur in Proteins," Proc. Symp. Falmouth, 1958, p. 127.
Klotz, I. M., Faller, J. L., and Urquhart, J. M. (1950), J. Phys. Chem. 54:18.
Knox, W. E. (1960), In "The Enzymes," Vol. 2 (Boyer, P., et al. eds.), p. 253, Academic, New York.
Koeppe, O. J., Boyer, P. D., and Stulberg, M. P. (1956), J. Biol. Chem. 219:569.
Kohno, K. (1966), J. Vitaminol. 12:137.
Koike, M., and Reed, L. J. (1960), J. Biol. Chem. 235:1931.
Kolthoff, I. M., and Anastasi, A., Tan, B. H. (1958), J. Amer. Chem. Soc. 80:3235.
Kolthoff, I. M., and Harris, W. E. (1946), Ind. Eng. Chem. Anal. Ed. 18:161.
Kolthoff, I. M., and Stricks, W. (1950), J. Amer. Chem. Soc. 72:1952.
Kolthoff, I. M., and Stricks, W. (1951a), Anal. Chem. 23:763.
Kolthoff, I. M., and Stricks, W. (1951b), J. Amer. Chem. Soc. 73:1728.
Kolthoff, I. M., Stricks, W., and Morren, L. (1954), Anal. Chem. 26:366.
Kolthoff, I. M., Stricks, W., and Kapoor, R. C. (1955a), J. Amer. Chem. Soc. 77:4733.
Kolthoff, I. M., Stricks, W., and Tanaka, N. (1955b), J. Amer. Chem. Soc. 77:4739.
Kolthoff, I. M., Matsuoka, M., Tan, B. H., and Shore, W. S. (1965a), Biochemistry 4:2389.
Kolthoff, I. M., Shore, W. S., Tan, B. H., and Matsuoka, M. (1965b), Anal. Biochem. 12:497.
Korman, S., and Clarke, H. T. (1956), J. Biol. Chem. 221:113, 133.
Kosower, E. M. (1956), J. Amer. Chem. Soc. 78:3497.
Kotaki, A., Harada, M., and Yagi, K. (1964), J. Biochem. Tokyo 55:553.
Kress, L. F., Bono, V. H., and Noda, L. (1966), J. Biol. Chem. 241:2293.
Kress, L. F., Wilson, K. A., and Laskowski, M. (1968), J. Biol. Chem. 243:1758.
Krimsky, I., and Racker, E. (1952), J. Biol. Chem. 198:721.
Krimsky, I., and Racker, E. (1954), Fed. Proc. 13:245.
Krimsky, I., and Racker, E. (1955), Science 122:319.
Krimsky, I., and Racker, E. (1963), Biochemistry 2:512.
Krull, L. H., and Friedman, M. (1967), Biochem. Biophys. Res. Commun. 29:373.
Krull, L. H., Gibbs, D. E., and Friedman, M. (1971), Anal. Biochem. 40:80.
Kuhn, N. J., and Lynen, F. (1965), Biochem. J. 94:240.
Kuhn, R., and Hammer, I. (1951), Ber. 84:91.
Kumar, K. S. V. S., Walsh, K. A., Bargetzi, J.-P., and Neurath, H. (1963), In "Aspects of Protein Structure" (Ramachandran, G. N., ed.) p. 319, Academic, New York.
Kuo, T., and De Luca, M. (1969), Biochemistry 8:4762.
Kuramitsu, H. K. (1968), J. Biol. Chem. 243:1016.
Lack, L. (1961), J. Biol. Chem. 236:2835.
Lake, A. W., and Lowe, G. (1966), Biochem. J. 101:402.
Laland, S. G., and Zimmer, T.-L. (1973), Essays Biochem. 9:31.
Lamfrom, H., and Nielsen, S. O. (1957), J. Amer. Chem. Soc. 79:1966.

Lamfrom, H., and Nielsen, S. O. (1958), Trav. Lab. Carlsberg, Ser. Chim. 30:349.
Larrabee, A. R., McDaniel, E. G., Bakerman, H. A., and Vagelos, P. R. (1965), Proc. Nat. Acad. Sci. U. S. 54:267.
Lawrence, F. K., and Laskowski, M. (1967), J. Biol. Chem. 242:4925.
Lawrence, P. J. (1969), Biochemistry 8:1271.
Lazarus, N. R., Derechin, M., and Barnard, E. A. (1968), Biochemistry 7:2390.
Leach, S. J. (1959), Biochim. Biophys. Acta 33:264.
Leach, S. J. (1960), Aust. J. Chem. 13:520.
Leach, S. J. (1966), In "Analytical Methods of Protein Chemistry" (Alexander, P., and Lundgren, H. P., ed.), Vol. 4, p. 3, Pergamon, Oxford.
Leach, S. J., and O'Donnell, I. J. (1961), Biochem. J. 79:287.
Leach, S. J., Swan, J. M., and Holt, L. S. (1963), Biochim. Biophys. Acta 78:196.
Leach, S. J., Meschers, A., and Swanepoel, O. A. (1965), Biochemistry 4:23.
Leach, S. J., Meschers, A., and Springell, P. H. (1966), Anal. Biochem. 15:18.
Lee, C. C., and Samuels, E. R. (1964), Can. J. Chem. 42:168.
Lee, P. P., and Westheimer, F. H. (1966), Biochemistry 5:834.
Leitzmann, C., Wu, J.-Y., and Boyer, P. D. (1970), Biochemistry 9:2338.
Leslie, J. (1967), Arch. Biochem. Biophys. 121:463.
Leslie, J., and Varricchio, F. (1968), Can. J. Biochem. 46:625.
Leslie, J., Butler, L. G., and Gorin, G. (1962a), Arch. Biochem. Biophys. 99:86.
Leslie, J., Williams, D. L., and Gorin, G. (1962b), Anal. Biochem. 3:257.
Levin, E. D. (1965), "The Scission of the Disulfide Bonds of Pepsin," Candidate's thesis, Moscow.
Levinthal, C., Signer, E. R., and Fetherolf, K. (1962), Proc. Nat. Acad. Sci. U. S. 48:1230.
Levison, M. E., Josephson, A. S., and Kirschenbaum, D. M. (1969), Experientia 25:126.
Li, T.-K., and Vallee, B. L. (1963), Biochem. Biophys. Res. Commun. 12:44.
Li, T.-K., and Vallee, B. L. (1964), Biochemistry 3:869.
Li, T.-K., and Vallee, B. L. (1965), Biochemistry 4:1195.
Lienhard, G. E., and Jencks, W. P. (1966), J. Amer. Chem. Soc. 88:3982.
Light, A., and Sinha, N. K. (1967), J. Biol. Chem. 242:1358.
Light, A., Frater, R., Kimmel, J. R., and Smith, E. L. (1964), Proc. Nat. Acad. Sci. U. S. 52:1276.
Light, A., Hardwick, B. C., Hatfield, L. M., and Sondack, D. L. (1969), J. Biol. Chem. 244:6289.
Liljas, A., Kannan, K. K., Bergstén, P.-C., Warra, I., Fridborg, K., Strandberg, B., Carlbom, U., Järup, L., Lövgren, S., and Petef, M. (1972), Nature (London), New Biol. 235:131.
Linderström-Lang, K., and Jacobsen, C. F. (1940), C. R. Trav. Lab. Carsberg 23:289.
Linderström-Lang, K., and Jacobsen, C. F. (1941), J. Biol. Chem. 137:443.
Lindley, H. (1956), Nature (London) 178:647.
Lindley, H. (1959), In "Sulfur in Proteins," Proc. Symp. Falmouth, 1958, p. 33.
Lindley, H. (1960), Biochem. J. 74:577.
Lindley, H. (1962), Biochem. J. 82:418.
Lindley, H., and Haylett, T. (1968), Biochem. J. 108:701.
Lindquist, R. N., and Cordes, E. H. (1968), J. Biol. Chem. 243:5837.

Link, T. P., and Stark, G. R. (1968), J. Biol. Chem. 243:1082.
Lipmann, F. (1945), J. Biol. Chem. 160:173.
Lipmann, F. (1946), Advan. Enzymol. 6:231.
Lipmann, F. (1948-1949), Harvey Lectures 44:99.
Lipmann, F. (1953), Fed. Proc. 12:673.
Lipmann, F., Kaplan, N. O., Novelli, G. D., Tuttle, L. C., and Guirard, B. M. (1947), J. Biol. Chem. 167:869.
Lipmann, F., Gevers, W., Kleinkauf, H., and Roskoski, R. (1970), Int. Cong. Biochem. 8th, Abst., p. 257.
Lipmann, F., Gevers, W., Kleinkauf, H., and Roskoski, R. (1971), Advan. Enzymol. 35:1.
Lipscomb, W. N., Hartsuck, J. A., Reeke, G. N., Quiocho, F. A., Bethge, P. H., Ludwig, M.L., Steitz, T. A., Muirhead, H., and Coppola, J. C. (1968), Brookhaven Symp. Biol. 21:24.
Little, C., and O'Brien, P. J. (1967a), Arch. Biochem. Biophys. 122:406.
Little, C., and O'Brien, P. J. (1967b), Biochem. J. 102:10P.
Little, C., Sanner, T., and Pihl, A. (1969), Eur. J. Biochem. 10:533.
Little, C., Sanner, T., and Pihl, A. (1969), Eur. J. Biochem. 8:229.
Little, G., and Brocklehurst, K. (1972), Biochem. J. 128:475.
Liu, T.-Y. (1967), J. Biol. Chem. 242:4029.
Liu, T.-Y., Stein, W. H., Moore, S., and Elliott, S. D. (1965), J. Biol. Chem. 240:1143.
Liu, W.-K., and Meienhofer, J. (1968), Biochem. Biophys. Res. Commun. 31:467.
Liu, W.-K., Trzeciak, H., Schüssler, H., and Meienhofer, J. (1971), Biochemistry 10:2849.
Lohmann, K. (1932), Biochem. Z. 254:332.
Lotspeich, W. D., and Peters, R. A. (1951), Biochem. J. 49:704.
Lovenberg, W., and McCarthy, K. (1968), Biochem. Biophys. Res. Commun. 30:453.
Lovenberg, W., and Williams, W. M. (1969), Biochemistry 8:141.
Lovenberg, W., Buchanan, B. B., and Rabinowitz, J. C. (1963), J. Biol. Chem. 238:3899.
Lowe, G., and Williams, A. (1964), Proc. Chem. Soc. 140.
Lowe, G., and Williams, A. (1965), Biochem. J. 96:189, 194, 199.
Lowe, G., and Yuthavong, Y. (1971), Biochem. J. 124:107, 117.
Lucas, E. C., and Williams, A. (1969), Biochemistry 8:5125.
Lucas, F. (1966), Nature (London) 210:952.
Ludescher, U., and Schwyzer, R. (1971), Helv. Chim. Acta 54:1637.
Lumper, L., and Zahn, H. (1965), Advan. Enzymol. 27:199.
Lundsgaard, E. (1930a), Biochem. Z. 217:162.
Lundsgaard, E. (1930b), Biochem. Z. 220:1, 8.
Luse, R. A., and McLaren, A. D. (1963), Photochem. Photobiol. 2:343.
Lyman, C. L., and Barron, E. S. G. (1937), J. Biol. Chem. 121:275.
Lynen, F. (1953), Fed. Proc. 12:683.
Lynen, F. (1961), Fed. Proc. 20:941.
Lynen, F. (1970), Biochem. Soc. Symp. No. 31, p. 1.
Lynen, F., and Reichert, E. (1951), Angew. Chem. 63:47.
Lynen, F., Reichert, E., and Rueff, L. (1951), Justus Liebig's Ann. Chem. 574:1.
Lyons, W. (1948), Nature (London) 162:1004.
Mackay, D. (1962), Arch. Biochem. Biophys. 99:93.

Maclaren, J. A. (1962), Aust. J. Chem. 15:824.
Maclaren, J. A., and Kirkpatrick, A. (1968), J. Soc. Dyers Colour. 84:564.
Maclaren, J. A., and Sweetman, B. J. (1966), Austr. J. Chem. 19:2355.
Maclaren, J. A., Savige, W. E., and Sweetman, B. J. (1965), Aust. J. Chem. 18:1655.
Maclaren, J. A., Kilpatrick, D. J., and Kirkpatrick, A. (1968), Aust. J. Biol. Sci. 21:805.
Macquarrie, R. A., and Bernhard, S. A. (1971), Biochemistry 10:2456.
Madsen, N. B. (1956), J. Biol. Chem. 223:1067.
Madsen, N. B. (1963) In "Metabolic Inhibitors," Vol. 2 (Hochster, R. M., and Quastel, J. H., eds.), p. 119, Academic, New York.
Madsen, N. B., and Cori, C. F. (1956), J. Biol. Chem. 223:1055.
Madsen, N. B., and Gurd, F. R. N. (1956), J. Biol. Chem. 223:1075.
Maeda, H., Glaser, C. B., and Meienhofer, J. (1970), Biochem. Biophys. Res. Commun. 39:1211.
Majerus, P. W., and Vagelos, P. R. (1965), Fed. Proc. 24:290.
Majerus, P. W., Alberts, A. W., and Vagelos, P. R. (1964), Proc. Nat. Acad. Sci. U. S. 51:1231.
Majerus, P. W., Alberts, A. W., and Vagelos, P. R. (1965a), Proc. Nat. Acad. Sci. U. S. 53:410.
Majerus, P. W., Alberts, A. W., and Vagelos, P. R. (1965b), J. Biol. Chem. 240:4723.
Malkin, R., and Rabinowitz, J. C. (1966), Biochemistry 5:1262.
Malkinson, A. M., and Hardman, J. K. (1969), Biochemistry 8:2769.
Manojlović-Muir, L. M. (1969), Nature (London) 224:686.
Markus, G. (1960), Fed. Proc. 19:340.
Markus, G. (1964), J. Biol. Chem. 239:4163.
Markus, G., and Karush, F. (1957), J. Amer. Chem. Soc. 79:134, 3264.
Markus, G., Grossberg, A. L., and Pressman, D. (1962), Arch. Biochem. Biophys. 96:63.
Marler, E., Nelson, C. A., and Tanford, C. (1964), Biochemistry 3:279.
Martin, R. B., and Edsall, J. T. (1958), Bull. Soc. Chim. Biol. 40:1763.
Martin, R. B., and Hedrick, R. I. (1962), J. Amer. Chem. Soc. 84:106.
Masri, M. S., Windle, J. J., and Friedman, M. (1972), Biochem. Biophys. Res. Commun. 47:1408.
Massey, V. (1960), J. Biol. Chem. 235:PC47.
Massey, V. (1963), In "The Enzymes," Vol. 7 (Boyer, P., et al., eds.), p. 275, Academic, New York.
Massey, V., and Edmondson, D. (1970), J. Biol. Chem. 245:6595.
Massey, V., and Gibson, Q. H. (1962), Int. Cong. Biochem. 5th, 5:157.
Massey, V., and Veeger, C. (1960), Biochim. Biophys. Acta 40:184.
Massey, V., and Veeger, C. (1961), Biochim. Biophys. Acta 48:33.
Massey, V., and Williams, C. H. (1965), J. Biol. Chem. 240:4470.
Massey, V., Gibson, Q. H., and Veeger, C. (1960), Biochem. J. 77:341.
Massey, V., Hofmann, T., and Palmer, G. (1962), J. Biol. Chem. 237:3820.
Massey, V., Brumby, P. E., Komai, H., and Palmer, G. (1969), J. Biol. Chem. 244:1682.
Mastin, H., and Schryver, S. B. (1926), Biochem. J. 20:1177.
Mathew, E., Agnello, C. F., and Park, J. H. (1965), J. Biol. Chem. 240:PC3232.
Mathew, E., Meriwether, B. P., and Park, J. H. (1967), J. Biol. Chem. 242:5024.
Matsumura, S., and Stumpf, P. K. (1968), Arch. Biochem. Biophys. 125:932.

Matsuo, Y. (1957), J. Amer. Chem. Soc. 79:2011.
Mayerle, J. J., Frankel, R. B., Holm, R. H., Ibers, J. A., Phillips, W. D., and Weiher, J. F. (1973), Proc. Nat. Acad. Sci. U.S. 70:2429.
McBride, O. W., and Harrington, W. F. (1965), J. Biol. Chem. 240:PC4545.
McBride, O. W., and Harrtington, W. F. (1967), Biochemistry 6:1484, 1499.
McKenzie, H. A., and Shaw, D. C. (1972), Nature (London) New Biol. 238:147.
McMurray, C. H., and Trenthan, D. R. (1969), Biochem. J. 115:913.
McPhee, J. R. (1956), Biochem. J. 64:22.
Meienhofer, J., Schnabel, E., Brenner, H., Brinkhoff, O., Zabel, R., Snoka, W., Keostermeyer, H., Brendenberg, D., Akuda, T., and Zahn, H. (1963), Z. Naturforsch. 18b:1120.
Mendel, L. B., and Blood, A. F. (1910), J. Biol. Chem. 8:177.
Michaelis, L., and Schubert, M. P. (1934), J. Biol. Chem. 106:331.
Middlebrook, W. R., and Phillips, H. (1942), Biochem. J. 36:428.
Miller, F., and Metzger, H. (1965), J. Biol. Chem. 240:4740.
Miller, R. W., and Massey, V. (1965), J. Biol. Chem. 240:1453.
Milligan, B., and Swan, J. M. (1962), Rev. Pure Appl. Chem. 12:72.
Milstein, C., and Pink, J. R. L. (1970), Prog. Biophys. Mol. Biol. 21:211.
Minkin, V. I., Osipov, O. A., and Zhdanov, Yu. A. (1968), "Dipole Moments in Organic Chemistry," p. 77, Khimiya, Leningrad [Engl. transl. Plenum, New York (1970)].
Mirsky, A. E. (1941), J. Gen. Physiol. 24:709, 725.
Mirsky, A. E., and Pauling, L. (1936), Proc. Nat. Acad. Sci. U. S. 22:439.
Mitchel, R. E. J., Chaiken, I. M., and Smith, E. L. (1970), J. Biol. Chem. 245:3485.
Mize, C. E., Thompson, T. E., and Langdon, R. G. (1962), J. Biol. Chem. 237:1596.
Mizobuchi, K., and Buchanan, J. M. (1966), Fed. Proc. 25:586.
Modig, H. (1968), Biochem. Pharmacol. 17:177.
Moore, E. C., Reichard, P., and Thelander, L. (1964), J. Biol. Chem. 239:3445.
Moore, S. (1963), J. Biol. Chem. 238:235.
Morino, Y., and Snell, E. E. (1967), J. Biol. Chem. 242:5591.
Mörner, K. A. H. (1899), Hoppe-Seyler's Z. Physiol. Chem. 28:595.
Morris, J. E., and Inman, F. P. (1968), Biochemistry 7:2851.
Murachi, T., and Takahashi, N. (1970), "Struct.-Funct. Relat. Proteolytic Enzymes," Proc. Int. Symp., p. 298, Munksgaard, Copenhagen.
Murphy, A. J., and Morales, M. F. (1970), Biochemistry 9:1528.
Nagai, S., and Black, S. (1968), J. Biol. Chem. 243:1942.
Nagy, J., and Straub, F. B. (1966), Acta Biochim. Biophys. 1:355.
Nagy, J., and Straub, F. B. (1969), Acta Biochim. Biophys. Acad. 4:15.
Naider, F., and Bohak, Z. (1972), Biochemistry 11:3208.
Nakagawa, Y., and Perlmann, G. E. (1970), Arch. Biochem. Biophys. 140:464.
Nakagawa, Y., and Perlmann, G. E. (1971), Arch. Biochem. Biophys. 144:59.
Nakai, N., and Hase, J. (1968), Chem. Pharm. Bull. Tokyo 16:2334, 2339.
Nakata, T., and Yagi, K. (1969), J. Biochem. Tokyo 66:409.
Narita, K., and Akao, M. (1965), J. Biochem. Tokyo 58:507.
Nawa, H., Brady, W. T., Koike, M., and Reed, L. J. (1960), J. Amer. Chem. Soc. 82:896.
Neims, A. H., Coffey, D. S., and Hellerman, L. (1966), J. Biol. Chem. 241:3036, 5941.

Nelander, B., and Sunner, S. (1972), J. Amer. Chem. Soc. 94:3576.
Némethy, G., and Scheraga, H. A. (1962), J. Phys. Chem. 66:1773.
Neuberg, C. (1902), Ber. 35:3161.
Neumann, H., Goldberger, R. F., and Sela, M. (1964), J. Biol. Chem. 239:1536.
Neumann, H., and Smith, R. A. (1967), Arch. Biochem. Biophys. 122:354.
Neumann, H., Steinberg, I. Z., Brown, J. R., Goldberger, R. F., and Sela, M. (1967), Eur. J. Biochem. 3:171.
Nezlin, R. S. (1966), "Biochemistry of Antibodies," Nauka, Moscow [Engl. trans. Plenum, New York (1970)].
Nickerson, W. J., Falcone, G., and Strauss, G. (1963), Biochemistry 2:537.
Nirenberg, M. W., and Jacoby, W. B. (1960), Proc. Nat. Acad. Sci. U. S. 46:206.
Nikkel, H. J., and Foster, J. F. (1971), Biochemistry 10:4479.
Noda, L. H., Kuby, S. A., and Lardy, H. A. (1953), J. Amer. Chem. Soc. 75:913.
Nolan, C., Margoliash, E., Peterson, J. D., and Steiner, D. F. (1971), J. Biol. Chem. 246:2780.
Nowak, T., and Himes, R. H. (1971), J. Biol. Chem. 246:1285.
O'Donnell, I. J., and Thompson, E. O. P. (1968), Aust. J. Biol. Sci. 21:385.
Ogilvie, J. W., Tildon, J. T., and Strauch, B. S. (1964), Biochemistry 3:754.
Okamoto, M., and Morino, Y. (1973), J. Biol. Chem. 248:82.
O'Kane, D.J., and Gunsalus, I. C. (1948), J. Bacteriol. 56:499.
Oliver, L., and Hartree, A. S. (1968), Biochem. J. 109:19.
Olomucki, M., and Diopon, J. (1972), Biochim. Biophys. Acta 263:213.
Olson, E. J., and Park, J. H. (1964), J. Biol. Chem. 239:2316.
Overberger, C. G., and Ferraro, J. J. (1962), J. Org. Chem. 27:3539.
Palacián, E., and Neet, K. E. (1970), Biochim. Biophys. Acta 212:158.
Palmer, G., and Massey, V. (1962), Biochim. Biophys. Acta 58:349.
Palmer, J. L., and Nisonoff, A. (1964), Biochemistry 3:863.
Park, J. H., Meriwether, B. P., Clodfelder, P., and Cunningham, L. W. (1960), Fed. Proc. 19:29.
Park, J. H., Meriwether, B. P., Clodfelder, P., and Cunningham, L. W. (1961), J. Biol. Chem. 236:136.
Park, J. H., Agnello, C. F., and Mathew, E. (1966), J. Biol. Chem. 241:769.
Parker, A. J., and Kharasch, N. (1959), Chem. Rev. 59:583.
Parker, A. J., and Kharasch, N. (1960), J. Amer. Chem. Soc. 82:3071.
Parker, D. J., and Allison, W. S. (1969), J. Biol. Chem. 244:180.
Pasynskii, A. G., and Chernyak, R. S. (1952), Biokhimiya 17:198.
Patchornik, A., and Degani, Y. (1971), J. Org. Chem. 36:2727.
Perham, R. N. (1969), Biochem. J. 111:17.
Perham, R. N., and Harris, J. I. (1963), J. Mol. Biol. 7:316.
Perkins, D. J. (1961), Biochem. J. 80:668.
Perlstein, M. T., Atassi, M. Z., and Cheng, S. H. (1971), Biochim. Biophys. Acta 236:174.
Perutz, M. F. (1965), J. Mol. Biol. 13:646.
Perutz, M. F., Muirhead, H., Cos, J. M., and Goaman, L. C. G. (1968), Nature (London) 219:131.
Peters, R. A. (1936), Nature (London) 138:327.
Peters, R. A. (1963), "Biochemical Lesions and Lethal Synthesis," Pergamon, Oxford.

Peters, R. A., Rydin, H., and Thompson, R. H. S. (1935), Biochem. J. 29:53, 61.
Peters, R. A., Sinclair, H. M., and Thompson, R. H. S. (1946), Biochem. J. 40:516.
Petushkova, E. V., and Bocharnikova, I. M. (1968), Biokhimiya 33:618 [Engl. transl. p. 504].
Pflumm, M. N., and Beychok, S. (1969), J. Biol. Chem. 244:3982.
Philipson, L. (1962), Biochim. Biophys. Acta 56:375.
Phillips, G. T., Nixon, J. E., Dorsey, J. A., Butterworth, P. H. W., Chesterton, C. J., and Porter, J. W. (1970), Arch. Biochem. Biophys. 138:380.
Piez, K. A. (1968), Ann. Rev. Biochem. 37:547.
Pihl, A., Lange, R. (1962), J. Biol. Chem. 237:1356.
Pihl, A., and Sanner, T. (1963a), Radiat. Res. 19:27.
Pihl, A., and Sanner, T. (1963b), Biochim. Biophys. Acta 78:537.
Pikkarainen, J., Rantanen, J., Vastamäki, M., Lampiaho, K., Kari, A., and Kulonen, E. (1968), Eur. J. Biochem. 4:555.
Pinkus, L. M., and Meister, A. (1972), J. Biol. Chem. 247:6119.
Pirie, N. W., and Pinhey, K. G. (1929), J. Biol. Chem. 84:321.
Plant, D., Tarbell, D. S., and Whiteman, C. (1955), J. Amer. Chem. Soc. 77:1572.
Plummer, T. H., and Hirs, C. H. W. (1964), J. Biol. Chem. 239:2530.
Poe, M., Phillips, W. D., McDonald, C. C., and Lovenberg, W. (1970), Proc. Nat. Acad. Sci. U. S. 65:797.
Poglazov, B. F., and Baev, A. A. (1961), Biokhimiya 26:535 [Engl. transl. p. 475].
Poglazov, B. F., Bilushi, V., and Baev, A. A. (1958), Biokhimiya 23:269.
Polgar, L. (1964), Acta Physiol. Acad. Sci. Hung. 25:1.
Polgar, L. (1966a), Experientia 22:232.
Polgar, L. (1966b), Biochim. Biophys. Acta 118:276.
Polgar, L. (1973), Eur. J. Biochem. 33:104.
Polyanovskii, O. L., and Telegdi, M. (1965), Biokhimiya 30:174 [Engl. transl. p. 148].
Pontremoli, S., Luppis, B., Traniello, S., Wood, W. A., and Horecker, B. L. (1965), J. Biol. Chem. 240:3469.
Pontremoli, S., Traniello, S., Enser, M., Shapiro, S., and Horecker, B. L. (1967), Proc. Nat. Acad. Sci. U. S. 58:286.
Porter, R. R. (1962), In "Symposium on Basic Problems in Neoplastic Diseases" (Gellhorn, A., and Hirschberg, E., Eds.), p. 177, Columbia Univ., New York.
Porter, R. R. (1967), Biochem. J. 105:417.
Prescott, D. J., Elovson, J., and Vagelos, P. R. (1969), J. Biol. Chem. 244:4517.
Price, P. A., Stein, W. H., and Moore, S. (1969), J. Biol. Chem. 244:929.
Pugh, E. L., and Wakil, S. J. (1965), J. Biol. Chem. 240:4727.
Putnam, F. W. (1953), In "Proteins," Vol. 1, Pt. B, p. 807, Academic, New York.
Putnam, F. W. (1969), Science 163:633.
Quiocho, F. E., and Thomas, J. W. (1973), Proc. Nat. Acad. Sci. U.S. 70:2858.
Rabenstein, D. L. (1973), J. Amer. Chem. Soc. 95:2797.
Rabin, B. R., and Trown, P. W. (1964), Nature (London) 202:1290.
Rabin, B. R., and Watts, D. C. (1960), Nature (London) 188:1163.
Rabin, B. R., and Whitehead, E. P. (1962), Nature (London) 196:658.
Racker, E. (1951), J. Biol. Chem. 190:685.
Racker, E. (1954), In "A Symposium on the Mechanism of Enzyme Actions" (McElroy, E. D., and Glass, B., eds.), p. 464, John Hopkins, Baltimore.

Racker, E. (1955), Physiol. Rev. 35:1.

Racker, E., and Krimsky, J. (1952), J. Biol. Chem. 198:731.

Racker, E., and Krimsky, J. (1958), Fed. Proc. 17:1135.

Raftery, M. A., and Cole, R. D. (1963), Biochem. Biophys. Res. Commun. 10:467.

Raftery, M. A., and Cole, R. D. (1966), J. Biol. Chem. 241:3457.

Ramponi, G., Cappugi, G., Treves, C., and Nassi, P. (1971), Biochemistry 10:2082.

Rapkine, L. (1933), C. R. Soc. Biol. 112:790, 1294.

Rapkine, L. (1936), J. Chim. Phys. 33:493.

Rapkine, L. (1938), Biochem. J. 32:1729.

Rapkine, L., Rapkine, S. M., and Trpinac, P. (1939), C. R. Acad. Sci. 209:253.

Ratner, S., and Clarke, H. T. (1937), J. Amer. Chem. Soc. 59:200.

Reed, L. J. (1960), In "The Enzymes," Vol. 3 (Boyer, P., et al., eds.), p. 195, Academic, New York.

Reed, L. J. (1966), In "Comprehensive Biochemistry," Vol. 14 (Florkin, M., and Stotz, E. H., eds.), p. 99, Elsevier, Amsterdam.

Reed, L. J. (1968), J. Vitaminol. 14:77.

Reed, L. J., De Busk, B. G., Gunsalus, I. C., and Hornberger, C. S. (1951), Science 114:93.

Reed, L. J., Gunsalus, I. C., Schnakenberg, G. H. F., Soper, Q. F., Boaz, H. E., Kern, S. F., and Parke, T. V. (1953), J. Amer. Chem. Soc. 75:1267.

Reed, L. J., Koike, M., Levitch, M. E., and Leach, F. R. (1958), J. Biol. Chem. 232:143.

Reeke, G. H., Hartsuck, J. A., Ludwig, M. L., Quiocho, F. A., Steitz, T. A., and Lipscomb, W. N. (1967), Proc. Nat. Acad. Sci. U. S. 58:2220.

Reichmann, M. E., and Hatt, D. L. (1961), Biochim. Biophys. Acta 49:153.

Reis, P. J., Tunks, D. A., Williams, O. B., and Williams, A. J. (1967), Aust. J. Biol. Sci. 20:153.

Rejnek, J., Appella, E., Mage, R. G., and Reisfeld, R. A. (1969), Biochemistry 8:2712.

Resnik, R. A. (1964), Nature (London) 203:880.

Richards, E. G., Snow, D. L., and McClare, C. W. F. (1966), Biochemistry 5:485.

Rickli, E. E., and Edsall, J. T. (1962), J. Biol. Chem. 237:PC258.

Riggs, A. F. (1953), J. Gen. Physiol. 36:1.

Riggs, A. F. (1959), In "Sulfur in Proteins," Proc. Symp. Falmouth, 1958, p. 173.

Riggs, A. F., and Wolbach, R. A. (1956), J. Gen. Physiol. 39:585.

Riordan, J. F., and Christen, P. (1968), Biochemistry 7:1525.

Risi, S., Dose, K., Rathinasamy, T. K., and Augenstein, L. (1967), Photochem. Photobiol. 6:423.

Robbins, F. M., and Fioriti, J. A. (1963), Nature (London) 200:577.

Robinson, G. W., Bradshaw, R. A., Kanarek, L., and Hill, R. L. (1967), J. Biol. Chem. 242:2709.

Robyt, J. F., Ackerman, R. J., and Chittenden, C. G. (1971), Arch. Biochem. Biophys. 147:262.

Rochat, C., Rochat, H., and Edman, P. (1970), Anal. Biochem. 37:259.

Rogers, K. S., Thompson, T. E., and Hellerman, L. (1962), Biochim. Biophys. Acta 64:202.

Rogers, K. S., Geiger, P. J., Thompson, T. E., and Hellerman, L. (1963), J. Biol. Chem. 238:481.

Rohrbach, M. S., Humphries, B. A., Yost, F. J., Rhodes, W. G., Boatman, S., Hiskey, R. G., and Harrison, J. H. (1973), Anal. Biochem. 52:127.
Ronchi, S., and Williams, C. H. (1972), J. Biol. Chem. 247:2083.
Rose, A. B., and Racker, E. (1962), J. Biol. Chem. 237:3279.
Rosenthal, N. A., and Oster, G. (1954), J. Soc. Cosmetic Chemists 5:286.
Rosenthal, N. A., and Oster, G. (1961), J. Amer. Chem. Soc. 83:4445.
Rost, J., and Rapoport, S. (1964), Nature (London) 201:185.
Rothfus, J. A. (1966), Anal. Biochem. 16:167.
Rothschild, H. A., and Barron, E. S. G. (1954), J. Biol. Chem. 209:511.
Rudinger, J., and Jost, K. (1964), Experientia 20:570.
Ryklan, L. R., and Schmidt, C. L. A. (1944), Arch. Biochem. 5:89.
Ryle, A. P., and Sanger, F. (1955), Biochem. J. 60:535.
Sachs, G. (1921), Chem. Ber. 54:1849.
Sajgó, M., and Telegdi, M. (1968), Acta Biochim. Biophys. 3:171.
Sakai, H. (1967), J. Biol. Chem. 242:1458.
Sakai, H. (1968), Anal. Biochem. 26:269.
Sakakibara, S., and Hase, S. (1968), Bull. Chem. Soc. Jap. 41:2816.
Salce, L., and Goodman, I. (1961), Fed. Proc. 20:387.
Salter, D. W., Norton, I. L., and Hartman, F. C. (1973), Biochemistry 12:1.
Sanadi, D. R., Langley, M., and Searls, R. L. (1959), J. Biol. Chem. 234:178.
Sanger, F. (1947), Nature (London) 160:295.
Sanger, F. (1949), Biohem. J. 44:126.
Sanger, F. (1953), Nature (London) 171:1025.
Sanner, T., and Pihl, A. (1962), Biochim. Biophys. Acta 62:171.
Sanner, T., and Pihl, A. (1963), J. Biol. Chem. 238:165.
Savige, W. E., and Maclaren, J. A. (1966), In "The Chemistry of Organic Sulfur Compounds," Vol. 2 (Kharasch, N., and Meyers, C. Y., eds.), p. 367, Pergamon, Oxford.
Saxena, V. P., and Wetlaufer, D. B. (1970), Biochemistry 9:5015.
Schellenberger, A. (1967), Angew. Chem. (Int. Ed. Engl.) 6:1024.
Scheraga, H. A. (1961), "Protein Structure," Academic, New York.
Schmidt, D. D., and Arens, A. (1968), Hoppe-Seyler's Z. Physiol. Chem. 349:1157.
Schmidt, U., Grafen, P., and Goedde, H. W. (1965), Angew. Chem. (Int. Ed. Engl.) 4:846.
Schöberl, A. (1933), Justus Liebig's Ann. Chem. 507:111.
Schöberl, A., and Ludwig, E. (1937), Chem. Ber. 70:1422.
Schöberl, A., and Rambacher, P. (1939), Justus Liebig's Ann. Chem. 538:84.
Schöberl, A., and Wagner, A. (1956), Proc. Int. Wool. Textile Res. Conf. Australia, 1955, Vol. C, p. 11.
Schöberl, A., Kawohl, M., and Hamm, R. (1951), Chem. Ber. 84:571.
Schneider, F., and Wenck, H. (1969), Hoppe-Seyler's Z. Physiol. Chem. 350:1521.
Schneider, J. F., and Westley, J. (1969), J. Biol. Chem. 244:5735.
Schram, E., Moore, S., and Bigwood, E. (1954), Biochem. J. 57:33.
Schroeder, W. A., and Kay, L. M. (1955), J. Amer. Chem. Soc. 77:3908.
Schroeder, W. A., Shelton, J. R., and Robberson, B. (1967), Biochim. Biophys. Acta 147:590.

Schubert, M. (1947), J. Amer. Chem. Soc. 69:712.
Schubert, M. P. (1932), J. Amer. Chem. Soc. 54:4077.
Schubert, M. P. (1935), J. Biol. Chem. 111:671.
Schubert, M. P. (1936a), J. Biol. Chem. 114:341.
Schubert, M. P. (1936b), J. Biol. Chem. 116:437.
Schubert, M. P. (1951), The First Symp. Chem. Biol. Correlations, Nat. Res. Council. Publ. p. 269.
Schwartz, A., and Oschmann, A. (1925), C. R. Sci. Biol. 92:169.
Schwartz, I. L., Rasmussen, H., Schoessler, M. A., Silver, L., and Fong, C. T. O. (1960), Proc. Nat. Acad. Sci. U. S. 46:1288.
Schwartz, I. L., Rasmussen, H., and Rudinger, J. (1964), Proc. Nat. Acad. Sci. U. S. 52:1044.
Schwyzer, R. (1953), Helv. Chim. Acta 36:414.
Scott, D. W., Finke, H. L., Gross, M. E., Guthrie, G. B., and Huffman, H. M. (1950), J. Amer. Chem. Soc. 72:2424.
Scott, D. W., Finke, H. L., McCullough, J. P., Gross, M. E., Pennington, R. E., and Waddington, G. (1952), J. Amer. Chem. Soc. 74:2478.
Scott, E. M., Duncan, I. W., and Edstrand, V. (1963), J. Biol. Chem. 238:3928.
Searls, R. L., Peters, J. M., and Sanadi, D. R. (1961), J. Biol. Chem. 236:2317.
Sedlak, J., and Lindsay, R. H. (1968), Anal. Biochem. 25:192.
Segal, H. L., and Boyer, P. D. (1953), J. Biol. Chem. 204:265.
Segal, H. L., and Gold, A. H. (1963), J. Biol. Chem. 238:PC2589.
Seibles, T. S., and Weil, L. (1967), In "Methods in Enzymology," Vol. 11 (Hirs, C. H. W., ed.), p. 204, Academic, New York.
Sela, M., White, F. H., and Anfinsen, C. B. (1959), Biochim. Biophys. Acta 31:417.
Shaltiel, S. (1967), Biochem. Biophys. Res. Commun. 29:178.
Shaltiel, S., and Soria, M. (1969), Biochemistry 8:4411.
Shapira, E., and Arnon, R. (1969), J. Biol. Chem. 244:1026.
Shifrin, S. (1964), Biochim. Biophys. Acta 81:205.
Shol'ts, Kh. F. (1964), Biokhimiya 29:557 [Engl. transl. p. 495].
Siebert, W., Fiore, C., and Dose, K. (1965), Z. Naturforsch. 20b:957.
Sieker, L. C., and Jencen, L. H. (1965), Biochem. Biophys. Res. Commun. 20:33.
Sillén, L. G., and Martell, A. E. (1964), "Stability Constants of Metal-Ion Complexes," Chemical Society, London.
Silverstein, E., and Sulebele, G. (1970), Biochemistry 9:274.
Simon, S. R., Arndt, D. J., and Konigsberg, W. H. (1971), J. Mol. Biol. 58:69.
Simoni, R. D., Griddle, R. S., and Stumpf, P. K. (1967), J. Biol. Chem. 242:573.
Simpson, R. B. (1961), J. Amer. Chem. Soc. 83:4711.
Singer, S. J. (1967), Advan. Protein Chem. 22:1.
Slobin, L. I., and Singer, S. J. (1968), J. Biol. Chem. 243:1777.
Sluyterman, L. A. (1957), Biochim. Biophys. Acta 25:402.
Sluyterman, L. A. (1966), Anal. Biochem. 14:317.
Sluyterman, L. A. E. (1967), Biochim. Biophys. Acta 139:439.
Sluyterman, L. A., and Wolthers, B. G (1969), Proc. Kon. Ned. Akad. Wetensch. Ser. B, 72:14.
Smillie, L. B., and Hartley, B. S. (1967), Biochem. J. 105:1125.
Smith, D. B., and Perutz, M. F. (1960), Nature (London) 188:406.

Smith, E. L. (1958), J. Biol. Chem. 233:1392.
Smith, E. L., and Kimmel, J. R. (1960), In "The Enzymes," Vol. 4 (Boyer, P. D., et al., eds.), p. 133, Academic, New York.
Smith, E. L., and Parker, M. J. (1958), J. Biol. Chem. 233:1387.
Smith, E. L., Davis, N. C. Adams, E., and Spackman, D. H. (1954), In "A Symposium on the Mechanism of Enzyme Action" (McElroy, W. D., and Glass, B., eds.), p. 291, John Hopkins, Baltimore.
Smith, E. L., Finkle, B. J., and Stockell, A. (1955a), Discuss. Faraday Soc. 20:96.
Smith, E. L., Kimmel, J. R., Brown, D. M., and Thompson, E. O. P. (1955b), J. Biol. Chem. 215:67.
Smith, E. L., Hill, R. L., and Kimmel, J. R. (1958), In "Symposium on Protein Structures" (Neuberger, A., ed.), p. 182, Methuen, London.
Smith, E. L., Kimmel, J. R., and Light, A. (1962), Int. Cong. Biochem. 5th, 4:33.
Smith, G. D., and Schachman, H. K. (1971), Biochemistry 10:4576.
Smith, H. A., Doughty, G., and Gorin, G. (1964), J. Org. Chem. 29:1484.
Smith, M. E., and Greenberg, D. M. (1957), J. Biol. Chem. 226:317.
Smyth, C., P. (1955), "Dielectric Behaviour and Structure, p. 244, McGraw-Hill, New York.
Smyth, D. G., Nagamatsu, A., and Fruton, J. S. (1960), J. Amer. Chem. Soc. 82:4600.
Smyth, D. G., Blumenfeld, O. O., and Konigsberg, W. (1964), Biochem. J. 91:589.
Smythe, C. V. (1936), J. Biol. Chem. 114:601.
Snell, J. M., and Weissberger, A. (1939), J. Amer. Chem. Soc. 61:450.
Snodgrass, P. J., Vallee, B. L., and Hoch, F. L. (1960), J. Biol. Chem. 235:504.
Snow, N. S. (1962), Biochem. J. 84:360.
Sogami, M., Petersen, H. A., and Foster, J. F. (1969), Biochemistry 8:49.
Sohler, S. M. R., Seibert, S. M. A., Kreke, C. W., and Cook, E. S. (1952), J. Biol. Chem. 198:281.
Sokolovsky, M., Harell, D., and Riordan, J. F. (1969), Biochemisty 8:4740.
Sondack, D. L., and Light, A. (1971), J. Biol. Chem. 246:1630.
Spackman, D. H. Stein, W. H., and Moore, S. (1958), Anal. Chem. 30:1190.
Spackman, D. H. Stein, W. H., and Moore, S. (1960), J. Biol. Chem. 235:648.
Speakman, J. B. (1933), Nature (London) 132:930.
Speakman, J. B. (1936), J. Soc. Dyers Colour. 52:335.
Speakman, J. B. (1963), Textile Rec. 81:46.
Spencer, R. L., and Wold, F. (1969), Anal. Biochem. 32:185.
Spencer, R. P., and Knox, W. E. (1962), Arch. Biochem. Biophys. 96:115.
Sperling, R., Burstein, Y., and Steinberg, I. Z. (1969), Biochemistry 8:3810.
Spolter, P. D., and Vogel, J. M. (1968), Biochim. Biophys. Acta 167:525.
Springell, P. H., Gillespie, J. M., Inglis, A. S., and Mclaren, J. A. (1964), Biochem. J. 91:17.
Stapleton, I. W., and Swan, J. M. (1960), Aust. J. Chem. 13:416.
Stark, G. R. (1964), J. Biol. Chem. 239:1411.
Stark, G. R. (1967), In "Methods in Enzymology," Vol. 11 (Hirs, C. H. W., ed), p. 590, Academic, New York.
Stark, G. R., and Crawford, L. V. (1972), Nature (London), New Biol. 237:146.
Stark, G. R., and Stein, W. H. (1964), J. Biol. Chem. 239:3755.
Stark, G. R., Stein, W. H., and Moore, S. (1960), J. Biol. Chem. 235:3177.

Stefanini, S., Chiancone, E., McMurray, C. H., and Antonini, E. (1972), Arch. Biochem. Biophys. 151:28.
Stein, W. H. (1964), Fed. Proc. 23:599.
Stein, W. H., and Moore, S. (1962), Int. Cong. Biochem. 5th, 4:33.
Steiner, D. F., and Clark, J. L. (1968), Proc. Nat. Acad. Sci. U. S. 60:622.
Steiner, D. F., and Oyer, P. E. (1967), Proc. Nat. Acad. Sci. U. S. 57:473.
Steiner, D. F., Cunningham, D., Spigelman, L., and Aten, B. (1967), Science, 157:697.
Steiner, R. F., De Lorenzo, F., and Anfinsen, C. B. (1965), J. Biol. Chem. 240:4648.
Steinrauf, L. K., Peterson, J., and Jensen, L. H. (1958), J. Amer. Chem. Soc. 80:3835.
Stern, J. R. (1956), J. Biol. Chem. 221:33.
Stockell, A., and Smith, E. L. (1957), J. Biol. Chem. 227:1.
Stocken, L. A., and Thompson, R. H. S. (1946), Biochem. J. 40:529, 535.
Stoffel, W. (1973), Biochem. Soc. Trans. 1:336.
Stockstad, E. L. R., Hoffman, C. E., Regan, M. A., Fordham, D., and Jukes, T. H. (1949), Arch. Biochem. 20:75.
Stoppani, A. O. M., and Brignone, J. A. (1957), Arch. Biochem. Biophys. 68:432.
Stoppani, A. O. M., Actis, A. S., Deferrari, S. O., and Gonzalez, E. L. (1953), Biochem. J. 54:378.
Straub, F. B. (1964), Advan. Enzynol. 26:89.
Strem, J., Krishna-Prasad, Y. S. R., and Schellman, J. A. (1961), Tetrahedron 13:176.
Stricks, W., and Kolthoff, I. M. (1956), J. Amer. Chem. Soc. 78:2085.
Stricks, W., Kolthoff, I. M., and Heyndrickx, A. (1954a), J. Amer. Chem. Soc. 76:1515.
Stricks, W., Kolthoff, I. M., and Tanaka, N. (1954b), Anal. Chem. 26:299.
Strittmatter, P., and Ball, E. G. (1955), J. Biol. Chem. 213:445.
Sunshine, G. H., Williams, D. J., and Rabin, B. R. (1971), Nature (London) New Bid. 230:133.
Suzuki, I., and Silver, M. (1966), Biochim. Biophys. Acta 122:22.
Suzuki, K. (1967), Biochemistry 6:1335.
Suziki, K., and Reed, L. J. (1963). J. Biol. Chem. 238:4021.
Swan, J. M. (1957a), Nature (London) 179:965.
Swan, J. M. (1957b), Nature (London) 180:643.
Swan, J. M. (1959), In "Sulfur in Proteins," Proc. Symp. Falmouth, 1958, p. 3.
Swan, J. M. (1961), Aust. J. Chem. 14:69.
Swanepoel, O. A. (1963), J. S. African Chem. Inst. 16:48.
Swanepoel, O. A., and van Rensburg, N. J. J. (1965), Photochem. Photobiol. 4:833.
Swanepoel, O. A., Mellet, P., and Scanes, S. G. (1969), Arch. Biochem. Biophys. 129:26.
Swart, L. S., and Haylett, T. (1971), Biochem. J. 123:201.
Swart, L. S., and Haylett, T. (1973), Biochem. J. 133;641.
Sweetman, B. J., and Maclaren, J. A. (1966), Aust. J. Chem. 19:2347, 2355.
Swenson, A. D., and Boyer, P. D. (1957), J. Amer. Chem. Soc. 79:2174.
Swoboda, G., and Hasselbach, W. (1973), Hoppe-Selyer's Z. Physiol. Chem. 354:1611.
Szabolcsi, G. (1958), Acta. Physiol. Acad. Sci. Hung. 13:213.
Szabolcsi, G., Biszku, E. (1961), Biochim. Biophys. Acta 48:335.
Szabolcsi, G., Biszku, E., and Sajgö, M. (1960), Acta Physiol. Acad. Sci. Hung. 17:183.
Szabolcsi, G., Biszku, E., and Szörenyi, E. (1959), Biochim. Biophys. Acta 35:237.
Szabolcsi, G., Boross, L., and Biszku, E. (1964), Acta Physiol. Acad. Sci. Hung. 25:149.

Szabolcsi, G., Biszku, E., Sajgö, M., Zánodszky, P., Friedrich, P., and Szajani, B. (1970), Int. Cong. Biochem. 8th, Abst. p. 110.
Szajáni, B., Sajgó, M., Biszku, E., Friedrich, P., and Szabolcsi, G. (1970), Eur. J. Biochem. 15:171.
Szendrö, P., Lampert, U., and Wrede, F. (1933), Hoppe-Seyler's Z. Physiol. Chem. 222:16.
Takagi, T., and Isemura, T. (1965), J. Biochem. Tokyo 57:89.
Takahashi, K., Stein, W. H., and Moore, S. (1967), J. Biol. Chem. 242:4682.
Tamburro, A. M., Boccu, E., and Celotti, L. (1970), Int. J. Protein Res. 2:157.
Tanaka, M., Nakashima, T., Benson, A., Mower, H., and Yasunobu, K. T. (1966), Biochemistry 5:1666.
Tanaka, N., Kolthoff, I. M., and Stricks, W. (1955), J. Amer. Chem. Soc. 77:2004.
Tanford, C. (1964), J. Amer. Chem. Soc. 86:2050.
Tanford, C., Kawahara, K., Lapanje, S., Hooker, T. M., Zarlengo, M. H., Salahuddin, A., Aune, K. C., and Takagi, T. (1967), J. Amer. Chem. Soc. 89:5023.
Taniguchi, N. (1971), Anal. Biochem. 40:200.
Tarbell, D. S. (1961), In "Organic Sulfur Compounds" (Kharasch, N., ed.), p. 97, Pergamon, Oxford.
Tarbell, D. S., and Harnish, D. P. (1951), Chem. Rev. 49:11.
Tashiro, Y., and Otsuki, E. (1970), Biochim. Biophys. Acta 214:265.
Taylor, J. F., Antonini, E., and Wyman, J. (1963), J. Biol. Chem. 238:2660.
Taylor, J. F., Antonini, E., Brunori, M., and Wyman, J. (1966), J. Biol. Chem. 241:241.
Telegdi, M., and Straub, F. B. (1973), Biochim. Biophys. Acta 321:210.
Thelander, L. (1967), J. Biol. Chem. 242:852.
Thelander, L. (1968), Eur. J. Biochem. 4:407.
Thelander, L. (1970), J. Biol. Chem. 245:6026.
Theorell, H., and Bonnichsen, R. (1951), Acta Chem. Scand. 5:1105.
Thompson, E. O. P., and O'Donnell, I. J. (1961), Biochim. Biophys. Acta 53:447.
Thompson, E. O. P., and O'Donnell, I. J. (1962), Aust. J. Biol. Sci. 15:757.
Thompson, E. O. P., and O'Donnell, I. J. (1967), Aust. J. Biol. Sci. 20:1001.
Toda, H., and Narita, K. (1968), J. Biochem. Tokyo 63:302.
Toda, H., Kato, I., and Narita, K. (1968), J. Biochem. Tokyo 63:295.
Toennies, G., and Kolb, J. J. (1945), J. Amer. Chem. Soc. 67:849.
Torchinskii, Yu. M. (1958), Byul. Éksperim. Biol. Med. 12:108.
Torchinskii, Yu. M. (1959), Ukrainsk. Biokhim. Zh., 31:589.
Torchinskii, Yu. M. (1961), Usp. Sovrem. Biol. 51(3):261.
Torchinskii, Yu. M. (1963), Usp. Sovrem. Biol. 55(2):161.
Torchinskii, Yu. M. (1964), Biokhimiya, 29:534 [Engl. transl. p. 458].
Torchinskii, Yu. M. (1968), Usp. Sovrem. Biol. 66(1)(4):13.
Torchinskii, Yu. M. (1971), unpublished data.
Torchinskii, Yu. M., and Sinitsyna, N. I. (1970), Molekul. Biol. 4:256. [Engl. transl. p. 205].
Torchinskii, Yu. M. Zufarova, R. A., Agalarova, M. B., and Severin, E. S. (1972) FEBS Lett. 28:302.
Tomkins, G. M. (1956), J. Biol. Chem. 218:437.
Toro-Goyco, E., Maretzki, A., and Matos, M. L. (1968), Arch. Biochem. Biophys. 126:91.
Traktatellis, A. C., and Schwartz, G. P. (1970), Nature (London) 225:548.

Trentham, D. R. (1968), Biochem. J. 109:603.
Trentham, D. R. (1971), Biochem. J. 122:71.
Trundle, D., and Cunningham, L. W. (1969), Biochemistry 8:1919.
Trung, A. K., and Yip, C. C. (1969), Proc. Nat. Acad. Sci. U. S. 63:442.
Tuppy, H. (1959), In "Sulfur in Proteins," Proc. Symp. Falmouth, 1958, p. 141.
Tuppy, H., and Paleus, S. (1955), Acta Chem. Scand. 9:353.
Turner, J. E., Kennedy, M. B., and Haurowitz, F. (1959), In "Sulfur in Proteins," Proc. Symp. Falmouth, 1958, p. 25.
Turpaev, T. M. (1954), Dokl. Akad. Nauk SSSR, 94:973.
Urnés, P., and Doty, P. (1961), Advan. Protein, Chem. 16:401.
Urry, D. W., Quadrifoglio, F., Walter, R., and Schwartz, I. L. (1968), Proc. Nat. Acad. Sci. U. S. 60:967.
Utsumi, S., and Karush, F. (1964), Biochemistry 3:1329.
Vagelos, P. R., Majerus, P. W., Alberts, A. W., Larrabee, A. R., and Ailhaud, G. P., (1966), Fed. Proc. 25:1485.
Vallee, B. L. (1962), Int. Cong. Biochem. 5th, 4:162.
Vallee, B. L., Coombs, T. L., and Hoch, F. L. (1960), J. Biol. Chem. 235:PC45.
Vallee, B. L., Riordan, J. F., and Coleman, J. E. (1963), Proc. Nat. Acad. Sci. U. S. 49:109.
Van den Bosch, H., and Vagelos, P. R. (1970), Biochim. Biophys. Acta 218:233.
Vanaman, T. C., and Stark, G. R. (1970), J. Biol. Chem. 245:3565.
Vanaman, T. C., Wakil, S. J., and Hill, R. L. (1968), J. Biol. Chem. 243:6420.
Van Eyes, J., and Kaplan, N. O. (1957), J. Biol. Chem. 228:305.
Van Rensburg, N. J. J., and Swanepoel, O. A. (1967), Arch. Biochem. Biophys. 118:531; 121:729.
Varandani, P. T. (1967), Biochim. Biophys. Acta 132:10.
Varandani, P. T., and Plumley, H. (1968), Biochim. Biophys. Acta 151:273.
Vas, M., and Boross, L. (1970), Acta Biochim. Biophys. 5:203, 215.
Vasil'eva, L. E., and Il'in, V. S. (1970), Biokhimiya, 35:458 [Engl. transl. p. 415].
Velick, S. F. (1953), J. Biol. Chem. 203:563.
Velick, S. F. (1958), J. Biol. Chem. 233:1455.
Velick, S. F., and Hayes, J. E. (1953), J. Biol. Chem. 203:545.
Velluz, L., and Legrand, M. (1965), Angew. Chem. (Int. Ed. Engl.) 4:838.
Venetianer, P., and Straub, F. B. (1963a), Biochim. Biophys. Acta 67:166.
Venetianer, P., and Straub, F. B. (1963b), Acta Physiol. Acad. Sci. Hung. 24:41.
Visser, J., and Veeger, C. (1968), Biochim. Biophys. Acta 159:265.
Vithayathil, P. J., and Richards, F. M. (1960), J. Biol. Chem. 235:2343.
Wakil, S. J., Pugh, E. L., and Saner, F. (1964), Proc. Nat. Acad. Si. U. S. 52:106.
Waldmann-Meyer, H. (1960), J. Biol. Chem. 235:3337.
Waley, S. G. (1966), Advan. Protein Chem. 21:1.
Walker, J. B. (1957), J. Biol. Chem. 224:57.
Walker, J. B. (1958), J. Biol. Chem. 231:1.
Walker, W. H., Kearney, E. B., Seng, R. L., and Singer, T. P. (1971), Eur. J. Biochem. 24:328.
Wallenfels, K., and Eisele, B. (1968), Eur. J. Biochem. 3:267.
Wallenfels, K., and Eisele, B. (1970), Biochem. Soc. Symp. No. 31, p. 21.
Wallenfels, K., and Streffer, C. (1966), Biochem. Z. 346:119.

Walsh, K. A., Kumar, K. S. V. S., Bargetzi, J. P., and Neurath, H. (1962), Proc. Nat. Acad. Sci. U. S. 48:1443.
Walsh, K. A., Ericsson, L. H., Bradshaw, R. A., and Neurath, H. (1970), Biochemistry 9:219.
Walsh, K. A., McDonald, R. M., and Bradshaw, R. A. (1970), Anal. Biochem. 35:193.
Walter, R., and Chan, W. Y. (1967), J. Amer. Chem. Soc. 89:3892.
Walter, R., and du Vigneaud, V. (1965), J. Amer. Chem. Soc. 87:4192.
Walter, R., and du Vigneaud, V. (1966), J. Amer. Chem. Soc. 88:1331.
Walter, R. Rudinger, J., and Schwartz, I. L. (1967), Amer. J. Med. 42:653.
Walton, E., Wagner, A. F., Bachelor, F. W., Peterson, L. H., Holly, F. W., and Folkers, K. (1955), J. Amer. Chem. Soc. 77:5144.
Warburg, O., and Christian, W. (1939), Biochem. Z. 303:40.
Warburg, O., Klotzsch, H., and Gawehn, K. (1954), Z. Naturforsch. 9b:391.
Warburg, O., Gawehn, K., and Geissler, A. W. (1957), Z. Naturforsch. 12b:47.
Ward, W. H., and Lundgren, H. P. (1954), Advan. Protein Chem. 9:243.
Wassarman, P. M., and Major, J. P. (1969), Biochemistry 8:1076.
Watenpaugh, K. D., Sieker, L. C., Herriott, J. R., and Jensen, L. H. (1972), Cold Spring Harbor Symp. Quant. Biol. 36:359.
Wallis, M. (1972), FEBS Lett. 21:118.
Watts, D. C., and Rabin, B. R. (1962), Biochem. J. 85:507.
Watts, D. C., Rabin, B. R., and Crook, E. M. (1961), Biochim. Biophys. Acta 48:380.
Webb, J. L. (1966), "Enzyme and Metabolic Inhibitors," Vol. 2, Academic, New York.
Weber, U., Hartter, P., and Flohé, L. (1970), Hoppe-Seyler's Z. Physiol. Chem. 351:1389.
Weigmann, H. D. (1968), J. Polymer. Sci. Pt. A-1, 6:2237.
Weigmann, H. D., and Rebenfeld, L. (1968), In "The Chemistry of Sulfides," Proc. Conf. 1966, p. 185.
Weil, L., and Seibles, T. S. (1961), Arch. Biochem. Biophys. 95:470.
Weiner, H., Batt, C. W., and Koshland, D. E. (1966), J. Biol. Chem. 241:2687.
Weinryb, I. (1968), Arch. Biochem. Biophys. 124:285.
Weiss, J. (1937), Chem. Ind. (London) 56:685.
Weitzman, P. D. J. (1965), Biochim. Biophys. Acta 107:146.
Wells, W. W., Schultz, J., and Lynen, F. (1967), Biochem. Z. 346:474.
Weltman, J. K., Szaro, R. P., Frackleton, A. R., Dowben, R. M., Bunting, J. R., and Cathou, R. E. (1973), J. Biol. Chem. 248:3173.
Weshhead, E. W., Butler, L., and Boyer, P. D. (1963), Biochemistry 2:927.
Whitaker, J. R. (1962), J. Amer. Chem. Soc. 84:1900.
Whitaker, J. R., and Bender, M. L. (1965), J. Amer. Chem. Soc. 87:2728.
White, F. H. (1960), J. Biol. Chem. 235:383.
White, F. H. (1961), J. Biol. Chem. 236:1353.
White, F. H. (1967), In "Methods in Enzymology," Vol. 11 (Hirs, C. H. W., ed.), p. 481, Academic, New York.
Whitehead, E. P., and Rabin, B. R. (1964), Biochem. J. 90:532.
Wieland, T., and Bokelman, E. (1952), Justus Liebig's Ann. Chem. 576:20.
Wieland, T., and Horning, H. (1956), Justus Liebig's Ann. Chem. 600:12.
Willecke, K., Ritter, E., and Lynen, F. (1969), Eur. J. Biochem. 8:503.
Wilhelm, J. M., and Haselkorn, R. (1970), Proc. Nat. Acad. Sci. U. S. 65:388.

Williams, V. R., and Libano, W. Y. (1966), Biochem. Biophys. Acta 118:144.
Williamson, A. R., and Askonas, B. A. (1968), Biochem. J. 107:823.
Winstead, J. A. (1967), Radiation Res. 30:832.
Wintersberger, E. (1965), Biochemistry 4:1533.
Wintersberger, E., Neurath, H., Coombs, T. L., and Vallee, B. L. (1965), Biochemistry, 4:1526.
Wit, J. G., and Leeuwangh, P. (1969), Biochim. Biophys. Acta 177:329.
Witter, A., and Tuppy, H. (1960), Biochim. Biophys. Acta 45:429.
Wolff, J., and Covelli, I. (1969), Eur. J. Biochem. 9:371.
Wolfram, L. J. (1965), Nature (London), 206:304.
Wollaston, W. H. (1810), Phil. Trans. Roy. Soc. 223.
Wong, K.-P., and Foster, J. F. (1969), Biochemistry 8:4104.
Wong, R. C., and Liener, I. E. (1964), Biochem. Biophys. Res. Commun. 17:470.
Woodin, T. S., and Segel, I. H. (1968), Biochim. Biophys. Acta 167:64.
Wright, W. B. (1958), Acta Crystallogr. 11:632.
Wronski, M. (1963), Monatsh. Chem. 94:197.
Wu, C. (1965), Biochim. Biophys. Acta 96:134.
Würz, H., and Haurowitz, F. (1961), J. Amer. Chem. Soc. 83:280.
Wyckoff, H. W., Tsernoglou, D., Hanson, A. W., Know, J. R., Lee, B., and Richards, F. M. (1970), J. Biol. Chem. 245:305.
Yakel, H. L., and Hughes, E. W. (1954), Acta Crystallogr. 7:291.
Yakovlev, V. A., and Torchinskii, Yu. M. (1958), Biokhimiya 23:755.
Yamashiro, D., Gillessen, D., and du Vigneaud, V. (1966), Biochemistry 5:3711.
Yanaihara, N., Yanaihara, C., Dupuis, G., Beacham, J., Camble, R., and Hofmann, K. (1969), J. Amer. Chem. Soc. 91:2184.
Yang, C. C. (1967), Biochim. Biophys. Acta 133:346.
Yankeelov, J. A., and Koshland, D. E. (1965), J. Biol. Chem. 240:1593.
Young, D. M., and Potts, J. T. (1963), J. Biol. Chem. 238:1995.
Yutani, K., Yutani, A., Imanishi, A., and Isemura, T. (1968), J. Biochem. Tokyo, 64:449.
Yutani, K., Yutani, A., and Isemura, T. (1969), J. Biochem. Tokyo 66:823.
Zahler, W. L., and Cleland, W. W. (1968), J. Biol. Chem. 243:716.
Zahn, H. (1951), Leder 2:8.
Zahn, H., and Golsch, E.(1963), Hoppe-Seyler's Z. Physiol. Chem. 330:38.
Zahn, H., and Lumper, L. (1968), Hoppe-Seyler's Z. Physiol. Chem. 349:485.
Zahn, H., and Trauman, K. (1954), Z. Naturforsch. 9B:518.
Zahn, H., Kunitz, F.-W., and Meichelbeck, H. (1961), "Structure de la Laine, Colloque Inst. Textile France, p. 227.
Zahn, H., Gutte, B., and Brinkhoff, O. (1965), Angew. Chem. (Int. Ed. Engl.) 4:528.
Zahn, H., Gutte, B., Pfeiffer, E. F., and Ammon, J. (1966), Ann. Chem. 691:225.
Zahn, H., Drechsl, E., and Puls, W. (1968), Hoppe-Seyler's Z. Physiol. Chem. 349:385.
Zahn, H., Brandenburg, D., and Gattner, H.-G. (1972), Diabetes 21:468.
Zanetti, G., and Forti, G. (1969), J. Biol. Chem. 244:4757.
Zanetti, G., and Williams, C. H. (1967), J. Biol. Chem. 242:5232.
Závodszky, P., Biszku, E., Abaturov, L. V., and Szabolcsi, G. (1972), Biochim. Biophys. Acta 7:1.
Ziegler, K. (1964), J. Biol. Chem. 239:PC2713.

Index